On-Wafer Calibration Techniques Enabling Accurate Characterization of High-Performance Silicon Devices at the mm-Wave Range and Beyond

RIVER PUBLISHERS SERIES IN ELECTRONIC MATERIALS AND DEVICES

Indexing: All books published in this series are submitted to the Web of Science Book Citation Index (BkCI), to SCOPUS, to CrossRef and to Google Scholar for evaluation and indexing.

The "River Publishers Series in Electronic Materials and Devices" is a series of comprehensive academic and professional books which focus on the theory and applications of advanced electronic materials and devices. The series focuses on topics ranging from the theory, modeling, devices, performance and reliability of electron and ion integrated circuit devices and interconnects, insulators, metals, organic materials, micro-plasmas, semiconductors, quantum-effect structures, vacuum devices, and emerging materials. Applications of devices in biomedical electronics, computation, communications, displays, MEMS, imaging, micro-actuators, nanoelectronics, optoelectronics, photovoltaics, power ICs and micro-sensors are also covered.

Books published in the series include research monographs, edited volumes, handbooks and textbooks. The books provide professionals, researchers, educators, and advanced students in the field with an invaluable insight into the latest research and developments.

Topics covered in the series include, but are by no means restricted to the following:

- Integrated circuit devices
- Interconnects
- Insulators
- Organic materials
- Semiconductors
- Qantum-effect structures
- Vacuum devices
- Biomedical electronics
- Displays and imaging
- MEMS
- Sensors and actuators
- Nanoelectronics
- Optoelectronics
- Photovoltaics
- Power ICs

For a list of other books in this series, visit www.riverpublishers.com

On-Wafer Calibration Techniques Enabling Accurate Characterization of High-Performance Silicon Devices at the mm-Wave Range and Beyond

Andrej Rumiantsev

MPI Corporation
Taiwan

River Publishers

Published, sold and distributed by:
River Publishers
Alsbjergvej 10
9260 Gistrup
Denmark

River Publishers
Lange Geer 44
2611 PW Delft
The Netherlands

Tel.: +45369953197
www.riverpublishers.com

ISBN: 978-87-7022-112-2 (Hardback)
978-87-7022-111-5 (Ebook)

Contents

Preface

This book is the culmination of more than a decade of my scientific research in the field of on-wafer RF calibration and measurements, collectively aimed at achieving a Dr.-Ing. Degree. My research was conducted in association with SUSS MicroTec Test Systems, Cascade Microtech and then later with Brandenburg University of Technologies BTU Cottbus-Senftenberg, Germany. The original manuscript was submitted to the Chair of Radio Frequency and Microwave Techniques BTU Cottbus-Senftenberg and my original Dr.-Ing. thesis was presented in April 2014. This version, while keeping the style and themes of the original publication, now includes updated and revised graphical materials to keep the content current.

Today within industry the drive is to replace conventional probe-tip RF calibration methods with *in-situ* 'device level' techniques. This approach has not only been shown to improve the overall test measurement accuracy, but it has also helped support advancements in new semiconductor design and processes going forward. Back in 2001, when I started working in the field of on-wafer RF calibration on Silicon, experts were still confronted with an overwhelming list of unsolved technical problems. The complexity and variety of these challenges involved cooperating with a wide range of specialists in the field of microwave and semiconductor technologies all over the world. The invaluable support and encouragement I received during this time provided numerous opportunities to collaborate with many experts in their field, many who have become good friends.

I would like to acknowledge the support I received from my former colleagues at SUSS MicroTec Test Systems (Dresden, Germany), Dr. Claus Dietrich (today with FormFactor, Germany) and Steffen Schott (today with Rosenberger, Germany). I would also like to thank Prof. Holger Heuermann from FH Aachen, Germany, who helped me to understand fundamentals of RF calibration mathematics and Prof. Michael Schroeter and Dr. Paulius Sakalas (CEDIC, TU Dresden, Germany) which the well-equipped laboratory became my second research office for years. It was also a great pleasure and honor for me to work with Nick Ridler (NPL, UK) on reconstructing the history of VNA calibration.

My first experiment of implementing the TMR (aka LRM+) calibration method on SiGe:C BiCMOS process was at IHP Microelectroncs (Frankfurt (O), Germany), with Dr. Rene Scholz, Dr. Falk Korndoerfer, and Dr. Biswanath Senapati (today with GLOBALFOUNDRIES). The results were extremely exciting but unvealed new difficulties in the quantitative verification of *in-situ* calibration accuracy.

Together with Phillip Corson and Susan Sweeney from IBM Microelectronics (Essex Junction, VT, USA, now GLOBALFOUNDRIES) and Dr. Uwe Arz (PTB, Braunschweig, Germany) we verified the characteristic impedance of coplanar lines for IBM RF CMOS process up to 110 GHz. This experiment enabled further quantitative verification of the transfer TMR method on SiGe BiCMOS process, that we completed with ST Microelectronic's (Grenoble, France) device modelling engineers Dr. Didier Celi, Nicolas Derriere and Frank Pourchon. Dr. Ed Godshalk (Maxim Integrated, Portland, OR, USA) helped me to explore methods of minimizing higher-order propagation modes in planar structures and probe-dependent calibration residual errors.

I would also like to acknowledge Dr. Franz Sischka from Agilent Eesof (now with SisConsult, Germany) for his valuable advices on verification of the on-wafer calibration accuracy from the device modelling perspective as well as for his critical feedback on the first revision of my manuscript.

I express my special appreciations to Ralf Doerner (Ferdinand-Braun-Institut Leibniz-Institut für Höchstfrequenztechnik, FBH, Berlin, Germany) who was my first mentor for on-wafer RF measurement techniques, a sharp critic but reliable supporter of my ideas. Ralf was an invaluable partner and contributor to the majority of my research, experiments and publications; and also became a very good friend over the years.

I would like to thank Prof. Matthias Rudolph from Brandenburg University of Technologies (BTU Cottbus-Senftenberg, Germany) who guided me at the last stage of my work, critically revised it and provided me an opportunity to defend it under his supervision.

And finally, I want to thank Dr. Stojan Kanev (previously with SUSS MicroTec Test Systems, now with MPI Corporation, Taiwan) who has both supported and pushed me to develop new and exciting RF on-wafer test solutions for industry.

MPI Corporation has greatly contributed to the publication of this book.

About MPI Corporation

Founded in 1995 and headquartered in Hsinchu, Taiwan, MPI Corporation is a global technology leader in Semiconductor, Light Emitting Diode (LED), Photo Detectors, Lasers, Materials Research, Aerospace, Automotive, Fiber Optic, Electronic Components and more. MPI's four main business sectors include Probe Card, Photonics Automation, Advanced Semiconductor Test and Thermal Divisions. MPI products range from various advanced probe card technologies, probers, testers, material handlers, inspection and thermal air systems. Many of these products are accompanied by state-of-the-art Calibration and Test & Measurement software suites. The diversification of product portfolio and industries allows a healthy environment for employee growth and retention. Cross pollination of divisional product technologies stimulates MPI product innovations that are meaningful to our precious customer base.

About MPI Advanced Semiconductor Test

MPI Advanced Semiconductor Test division offers a variety of manual, semi-automated and fully automated engineering probe systems, RF probes from 26 to 110 GHz, and unique RF calibration software QAlibria®. Major applications include RF & mmW, Device Characterization for modeling and process development, High Power, Wafer-Level Reliability, Failure Analysis, IC Engineering and Design Validation, Silicon Photonics, and more.

mpi-corporation.com/ast

List of Figures

List of Tables

List of Abbreviations

BEBC	Base-Emitter-Base-Collector
BEOL	Backend of Line
BiCIMOS	Bipolar Complementary Metal-Oxide-Semiconductor
BJT	Bipolar Junction Transistor
CBCBE	Collector-Base-Collector-Base-Emitter
CBE	Collector-Base-Emitter
CBEB	Collector-Base-Emitter-Base
CBEBC	Collector-Base-Emitter-Base-Collector
CMOS	Complementary Metal-Oxide-Semiconductor
CPW	Coplanar Waveguide
DUT	Device Under Test
FoM	Figure of Merits
GaAs	Gallium Arsinide
GaN	Gallium Nitride
GS	Ground-Signal
GSG	Ground-Signal-Ground
GSGSG	Ground-Signal-Ground-Signal-Ground
GSSG	Ground-Signal-Signal-Ground
HBT	Heterojunction Bipolar Transistor
IC	Integrated Circuit
InP	Indium Phosphide
ISS	Impedance Standard Substrate
ITRS	International Technology Roadmap for Semiconductors
LRM	Line-Reflect-Match
LRRM	Line- Reflect-Reflect-Match
PDK	Process Design Kit
RAM	Radiation Absorbing Material
RF	Radio Frequency
RRMT	Reflect-Reflect-Match-Thru
RTC	Resistor Thermal Coefficient
SG	Signal-Ground

SGS	Signal-Ground-Signal
Si	Silicon
SiGe	Silicon Germanium
SMU	Source Monitor Unit
SOLT	Short-Open-Load-Thru
SOLT	Short-Open-Load-Thru
SPICE	Simulation Program with Integrated Circuit Emphasis
TE	Transverse Electric (wave)
TEM	Transverse Electromagnetic (wave)
TM	Transverse Magnetic (wave)
TMR	Thru-Match-Reflect
TRL	Thru-Reflect-Line
UMSO	Unknown-Match-Sort-Open
VNA	Vector network analyzer

PART I

Development and Implementation

1

Motivation

1.1 Advances in Silicon Technologies for mm-Wave Applications

The continuously increasing demand for more content, services and security drives the development of high-speed wireless technologies, optical communication, automotive radar, imaging and sensing systems and many other mm-wave and THz applications. Commercialization, the need for highly integrated and cost-effective circuits and systems push the transition of mm-wave and sub-mm-wave devices and circuits from III-V materials to silicon that is more attractive for consumer market (e.g. [1]).

As Harame stated in [2], both the complementary metal oxide semiconductor (CMOS) and the bipolar CMOS (BiCMOS) technologies play an important role in the commercial RF and mm-wave applications. While the BiCMOS is favored for high-performance applications with low digital content, the price advantage for the high-volume production stimulates the further optimization of the RF CMOS for the needs of mm-wave applications (Figure 1.1).

Due to shrinking lateral device geometries and technology improvement, it became possible to achieve operation frequencies of half of THz for silicon-germanium SiGe:C heterojunction bipolar transistors (HBTs) [3]. Several ongoing European projects aimed to realize SiGe HBTs operating at a maximum frequency up to 0.7 THz as well as to integrate and to optimize back-end modules in an advanced 55 nm technology. This would enable a 0.5 THz 55 nm SiGe BiCMOS platform for design of RF, mm-wave and THz system-on-chip for commercial applications [4, 5].

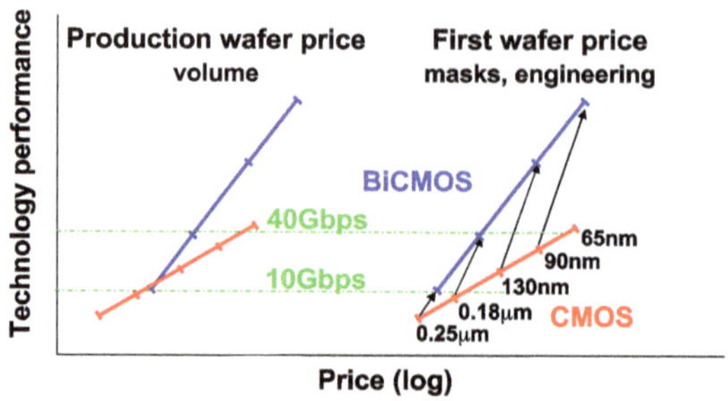

Figure 1.1 Performance vs. price for silicon technology generations. Picture from [2].

1.2 Challenges of RF Characterization of Advanced Devices

S-parameter measurement at mm-wave and sub-mm-wave frequencies plays a crucial role in the IC design debug. Most important, however, is the step of device characterization for development and optimization of device model parameters for new technologies. Accurate characterization of the intrinsic device in its entire operation frequency range becomes extremely important. It enables truthful extraction of the device model parameters radically decreasing IC design iterations and, as a result, the development time and cost. Thus, the need for accurate, reliable and easy-to-implement procedures for intrinsic device characterization at mm-wave frequencies and beyond drastically increases.

Calibrating S-parameter measurement systems is crucial even at relatively low frequencies of hundreds of megahertz: the wavelength of the test signal is of several orders of magnitude shorter than the equivalent length of the measurement signal path. The RF calibration is intended to remove systematic errors from the instrument hardware (and to take into account the presence of any accessories that may have been added to enable specific measurements to be performed) at the required frequencies for the measurements. For example, measurements may be required to be made in an on-wafer environment. In this case, first, cables need to be connected to the measurement instrument (vector network analyzer or VNA) front panel connectors, followed by coaxial adaptors, and finally, on-wafer probes. RF calibration will correct for the impact of these added components as well as correct the systematic errors in the instrument. This is why this type of calibration is often referred to as "error correction" (Figure 1.2).

The error correction procedure can also be imagined as a virtual shifting of the measurement reference plane (i.e. the "zero point" of the magnitude

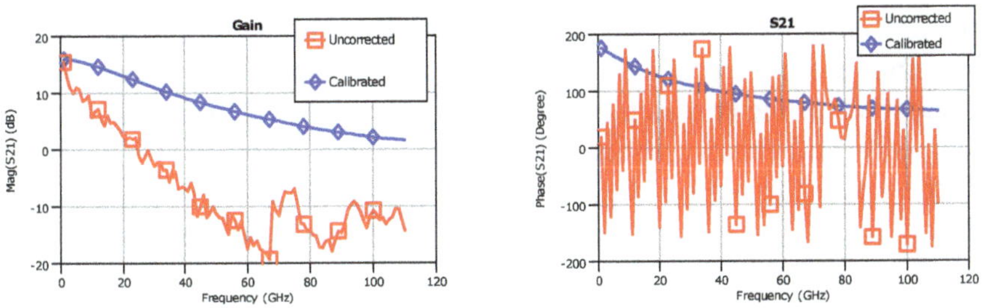

Figure 1.2 Uncorrected and calibrated measurements of gain of the same transistor.

and phase of the measurement signal) to the desired location. For example, the calibration procedure sets the measurement reference plane to the cable end for measurements in coaxial environment. The systematic measurement errors are defined using the set of coaxial calibration standards – the calibration kit. Conventionally, the characterization of the RF devices at the wafer level is done with respect to the probe-tip calibration. In this case, the measurement reference plane is set close to the end of the RF probes using planar calibration standards.

Vendors of RF probes provide probe-tip calibration standards and verification elements and guarantee accurate system calibration and calibration accuracy verification at the wafer level. A simple short-open-load-thru (SOLT) method that is available from the front panel of every VNA can successfully be used with the probe-tip standards at frequencies of up to 20 GHz at least [6]. Thus, calibration of the measurement system, the calibration validation as well as the sharing of the measurement data between different parties is relatively simple and is well established.

The RF probe-tip calibration provides the appropriate error correction for measurement of integrated circuits (ICs). The characterization of an intrinsic device, however, is associated with additional challenges: the parasitic impedance of the device contact pads, interconnect lines and via stack is of some order of magnitude greater than the input and output impedances of the device itself. The contact pad de-embedding procedure is a conventional solution of this problem. Usually, two special elements, the so-called de-embedding open and short dummies, are placed on the wafer close to the DUT. They are laid out so that they represent the parallel and the serial portion of the device parasitic impedances. Therefore, both parasitic components can be measured together with the DUT and then subtracted from the DUT measurement results in two steps: first its parallel portion followed by the serial portion. In other words, the pad de-embedding shifts the measurement reference plane from the probe tip further towards the device terminals.

The accuracy of the two-step de-embedding decreases with the frequency and becomes unsatisfactory above 40 GHz. As many publications already showed, the approximation of the device parasitic impedance over just two parallel and serial equivalent components is insufficient at mm-wave frequencies. The parasitic impedance becomes more complex in nature; the portion of its distributed component drastically increases. Though several advanced de-embedding methods already demonstrated reasonable accuracy improvement at mm-wave frequencies, they did not find a wide application on practice. Utilizing complex approaches for more accurate capturing of the Backend of Line (BEOL) parasitics, they require five, six or more dummy elements, measurement and post-processing steps. It significantly increases the complexity of the method and the difficulty of its practical implementation. The risk of errors is also increased, often leading to such common mistakes as under- and/or over-de-embedding. That is why the device model parameters are extracted for the 40 GHz (often even 20 GHz) measured and de-embedding using the two-step method, as International Technology Roadmap for Semiconductors (ITRS) describes in [7]. Such a compromise is justified by two factors:

1. It is a common understanding that the compact model parameters of a device such as, for instance, its junction capacitances should be constant over the wide frequency range;
2. The simplicity of the de-embedding step.

However, the extraction of another important device (and technology), Figure of Merits (FoM), suffers from this compromise: the unity current gain cut-off frequency and the unity power gain (the maximum oscillation frequency) (Figure 1.3). Both parameters depend on the device geometry and are extremely sensitive to the device parasitics. Because the solution for accurate characterization of the intrinsic device up to sub-THz frequencies (this is the range of performance of the modern advanced devices) is not yet available and are extrapolated from the data that are obtained much below 110 GHz.

It is still a subject of hard debates within the modeling engineer community which method of the extraction of and provides reliable results (e.g. [7, 9–12]). Different methods may show up to 30% variation of the extrapolated parameters. With any further progress step in the device performance, it becomes more difficult to justify the sub-THz values of that are

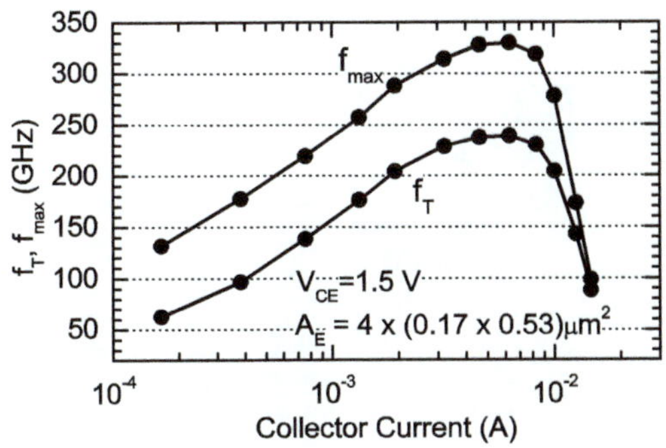

Figure 1.3 f_T and f_{MAX} of a high-speed SiGe BiCMOS HBT as a function of collector current extrapolated from h_{21} and the unilateral gain U at 40 GHz. Picture from [8].

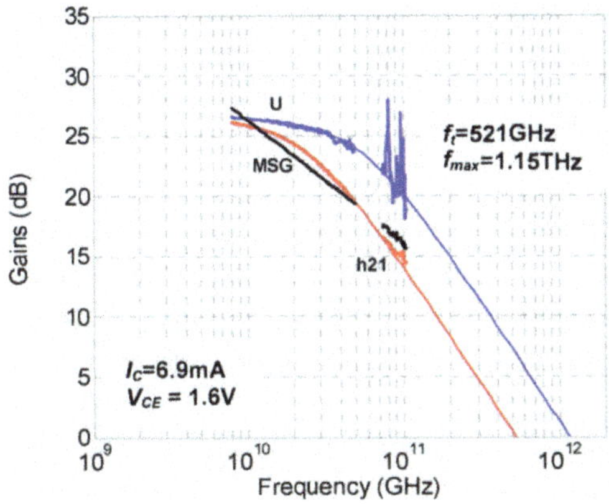

Figure 1.4 Extrapolated f_T and f_{MAX} of a 0.13 μm × 2 μm HBT from the 130 nm InP process. The device is measured up to 50 GHz and from 75 to 110 GHz. Picture from [13].

extracted from data measured below 65 GHz (e.g. [13, 14]). A good example of such a situation is shown in Figure 1.4. The values of the GHz THz were extracted from the 50 GHz data. Measurement results of the same device at frequencies from 75 to 110 GHz are less accurate and may lead to very different extraction results. Obviously, the need for accurate characterization of the intrinsic device at mm-wave and sub-mm-wave frequencies turned out to be more meaningful and critical for advanced technologies featuring sub-THz f_T and f_{MAX}.

Advanced mm-wave and sub-mm-wave III-Vs semiconductor technologies (e.g. Gallium Arsenide (GaAs), Gallium Nitride (GaN) and Indium Phosphide (InP)) do not utilize multiple metallization layers, typically only

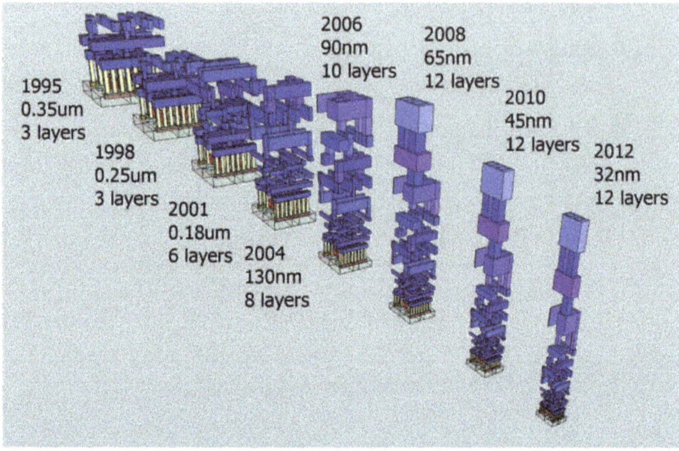

Figure 1.5 An inverter realized in different silicon technologies: from the 0.35 um to the 32 nm CMOS processes. The device layout is simulated based on the ITRS data. Picture is courtesy of Akira Tsuchiya (Kyoto University, Japan).

two (e.g. [15]). Technology steps mostly address vertical and horizontal scaling of a lateral device (e.g. [16]). That is why the measurement reference plane can be shifted from the probe tip close to intrinsic device terminals by either a simple two-step de-embedding or straightforward by implementing the *in-situ* calibration (e.g. [13, 17]).

A progress of the silicon technologies is more complex. Figure 1.5 shows the same simple inverter as being realized from the 0.35 um to 32 nm CMOS processes according to the ITRS data. Together with the downscaling of a lateral device, every next step of the silicon technologies adds more metal layers to the process BEOL. As a result, the extraction of reliable FoM using conventional methods becomes more and more challenging: the device performance level increases, its parasitic resistances and capacitances are reduced and the BEOL parasitics are drastically increased. The need for new calibration methods, that can sufficiently improve the characterization accuracy of the intrinsic device and, at the same time, will be simple for practical implementation, is drastically increasing.

1.3 Contribution of This Work

This work addressed the change of accurate mm-wave characterization of devices fabricated in advanced semiconductor technologies. It developed the *in-situ* calibration solution that is easy to be implemented for silicon technologies. The new technique was verified up to 110 GHz on three different processes: high-performance SiGe:C BiCMOS from IHP Microelectronics (Germany), BiCMOS9MMW from STMicroelectronics (France) and RF CMOS 8SF from IBM Microelectronics (USA). Practical results demonstrated that the proposed *in-situ* calibration significantly outperforms the convention method independently on the process specifics and complexity.

In this work, several RF calibration methods were analyzed, and it was shown that two of them were good candidates for the *in-situ* application: the multiline TRL[1] and the transfer TMR[2]. The accuracy of both methods was verified qualitatively and quantitatively. While the experimental data proved that both methods are comparable to each other from the measurement accuracy perspective, the multiline TRL was recommended for the technology development step, whereas the transfer TMR was suggested for the final production step. It was also shown how this *in-situ* calibration solution can be expanded to address the needs of multiport measurements.

A special attention was given to topics such as the following:

1. Specifics of wafer-level measurement and calibration caused by application of planar calibration elements as well as advantages and drawbacks of S-parameter calibration methods for the *in-situ* system calibration;
2. The optimal location of the *in-situ* calibration reference plane. The proposed solution addressed the needs of device modeling engineers, circuit designers and the characterization engineers, as well as the specific requirements rising at sub-mm-wave frequencies;
3. Design rules for the on-wafer calibration standards. The developed design recipe can be successfully applied for CMOS and BiCMOS silicon processed as well as on III-Vs semiconductors;
4. Definition of electrical properties of custom standards. Detailed analysis and the extensive number of experiments yielded to a list of simple recommendations for specifics of every calibration method. The variation of the standard properties caused by the instability of the fabrication process and the need for measurements at multiple temperatures was also addressed;
5. Verification of the calibration accuracy. The work presented results of the quantitative verification of the *in-situ* calibration accuracy using the

[1]Thru-Reflect-Line

[2]Thru-Match-Reflect

method proposed by the National Institute of Standards and Technology (NIST, USA). It also proposed a simple qualitative verification method that is simple for practical applications.

Some important aspects of the on-wafer S-parameter measurement assurance were presented as well. The discussion included the analysis of the calibration residual errors caused by the improper boundary conditions of coplanar calibration standards and the impact of the RF probe-tip design. In conclusion, some suggestions for further accuracy improvement of the proposed method were given.

2

Introduction

2.1 *S*-parameters of a Linear Network

Electrical behavior of a linear n-port device is usually described over its input and output impedance or admittance parameters (*S*- or *Y*-parameters, respectively). These parameters are defined as voltage V and current I relationship at input and output ports of the device connected to a generator and terminated to an open or a short circuit (Figure 2.1). With increase of operation frequency, the realization of an ideal open or short termination becomes more challenging. Moreover, many types of DUTs (e.g. filters, power amplifiers) cannot be measured in this way as shortcutting their output can change the electrical characteristics or even destroy the device. Instead, another set of parameters is commonly used to describe a behavior at high frequencies: the *scattering* or *S*-parameters. *S*-parameters define the relation between the incident, reflected and transmitted power waves at each device terminal under specified load conditions. *S*-parameters are significantly easier to be measured. If required, they can be converted to other parameters, such as Z, Y, T, H (e.g. [18]).

In [19], Kurokawa proposed a concept of incident a and b reflected power waves that simplified the analysis of microwave networks:

$$a_i = \frac{V_i + Z_i I_i}{2\sqrt{|\Re(Z_i)|}} \tag{2.1}$$

$$b_i = \frac{V_i - Z_i^* I_i}{2\sqrt{|\Re(Z_i)|}}, \tag{2.2}$$

where V_i and I_i are voltage and current flowing into the ith port of a junction, respectively. Z_i is the impedance of the ith port. Z_i^* is the complex conjugate of Z_i.

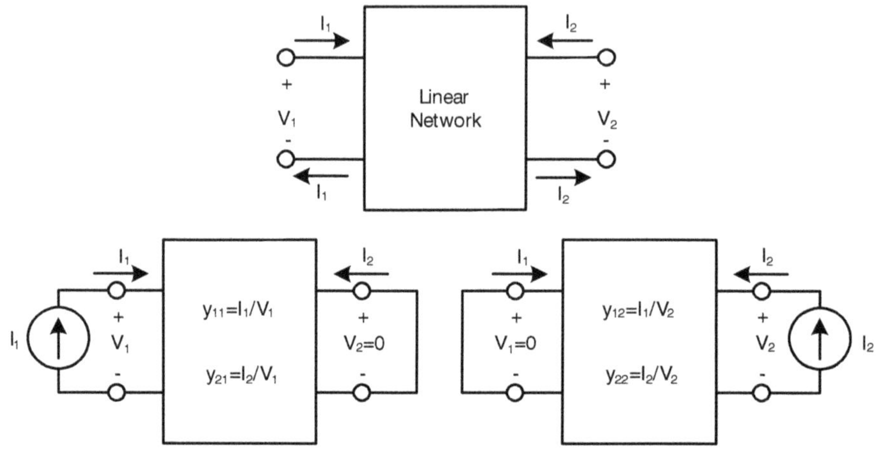

Figure 2.1 A linear network (top) and definition of its Y-parameters (bottom).

The ratio of a_i and b_i gives the power wave reflection coefficient:

$$s = \frac{b_i}{a_i} \tag{2.3}$$

For a linear n-port network with vectors a, b, v and i, the ith components are a_i, b_i, V_i and I_i at its ith port. So, a and b can be written in terms of v and i as

$$a = F(v + G_i) \quad b = F(v - G^{+}i), \tag{2.4}$$

where F and G are the diagonal matrices. Their ith diagonal components are $1/2\sqrt{|\Re Z_i|}$ and Z_i, respectively. G^{+} is the complex conjugate transposed matrix of G.

The linear relationship between v and i as well as b and a are given by

$$v = Z \cdot i \quad \text{and} \quad b = S \cdot a, \tag{2.5}$$

where Z is the impedance matrix and S is the power wave scattering matrix.

The i, jth element of the scattering matrix is

$$S_{ij} = \left.\frac{b_i}{a_j}\right|_{a_k = 0,\ k \neq j.} \tag{2.6}$$

An example for a two-port network is shown in Figure 2.2. S_{11} and S_{22} are its forward and reverse reflection parameters, while S_{21} is the transmission

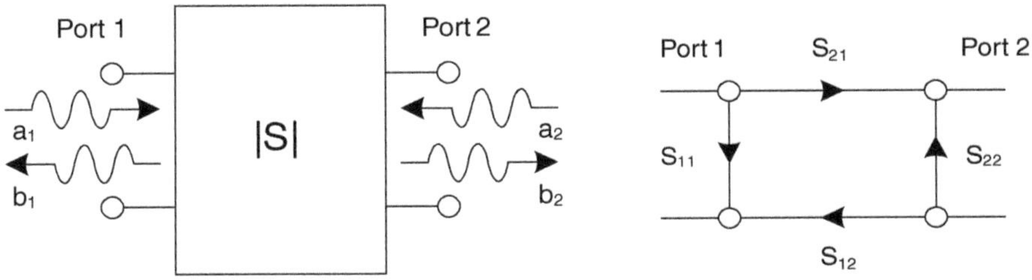

Figure 2.2 Definition of incident and reflected waves (left) and the signal flow graph of a two-port network (right).

parameter for forward (from port 1 to port 2) and S_{12} for reverse (from port 2 to port 1) directions, respectively. Often, S S-parameters of a network are presented as a flow graph (Figure 2.2, right).

2.2 Theory of a Uniform Waveguide Mode

Marks and Williams introduced in [20] an alternative set of parameters, the "pseudo-waves", that are more applicable for developing VNA calibration techniques at the wafer level. At the wafer level, the measurement reference impedance Z_{REF} and the characteristic impedance of the transmission media Z_0 often do not equal each other.

This section gives a short compendium of Marks' and Williams' General Waveguide Circuit Theory. It is essential for this work and will be regularly referred to in other sections.

2.2.1 Waveguide Voltage and Current

For given waveguide voltage v and waveguide current i, the transverse components of the total electrical and magnetic fields $\mathbf{E}$ and $\mathbf{H}$ can be represented by

$$\mathbf{E}_t = c_+ e^{-\gamma z}\mathbf{e}_t + c_- e^{+\gamma z}\mathbf{e}_t \equiv \frac{v(z)}{v_0} \cdot \mathbf{e}_t, \tag{2.7}$$

$$\mathbf{H}_t = c_+ e^{-\gamma z}\mathbf{h}_t + c_- e^{+\gamma z}\mathbf{h}_t \equiv \frac{i(z)}{i_0} \cdot \mathbf{h}_t, \tag{2.8}$$

where $\mathbf{e}_t$ and $\mathbf{h}_t$ are the transverse components of the electric and magnetic fields, respectively;

γ is the propagation constant composed of real α and imaginary β components as: $\gamma \equiv \alpha + j\beta$;

z is the longitudinal unit vector;

v_0 and i_0 are the normalization constants, so both v and v_0 have units of voltage and i and i_0 have units of current.

The fields $\mathbf{e}_t$ and $\mathbf{h}_t$ are in the normalized forward-propagating mode with the propagation constant γ, waveguide voltage $v(z) = v_0 e^{-\gamma z}$ and waveguide current $i(z) = i_0 e^{-\gamma z}$. For the normalized backward-propagating mode, the propagation constant is $-\gamma$, $u(z) = v_0 e^{+\gamma z}$ and $i(z) = -i_0 e^{+\gamma z}$.

2.2.2 Power

The net complex power $p(z)$ crossing a transverse plane S is given by

$$p(z) \equiv \int_S \mathbf{E}_t \times \mathbf{H}_t^* \mathbf{z} dS = \frac{v(z) i^*(z)}{v_0 i_0^*} p_0, \tag{2.9}$$

where

$$p_0 \equiv \int_S \mathbf{e}_t \times \mathbf{h}_t^* \mathbf{z} dS. \tag{2.10}$$

Imposing the constraint:

$$p_0 = v_0 i_0^* \tag{2.11}$$

the condition

$$p = vi^* \tag{2.12}$$

can be achieved with arbitrary choices of the normalization constants v_0 and i_0. Equations (2.9) and (2.12) can be simultaneously satisfied. When choosing v_0 or i_0 arbitrary, the other will be defined by Equation (2.11).

The average power flow $P(z)$ across the section S is given by the real part of $p(z)$ as

$$P(z) \equiv \Re(p(z)) = \Re\left(\mathbf{E}_t \times \mathbf{H}_t^* \mathbf{z} dS\right) = \Re(vi^*). \tag{2.13}$$

This leads to the conclusion that the power is *not* generally a linear combination of the forward and backward mode powers since it is given by a nonlinear expression. Consequently, the net real power P is, in general, *not* simply the difference of the power carried by the forward and backward modes.

2.2.3 Characteristic Impedance

The characteristic impedance Z_0 of the forward mode is defined by

$$Z_0 = \frac{v_0}{i_0} = \frac{|v_0|^2}{p_0^*} = \frac{p_0}{|i_0|^2}. \tag{2.14}$$

Definition (2.14) is an analogy to the electrical circuit theory. It is important to note that Z_0 is independent from the normalization of the modal fields **e** and **h**, which affect $|p_0|$. While the magnitude of Z_0 does depend on the choice of either v_0 or i_0, its phase is identical to that of p_0 and therefore independent of all normalizations. The phase of the characteristic impedance Z_0 is a fixed, inherent and unambiguous property of the mode.

Marks and Williams give an example of the special case of TE (transverse electric), TM (transverse magnetic) or TEM (transverse electromagnetic) modes in homogeneous matter to illustrate the close correspondence between their definition of Z_0 and the conventional definition of the characteristic impedance. Assuming that the waveguide impedance η is constant over the cross section, each mode has fields that satisfy

$$\mathbf{z} \times \mathbf{e}_t = \eta \mathbf{h}_t. \tag{2.15}$$

So, Z_0 is given by

$$Z_0 = \frac{|v_0|^2}{\int_S |\mathbf{e}_t|^2 \mathrm{d}S} \eta. \tag{2.16}$$

The last one proves that the phase of the characteristic impedance is equal to that of the wave impedance: the modal field $\mathbf{e}_t$ is normalized, the denominator is fixed and the magnitude of Z_0 depends only on v_0.

2.2.4 Equivalent Circuit Representation of Transmission Line

The four parameters that are used to describe a section of a transmission line are capacitance C, inductance L, resistance R and conductance G per unit length (Figure 2.3):

$$j\omega C + G \equiv \frac{\gamma}{Z_0}, \tag{2.17}$$

$$j\omega L + R = \gamma Z_0. \tag{2.18}$$

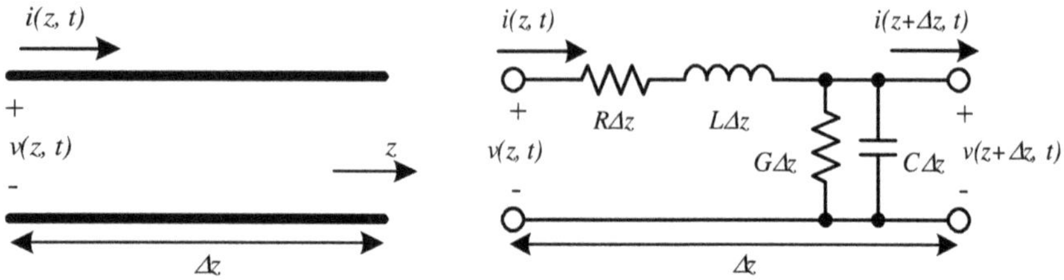

Figure 2.3 The voltage and current definitions and the equivalent circuit of an incremental length of a transmission line.

Also, these parameters often appear in equations satisfied by v and i (e.g. [18]):

$$\frac{\mathrm{d}v}{\mathrm{d}z} = (j\omega L + R)i \tag{2.19}$$

$$\frac{\mathrm{d}i}{\mathrm{d}z} = -(j\omega C + G)v. \tag{2.20}$$

From (2.17) and (2.18), the propagation constant γ and the characteristic impedance Z_0 of the line are defined as

$$\gamma = \sqrt{(j\omega L + R)(j\omega C + G)} \tag{2.21}$$

$$\gamma = \sqrt{(j\omega L + R)(j\omega C + G)}. \tag{2.22}$$

Both Equations (2.21) and (2.22) have two roots corresponding to the forward- and backward-propagation modes. Each mode has identical C, L, G and R, but opposite γ and Z_0. For the forward mode $\Re(Z_0) \geq 0$, and, hence, its γ can be calculated from either (2.17) or (2.18).

2.2.5 Effective Permittivity and Measurement of Characteristic Impedance

Effective permittivity $e_{r,eff}$ (also called effective relative dielectric constant) equates to the propagation constant γ of a TEM mode in a fictitious medium of permittivity $e_{r,eff}$ and permeability μ_0. It is defined as

$$e_{r,eff} \equiv -\left(\frac{c\gamma}{\omega}\right)^2, \tag{2.23}$$

where c is the speed of light in vacuum.

Using (2.21), we can define

$$e_{r,eff} = \frac{c^2}{\omega^2}\left(\omega^2 LC - RG - j\omega(LG+RC)\right). \tag{2.24}$$

In most cases, C, L, G and R are nonnegative, which leads to: $\Im(e_{r,eff}) \leq 0$. Typically, $\Re(e_{r,eff}) > 0$; however, it becomes negative in lossy lines at low frequencies when $RG > \omega^2 LC$.

Equation (2.22) can be re-written as

$$Z_0 = \frac{\sqrt{e_{r,eff}}}{c \cdot C\left(1 + \dfrac{G}{j\omega C}\right)} \tag{2.25}$$

which can be applied for the determination of Z_0 [21]. This relation will be used further to measure characteristic impedance of commercial and custom line standards.

2.3 Waveguide Circuit Theory

These results were essential for the development of Marks' and Williams' waveguide circuit theory. This section will first show how they defined the traveling and the pseudo-traveling waves for a single uniformed waveguide. Next, it will present the scattering and the pseudo-scattering parameters, impedance matrices and transformation of the reference impedance, as given in the General Waveguide Circuit Theory [20].

2.3.1 Traveling Wave

The forward and backward traveling waves a_0 and b_0 are derived by normalizing the forward and backward modes in (2.7) and (2.8):

$$a_0 \equiv \sqrt{\Re(p_0)}c_+e^{-\gamma z} = \frac{\sqrt{\Re(p_0)}}{2v_0}(v+iZ_0) \tag{2.26}$$

and

$$b_0 \equiv \sqrt{\Re(p_0)}c_-e^{+\gamma z} = \frac{\sqrt{\Re(p_0)}}{2v_0}(v-iZ_0) \tag{2.27}$$

for positive roots. Both a_0 and b_0 are independent of the magnitude of the modal field $\mathbf{e}_t$. The phases of a_0 and b_0 depend on the phase of $\mathbf{e}_t$ in the same way as the phases of c_+ and c_- do.

Assuming that $\Re(Z_0) \neq 0$, (2.26) and (2.27) can be re-written as

$$v(z) = \frac{v_0}{\sqrt{\Re(p_0)}}(a_0 + b_0) \tag{2.28}$$

$$i(z) = \frac{i_0}{\sqrt{\Re(p_0)}}(a_0 - b_0). \tag{2.29}$$

Therefore, the real power results from (2.13):

$$P(z) = |a_0|^2 - |b_0|^2 + 2\Im(a_0 b_0^*)\frac{\Im(Z_0)}{\Re(Z_0)}. \tag{2.30}$$

Equation (2.30) demonstrates that the real net power P crossing a reference plane is not equal to the difference of the powers carried by the forward and the backward waves. The only exception is when the characteristic impedance is real or when either a_0 or b_0 vanish.

The reflection coefficient Γ_0 is defined by

$$\Gamma_0(z) \equiv \frac{b_0(z)}{a_0(z)}. \tag{2.31}$$

The power can be expressed in terms of reflection coefficient as

$$P = |a_0|^2\left(1 - |\Gamma_0|^2 - 2\Im(\Gamma_0)\frac{\Im(Z_0)}{\Re(Z_0)}\right). \tag{2.32}$$

It is important to note that $|\Gamma_0|^2$ is not a power reflection coefficient and it may exceed 1 if Z_0 is not real.

2.3.2 Pseudo-Waves

For wafer-level measurements, it turned out to be more suitable to introduce *pseudo-waves* as an alternative set of parameters. Pseudo-waves are the mathematical artifacts defined for a specific reference impedance Z_{REF} which satisfies $\Re(Z_{REF}) \geq 0$. The complex pseudo-wave amplitudes a and b are defined as

$$a(Z_{REF}) \equiv \left(\frac{|v_0|}{v_0}\frac{\sqrt{\Re(Z_{REF})}}{2|Z_{REF}|}\right)(v + iZ_{REF}) \tag{2.33}$$

and

$$b(Z_{REF}) \equiv \left(\frac{|v_0|}{v_0}\frac{\sqrt{\Re(Z_{REF})}}{2|Z_{REF}|}\right)(v - iZ_{REF}). \tag{2.34}$$

Therefore, v and i can be described inversely as

$$v = \left(\frac{v_0}{|v_0|}\frac{|Z_{REF}|}{\sqrt{\Re(Z_{REF})}}\right)(a+b) \tag{2.35}$$

and

$$i = \frac{1}{Z_{REF}}\left(\frac{v_0}{|v_0|}\frac{|Z_{REF}|}{\sqrt{\Re(Z_{REF})}}\right)(a-b). \tag{2.36}$$

The root has to be positive for (2.33) through (2.36). This leads to the re-definition of the average power from (2.13) as

$$P = |a|^2 - |b|^2 + 2\Im(ab^*)\frac{\Im(Z_{REF})}{\Re(Z_{REF})}. \tag{2.37}$$

The pseudo-reflection coefficient Γ is defined by

$$\Gamma(Z_{REF}) \equiv \frac{b(Z_{REF})}{a(Z_{REF})} \tag{2.38}$$

and depends on Z_{REF}. Finally, the power P can be expressed over the reflection coefficient (from (2.32)):

$$P = |a|^2\left(1 - |\Gamma|^2 - 2\Im(\Gamma)\frac{\Im(Z_{REF})}{\Re(Z_{REF})}\right). \tag{2.39}$$

Obviously, the pseudo-waves are equal to the actual traveling waves when the reference impedance is set to be the characteristic impedance of the mode. However, this is not always convenient. So, in case of a lossy and dispersive planar transmission line, it makes more sense to use a constant reference impedance (such as $Z_{REF} = 50\ \Omega$) because the characteristic impedance varies significantly with the frequency [21, 22]. Thus, measurement results for $Z_{REF} = Z_0$ may be difficult to interpret.

Another convenient choice is to set Z_{REF} such that $b(Z_{REF})$ vanishes at a given point on the line. Thus, the pseudo-reflection coefficient cancels out at a particular z and for a particular termination. Many calibration methods use this approach and force the pseudo-reflection coefficients of some calibration standards (typically "match" or "load") to vanish. A good example is the conventional LRM [23]. It is often used in coaxial measurement environment and is hardly applicable at the wafer level for planar standards. This problem was recognized already at early stages of the wafer-level RF measurement technology and facilitated the development of dedicated methods (e.g. [24]). A special attention to this problem is given later in this work (see Chapters 3 and 4).

2.3.3 Scattering and Pseudo-Scattering Matrices

For a linear waveguide circuit with a typical multiport configuration, one can define the reference impedance Z^i_{REF} at each waveguide port i with Z^i_{REF} and Z^i_{REF} given by (2.33) and (2.34). Similar to (2.5), the vector of the outgoing pseudo-waves **b** is linearly related to the vector of incoming pseudo-waves **a** by a pseudo-scattering matrix **S**:

$$\mathbf{b} = \mathbf{S}\mathbf{a}. \tag{2.40}$$

Similarly, the scattering matrix $\mathbf{S}^0$ is defined as the relation of the vectors $\mathbf{a}^0$ and $\mathbf{b}^0$ (vectors of incoming a^i_0 and outgoing b^i_0 waves):

$$\mathbf{b}_0 = \mathbf{S}^0\mathbf{a}_0. \tag{2.41}$$

The pseudo-scattering matrix **S** is equal to the scattering parameter matrix $\mathbf{S}^0$ if $Z^i_{REF} = Z^i_0$ at each ith port.

For a one-port, the reflection coefficients Γ^0 and Γ are the single elements of scattering matrices $\mathbf{S}^0$ and **S**.

2.3.4 The Cascade Matrix

Often, it is more practical to operate with the so-called transmission "T" or cascade parameters. Similar to (2.40), the pseudo-cascade parameter matrix is given as

$$\begin{bmatrix} b_1(Z^i_{REF}) \\ a_1(Z^i_{REF}) \end{bmatrix} = T^{ij} \begin{bmatrix} a_2(Z^j_{REF}) \\ b_2(Z^j_{REF}) \end{bmatrix}, \tag{2.42}$$

where Z^i_{REF} and Z^j_{REF} are reference impedances of the ports i and j.

In analogy to (2.41), one can introduce the special notation T_0 to describe the cascade matrix:

$$\begin{bmatrix} b_{01} \\ a_{01} \end{bmatrix} = T^0 \begin{bmatrix} a_{02} \\ b_{02} \end{bmatrix}, \tag{2.43}$$

where $T = T^0$ if $Z^i_{REF} = Z^i_0$ for each ith port.

2.3.5 Change of Reference Impedance

The relationship between the pseudo-wave amplitudes for different reference impedances Z^n_{ref} and Z^m_{REF} results from Equations (2.33–2.36) as

$$\begin{bmatrix} a(Z^n_{REF}) \\ b(Z^n_{REF}) \end{bmatrix} = Q^{ij} \begin{bmatrix} a(Z^m_{REF}) \\ b(Z^m_{REF}) \end{bmatrix} \tag{2.44}$$

where

$$Q^{nm} \equiv \frac{1}{2Z^m_{REF}} \left| \frac{Z^m_{REF}}{Z^n_{ref}} \right| \sqrt{\frac{\Re(Z^n_{ref})}{\Re(Z^m_{REF})}} \begin{bmatrix} Z^m_{REF} + Z^n_{ref} & Z^m_{REF} - Z^n_{ref} \\ Z^m_{REF} - Z^n_{ref} & Z^m_{REF} + Z^n_{ref} \end{bmatrix}. \tag{2.45}$$

As stated in [20], (2.44) and (2.45) are an exact expression of the complex impedance transformation. Therefore, the pseudo-waves can be accurately referred to as impedance-transformed traveling waves.

Q^{nm} can be re-written for the complex impedance transformation of reflection coefficients:

$$Q^{nm} = \sqrt{\frac{1 - j\frac{\Im(Z^m_{REF})}{\Re(Z^m_{REF})}}{1 - j\frac{\Im(Z^n_{ref})}{\Re(Z^n_{ref})}}} \frac{1}{\sqrt{1 - \Gamma^2_{nm}}} \begin{bmatrix} 1 & \Gamma_{nm} \\ \Gamma_{nm} & 1 \end{bmatrix} \tag{2.46}$$

where

$$\Gamma_{nm} \equiv \frac{Z^m_{REF} - Z^n_{ref}}{Z^m_{REF} + Z^n_{ref}}. \tag{2.47}$$

Finally, the reflection coefficient is transformed by

$$\Gamma(Z^n_{ref}) = \frac{\Gamma_{nm} + \Gamma(Z^m_{REF})}{1 + \Gamma_{nm}\Gamma(Z^m_{REF})}. \tag{2.48}$$

Marks and Williams gave some useful examples of transferring reflection coefficients of ideal open, short and load standards. In fact, the open and short standards have a unique physical manifestation: their reflection coefficients are independent from the reference impedance.

2.3.6 Two-Port Reference Impedance Transformation

A simple way to transform the impedance for a two-port S-parameter matrix is to compute the associated cascade matrix T, transform it to the new reference impedance and convert results back to the S-parameter matrix. The important relationship of

$$\begin{bmatrix} b(Z^n_{ref}) \\ a(Z^n_{ref}) \end{bmatrix} = Q^{nq} \begin{bmatrix} b(Z^q_{ref}) \\ a(Z^q_{ref}) \end{bmatrix} \tag{2.49}$$

leads to

$$T^{pq} = Q^{pm} T^{mn} Q^{nq}. \tag{2.50}$$

The last equation shows the transformation path of the reference impedance of port 1 from Z^m_{REF} to Z^p_{ref} and for port 2 from Z^n_{ref} to Z^q_{ref}. This is the general case where both ports have different reference impedances. In most practical situations, both ports of a device use equal reference impedances. So, (2.50) simplifies to

$$T^{pp} = Q^{pm} T^{mm} Q^{mp} = \frac{1}{1-\Gamma^2_{pm}} \begin{bmatrix} 1 & \Gamma_{pm} \\ \Gamma_{pm} & 1 \end{bmatrix} T^{mm} \begin{bmatrix} 1 & -\Gamma_{pm} \\ -\Gamma_{pm} & 1 \end{bmatrix}. \tag{2.51}$$

The last equation is used very often when performing the transformation of wafer-level calibration results from calibration reference impedance to measurement reference impedance (see Chapters 3–5).

2.3.7 Load Impedance

The load impedance is defined as

$$Z_{LOAD} \equiv \frac{v}{i} \tag{2.52}$$

at the reference plane at which the electromagnetic waves propagates only as a single mode. The load impedance is independent from the reference impedance, similar to v and i. However, [20] states that Z_{LOAD} is not a unique property of a one-port itself. Unlike the low-frequency circuit theory, Z_{LOAD} depends on the fields of the mode upon it. Loading the same one-port on different waveguides or on different modes of the same waveguide may give very different results for Z_{LOAD}.

From previously defined Equations (2.34) and (2.52),

$$b(Z_{LOAD}) = 0 \quad \text{and} \quad \Gamma(Z_{LOAD}) = 0 \tag{2.53}$$

when the reference impedance Z_{REF} is equal to the load impedance Z_{LOAD}. In this case, the impedance looks like a perfect match and its reflection coefficient vanishes. The reverse statement is relevant for the system calibration and, unfortunately, is often misunderstood: assuming that the reflection coefficient cancels out when a certain load is connected to the waveguide, the reference impedance is effectively set to $Z_{REF} = Z_{LOAD}$. This may contradict the initial goal, especially if the load impedance is complex. Figure 2.4 impressively demonstrates such a case. Here, the reflection coefficient of the same open element (probes in air) is measured with respect to three different LRM calibration runs. Calibrations were made using the same set of standards but with different overlaps of the probes on the planar load of 0, 25 and 50 μm. This led to a broad variation of the equivalent reactance of the load standard and, finally, to different calibration reference impedances [25]. A similar case is when calibrating with asymmetrical load standard that is shown in Chapter 6.

From (2.53), (2.47) and (2.48), we get

$$\Gamma(Z_{REF}) = \frac{Z_{LOAD} - Z_{REF}}{Z_{LOAD} + Z_{REF}} \tag{2.54}$$

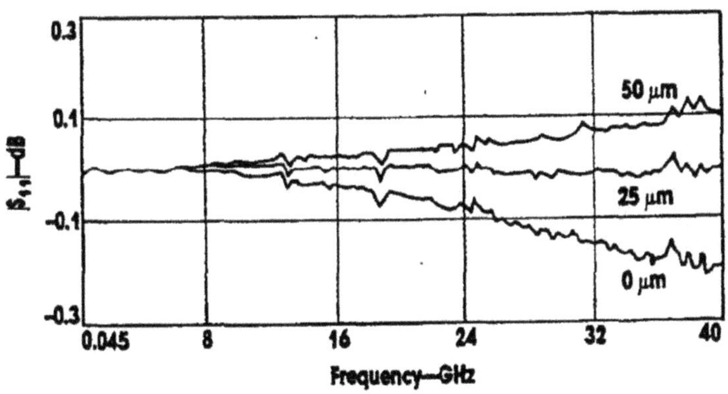

Figure 2.4 Reflection coefficient of the planar open standard corrected by probe-tip LRM calibration. The probe placement on the planar load standard affects the LRM calibration accuracy (reference impedance). Picture from [25].

or, when solved for Z_{LOAD}:

$$Z_{LOAD} = Z_{REF}\frac{1+\Gamma(Z_{REF})}{1-\Gamma(Z_{REF})}. \tag{2.55}$$

These two equations represent the general case. They are turning into the widely used original waveguide circuit theory equations (e.g. from [18]) when the reference impedance Z_{REF} is equal to the characteristic impedance Z_0.

Equation (2.55) was used to calculate the equivalent impedance of the custom load standard at the middle of a custom thru standard with the characteristic impedance Z_{LINE} different from the desired calibration reference impedance (see Chapters 3 and 4).

2.4 *S*-parameter Measurement

The measurement of RF and microwave characteristics of a network is a crucial part of the development, optimization and production cycles of high-frequency components and systems. The characterization of a linear DUT in the frequency domain is carried out by the vector network analyzer (VNA). Vector network analysis technique accounts over 50 years of history [26]. Starting from simple test sets assembled and manually operated by RF engineers in the late 1950s, it has been developed to highly sophisticated automated measurement systems that cover frequency ranges from a few tens of kHz up to THz (e.g. [27–29]).

The basic measurement concept implemented in modern VNA remains the same for all frequencies: it measures the incident a_i, reflected and transmitted b_i wave quantities at the ith port of the DUT (Figure 2.5).

In a general way, VNA allows measurement and computation of *S*-parameters of the DUT in a system reference impedance, typically at 50 Ω. The incident signals are generated by the internal source, while the reflected and transmitted signals are measured by receivers. The VNA has an internal switch to redirect the incident signal (positions I and II, Figure 2.5). So, the DUT can be automatically measured in both forward and reverse directions without reconnecting. The *S*-parameter matrix of the DUT is computed as

$$\begin{bmatrix} b_1 \\ b_2 \end{bmatrix} = \begin{bmatrix} S_{11} & S_{12} \\ S_{21} & S_{22} \end{bmatrix} \begin{bmatrix} a_1 \\ a_2 \end{bmatrix} \tag{2.56}$$

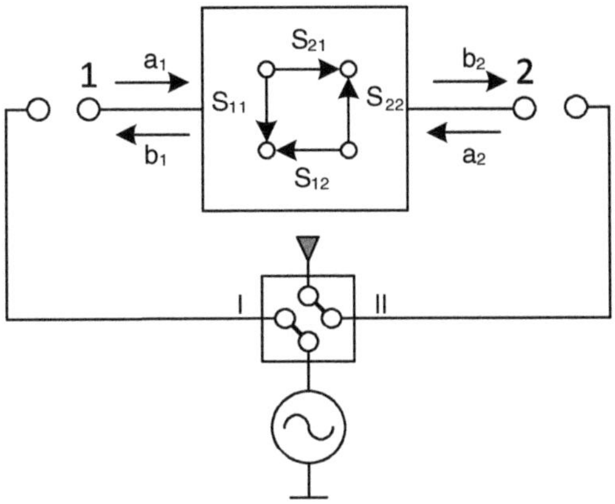

Figure 2.5 A basic measurement concept of S-parameters realized in modern VNA.

where

$$S_{11} = \left.\frac{b_1}{a_1}\right|_{a_2 = 0} \quad S_{21} = \left.\frac{b_2}{a_1}\right|_{a_2 = 0} \tag{2.57}$$

for forward direction (when $a_2 = 0$), and

$$S_{12} = \left.\frac{b_1}{a_2}\right|_{a_1 = 0} \quad S_{22} = \left.\frac{b_2}{a_2}\right|_{a_1 = 0} \tag{2.58}$$

for reverse direction (when $a_1 = 0$).

Modern VNAs realize different receiver concepts (reference channel and double reflectometer), have two and more measurement ports and can apply single-ended and true differential incident signals to the DUT (e.g. [30]).

2.5 Systematic Measurement Errors and *S*-parameter Calibration

In [31], I presented a detailed investigation of the history and the state-of-the-art VNA measurements and S-parameter calibration techniques (Attachment 12). My co-author, Nick Ridler from NPL (UK) added metrological aspects, such as the definition of measurement traceability, measurement assurance as well as clarification of the term “calibration” that, in fact, has multiple meanings for vector network analysis.

This research helped to form a strategy of how to approach the specific problem of the on-wafer calibration on silicon. It became clear that the calibration method should be based on a seven-term error model and should include a self-calibration step. The reflection coefficient of a high-reflective element (short or open) should be chosen as a free parameter as it is very hard to simulate or measure it for custom standards at mm-wave frequencies.

2.6 *S*-parameter Measurements at the Wafer Level

Characterizing the un-packaged DUT at the wafer-level requires a complex measurement system with several components: VNA, source monitor unit (SMU), the probe station, cables and wafer probes. The probe station provides the positioning of the test wafer with DUT, while the wafer probes convert the measurement signal from the 3D transmission media (RF cables or rectangular waveguides) to the quasi-2D coplanar DUT interface (Figure 2.6).

2.6.1 Design of Modern Wafer Probes

There are several configurations of probe tips: Ground-Signal (GS or SG), Ground-Signal-Ground (GSG) for single-ended measurements as well as various combinations of signal and ground contacts for dual/differential measurements (such as GSGSG, GSSG, SGS, etc.).

The wafer probe itself should carefully match the characteristic impedances Z_0 of the transmission media and provide proper conversion of the electromagnetic energy between different propagation modes. The conversion of the EM-field pattern is maintained by several RF transition

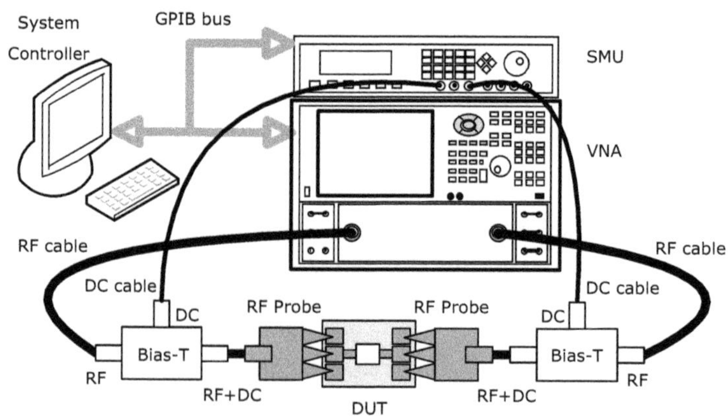

Figure 2.6 A schematic representation of a measurement system for device characterization at the wafer level.

steps within a single probe assembly. A conventional RF probe consists of the following parts:

1. Test instrumentation interface (coaxial or waveguide);
2. Transition from the test interface to the micro-coax cable;
3. Transition from micro-coax cable to a planar waveguide, such as CPW or microstrip;
4. Coplanar interface to a DUT on the wafer (or probe tip).

Several probes either combine steps 3 and 4 or do not use the micro-coax cable (Figure 2.7). A coaxial connector is a commonly used test system interface of RF probes below 65 GHz. Both coaxial and waveguide connections are possible interfaces for the frequency range from 50 to 110 GHz. Broadband measurement systems covering a frequency range from DC to 110 GHz in a single sweep utilize the smallest size (1 mm) coaxial connector [27–29]. Rectangular waveguides of different dimensions interface to the measurement system from above 110 GHz.

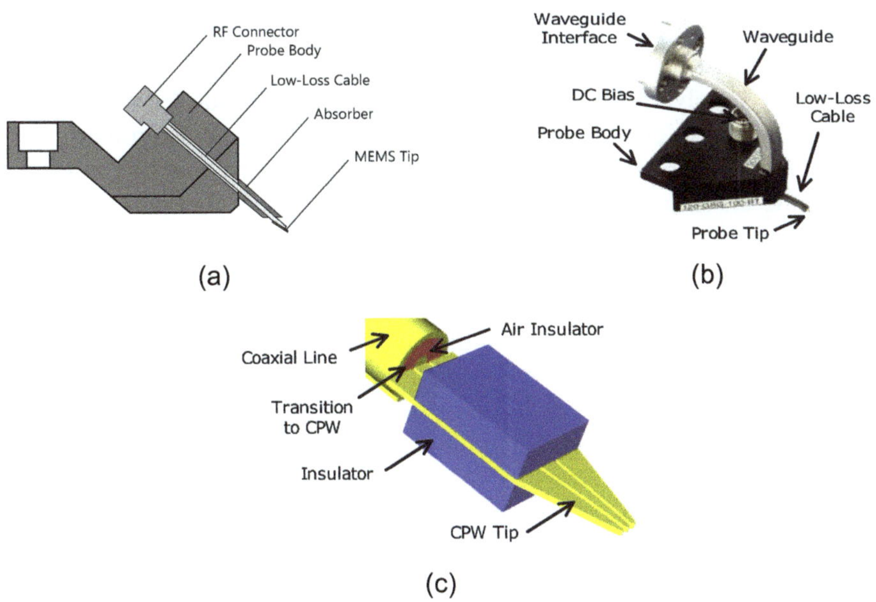

Figure 2.7 A concept of RF probe: based on a micro-coax cable for coaxial (Picoprobe, ACP, TITAN, top left) and waveguide (Picoprobe, ACP, Infinity) interface (top right) and direct transition from the coaxial to the coplanar line (|Z|-Probe) (bottom).

Recently, the 0.8 mm coaxial connector was introduced for a measurement system covering a band from 70 kHz to 145 GHz in a single sweep. The 0.8 mm connector-based RF probe is also underway [32].

At mm-wave frequencies, the rectangular waveguide TE_{10} EM fields can be converted into coplanar TEM-type EM fields either directly or in two steps. The two-step method applies a short section of a micro-coax cable and is common for RF probes up to 325 GHz. The use of a micro-coax cable is advantageous from the mechanical point of view. However, its relatively high insertion loss together with the impact of additional transition step reduces the overall probe electrical performance. The direct transition to a microstrip membrane tip [33] or a micro-machined silicon CPW [34] makes the micro-coax cable dispensable.

There are several approaches on how to design probe tips and how to attach them to the rest of the probe (Figure 2.8):

1. The signal conductor of the micro-coax cable is shaped out to form the signal tip. Ground blades of the tip are soldered from both sides of the cable (Picoprobe, GGB);
2. The air-isolated CPW tip is attached to the micro-coax cable (ACP, Cascade Microtech and TITAN Probes, MPI Corporation);
3. The flexible polyamide microstrip line, that ends with the CPW tips, is attached to the micro-coax cable (Infinity Probe, Cascade Microtech);
4. Direct transmission (without micro-coax cable) from the coaxial connector to the air-isolated CPW contacts (|Z|-Probe, Rosenberger-Cascade Microtech);
5. Direct transmission from the rectangular waveguide to a polyamide microstip line (500 GHz Infinity Probe, Cascade Microtech);
6. Direct transition from the rectangular waveguide to a micro-machined silicon CPW contact structure (DMPI, Figure 2.9).
7. This variety of design concepts roots in a tradeoff between mechanical and electrical requirements for probing on different contact pad materials. The interfaces, in particular the probe tip, bring discontinuities into the measurement signal path. As per [20], such a discontinuity causes the generation of higher-order propagation modes *per se*. Thus, wafer probes and DUT launches must support only a single quasi-TEM propagation mode and should exclude higher-order modes or exhibit a significantly higher impedance to them (e.g. [35]).

The impact of the probe design on the calibration residual errors and the measurement accuracy is discussed later in Chapter 9.

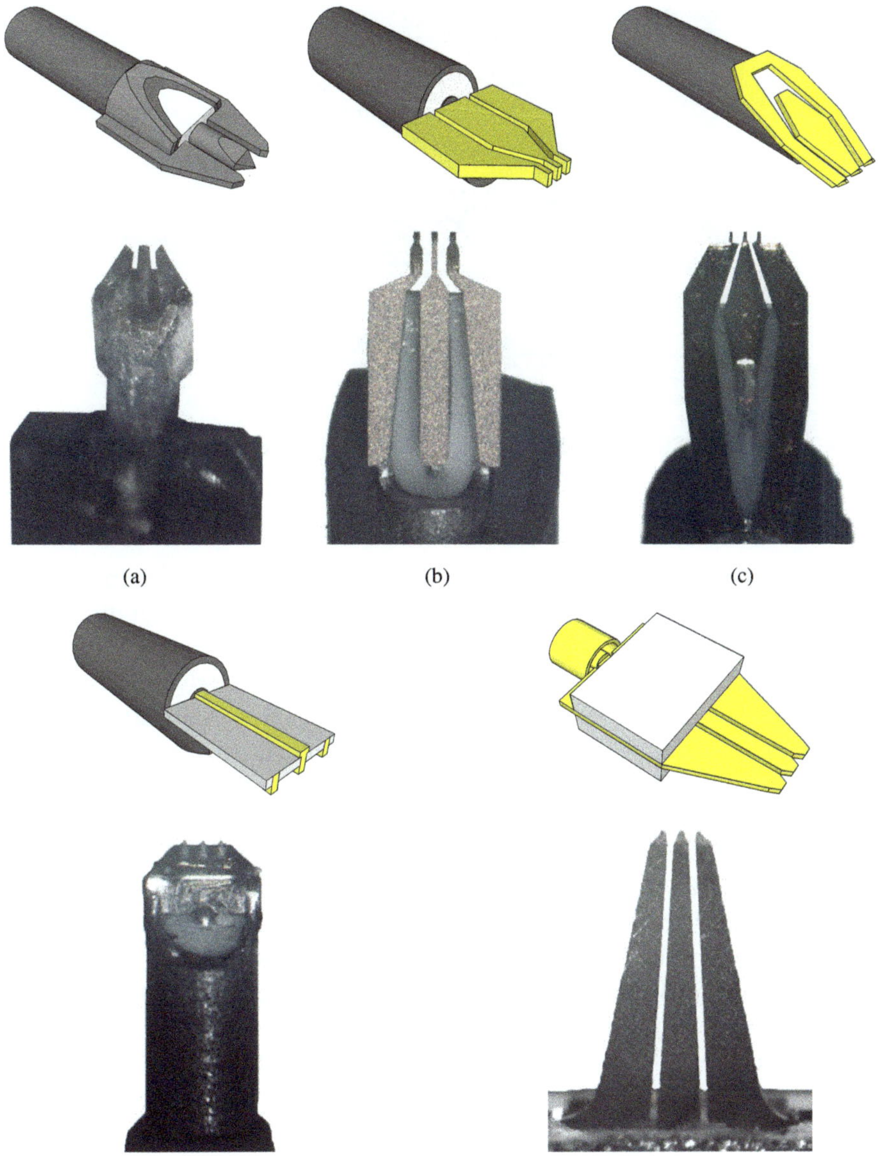

(a) (b) (c)

(d) (e)

Figure 2.8 Design concept and photograph of various Ground-Signal-Ground probe tips, bottom view, from left to right (top): (a) Picoprobe; (b) ACP; (c) TITAN Probe. On the bottom: (d) Infinity Probe; (e) |Z|-Probe. All photographs are of the same magnification.

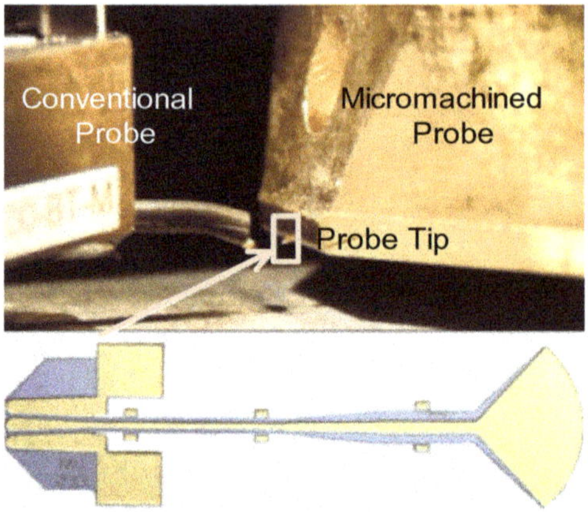

Figure 2.9 Ground-Signal-Ground probe tip of the DMPI probe. Picture from [16].

2.6.2 DUT Contact Pads and BEOL Parasitics

The size of an elementary device (i.e. a transistor) of advanced mm-wave technologies is just a few micrometers. It is not possible to contact it directly even using RF probes with the smallest available pitch (distance between signal and ground) of 50 μm. Moreover, active devices from advanced semiconductor processes are covered by many layers of BEOL metallization. The BEOL "device infrastructure" includes the probe contact pads, interconnect lines and via stacks. The impact of the BEOL parasitic impedance became a substantial problem for high-performance devices at higher measurement frequencies: the parasitics are several orders of magnitude larger than the intrinsic parameters of the device (see also Figure 1.5).

Therefore, the "intrinsic device" must be de-embedded from the surrounding parasitics of the BEOL and pads of a "total device". De-embedding is the second step of the calibration process. It uses special structures that are often called de-embedding "dummies". They are laid out similar to the device and, therefore, represent the DUT BEOL parasitics. Dummies are measured in the same series as the DUT. The measured parasitics are subtracted from the "total device" yielding the characteristics of the intrinsic device.

There are different ways and complexity orders of de-embedding depending upon the measurement application, the design of the DUT, measurement parameters and the frequency range. It can be applied to passive (inductors, capacitors and resistors) as well as active devices (transistors, diodes, etc.). Because substrate parameters (such as resistivity,

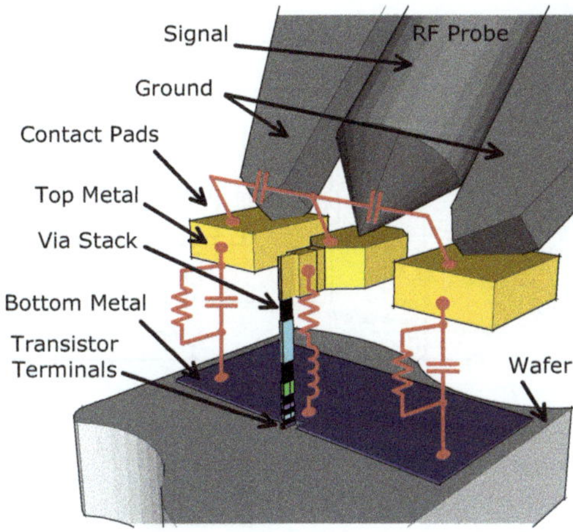

Figure 2.10 An example of a Ground-Signal-Ground pads and probe, and simplified presentation of the BEOL parasitic circuit elements of a DUT.

oxide thickness, etc.) vary across the wafer, dummy structures should be located near to the DUT.

The increase of the complexity of the semiconductor process, the operation frequency of the device and the demand for improvement of the accuracy of device characterization and modeling pushed the development of the de-embedding procedures. In 1987, Van Wijnen introduced the two-step open short de-embedding procedure for characterization of Si BJTs up to 18 GHz. Since then, semiconductor processes, device maximum operation frequencies and, consequently, de-embedding methodology have made a substantial progress. Today, many advanced de-embedding procedures have been proposed, such as the lumped-element-based five-step method (proven up to 50 GHz, [36]), a general 4-port de-embedding (proven up to 110 GHz, [37]), the scalable 6-step method (proven up to 110 GHz, [38]) or the transmission line based method (proven up to 170 GHz, [39]). Figure 2.11 compares how DUT backend parasitics are captured by different de-embedding algorithms over different equivalent circuits.

With the increase of measurement frequencies and the decrease of the size of an elementary device, de-embedding of the BEOL parasitic effects became a very complex process. It requires comprehensive equivalent circuits, more dummy elements as well as multiple measurement and calculation steps. Numerous additional sources of errors occur leading to common mistakes, such as under- or over-de-embedding.

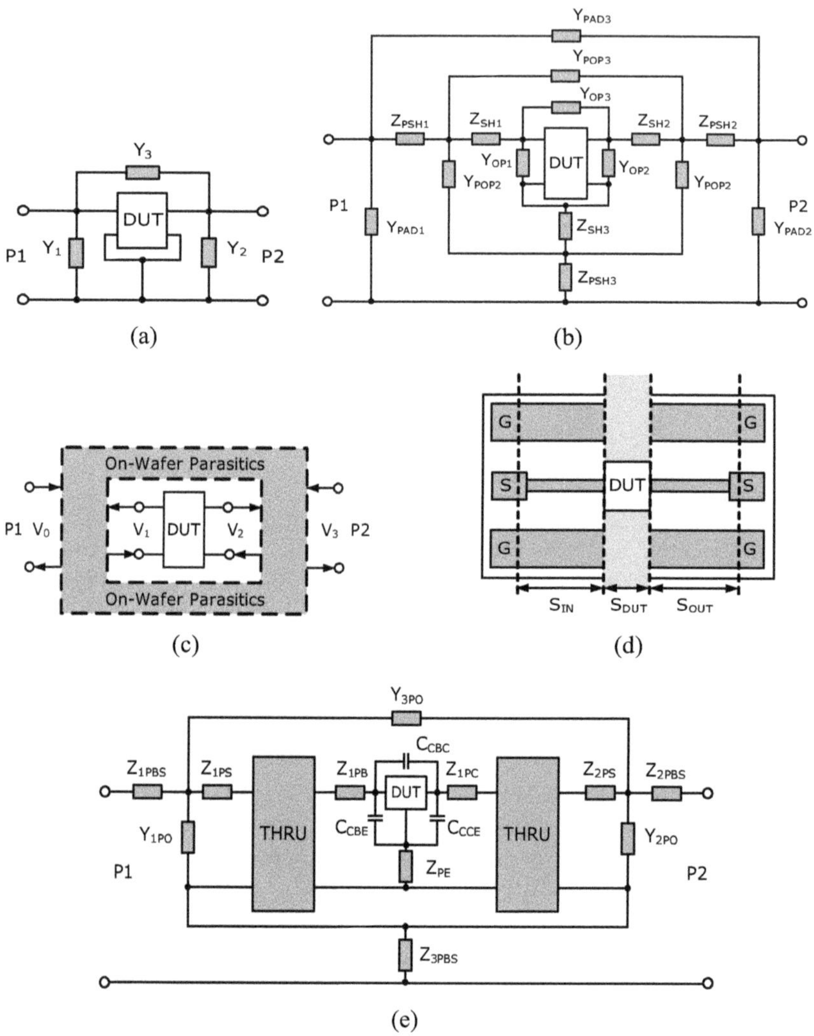

Figure 2.11 Representation of the BEOL parasitics by different de-embedding methods: two-step (a), five-step (b), general 4-four-port (c), transmission line-based (d) and the scalable six-step (e).

In Chapter 7, it will be shown that the implementation of the *in-situ* calibration approach captures the largest portion of the BEOL parasitics within one calibration step. The remaining parasitics are of lumped nature. They can easily be described by a simple model and calculated out using a two-step de-embedding method even at mm-wave frequencies.

2.6.3 Calibration Standards

For the wafer-level measurements, the "calibration standard" term is applied to an element with fully (or partly) known electrical characteristics, such as impedance Z of reflection elements and characteristic impedance Z_0 and propagation constant γ, for transmission (line) elements. These elements are often realized in the coplanar waveguide (CPW) design and accumulated on an alumina substrate, also called "calibration substrate" or impedance standard substrate (ISS; Figures 2.12 and 2.13). Such standards are used for the probe calibration process when the measurement reference plane is set near to the probe-tip end (Figure 2.14).

An alternative approach is to fabricate calibration standards together with the DUT on the same media. They share the same design with the DUT (microstrip or CPW) and are located close to it. Wafer-embedded reference elements are usually called "on-wafer" standards to emphasize the difference to the ISS standards, which are fabricated on a different media, or "off-wafer".

On-wafer calibration is a common practice for III-Vs semiconductor technologies. Implementing on-wafer standards on silicon is challenging and less common due to high substrate loss and its frequency-dependent conductivity. Figure 2.15 gives an example of the on-wafer TRL kit implemented on 65 nm CMOS process for characterization of a power amplifier from 140 up to 220 GHz [40].

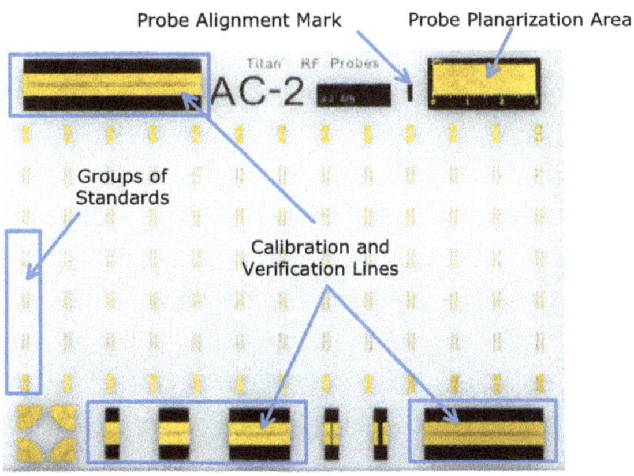

Figure 2.12 Example of commercially available calibration substrate that includes lumped-element impedance standards, calibration and verification lines, probe alignment marks as well as an area for probe planarization.

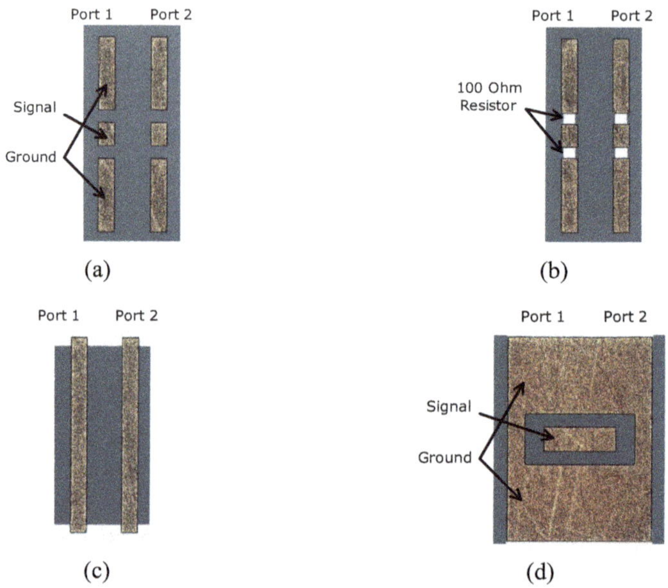

Figure 2.13 Example of commercially available off-wafer coplanar calibration standards (from left to right): paired opens, paired loads, paired shorts and the thru. Such standards are used for most popular wafer-level calibration procedures.

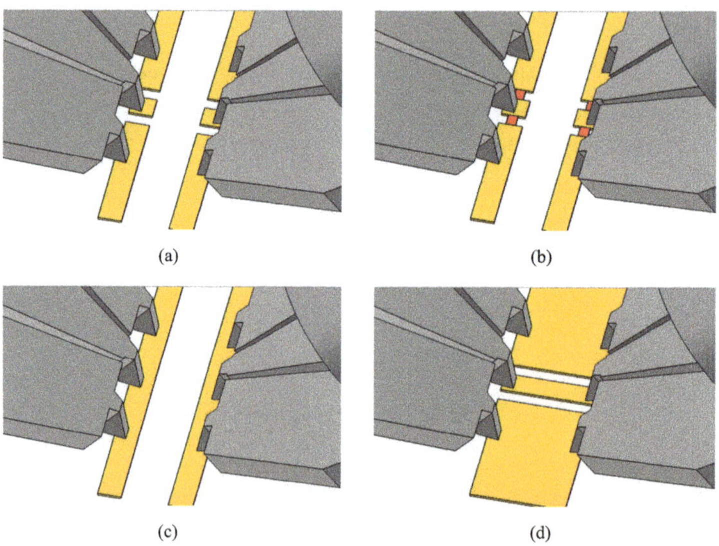

Figure 2.14 Example of the probe-tip calibration. RF probe contacting the CPW commercial standards: open (a), load (b), short (c) and line/thru (d).

Chapters 4–6 discuss the on-wafer implementation of the multiline TRL and the transfer TMR calibration methods for advanced CMOS and BiCMOS technologies. It is also shown that the electrical parameters of commercial calibration standards strongly depend on the material the calibration substrate was placed on (Chapter 9).

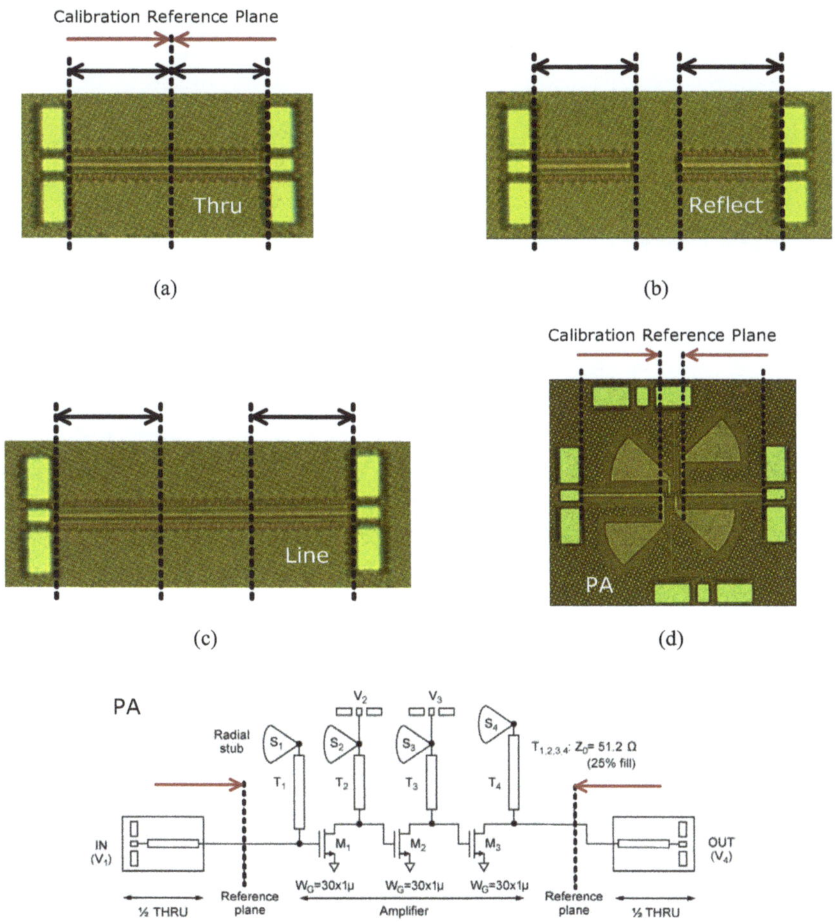

(e)

Figure 2.15 Example of the on-wafer TRL calibration kit design for measurements of CMOS PA: the thru (a), the reflect (b) and the line (c) calibration standards, the test power amplifier (d) and (e). Pictures are from [40].

3

Development of Calibration Solutions

3.1 Requirements

On-wafer RF measurement requires the shift of the S-parameter measurement reference plane either to the probe-tip end or to a certain virtual on-wafer location close to the terminals of the intrinsic device.

Limitations of planar calibration standards require special wafer-level calibration procedures of the VNA. Conventional methods that are developed for coaxial and/or waveguide environment may lead to significant calibration and measurement errors at the wafer level, especially at higher frequencies [41]. The need for accurate and consistent measurement results pushed engineers to develop dedicated wafer-level calibration techniques, such as LRRM [24], the NIST multiline TRL [42] and the translate LRM [43].

Silicon application, however, raises additional calibration challenges. The key requirements to on-wafer calibration on silicon are as follows

1. Accuracy. Sensitivity of a calibration method to non-ideal characteristics of custom standards limits its accuracy. In contrast to the probe-tip calibration, on-wafer calibration is an in-line step. It is hard or impossible to trim standards to the desired performance in the same way as the ISS elements. Calibration methods should be insensitive to non-ideal standards.
2. Minimal test chip size. Standards are embedded on every test chip together with devices. In contrast to the III–V's, silicon is mostly used for RF consumer applications. Minimal size of calibration chips is mandatory to save the on-wafer real estate and reduce test and manufacturing cost of the final product.
3. Same geometry with the DUT. The design of standards should match the geometry of the DUT (for elementary devices).

4. Reduction of the cost of test. Between the calibration and device measurement steps, there should be no operator interaction, like re-positioning of probes. Calibration should be run automatically and without a need for motorized RF positioners.

3.2 Calibration of a Two-Port VNA

Analyzing calibration theory and specifics of advanced silicon processes led to the conclusion that the transfer TMR self-calibration method was the best candidate for application at the device production environment, while the multiline TRL could serve as the benchmark, e.g. for the process development and integration cycle. Conventional LRM sets the calibration reference impedance to the impedance of the load standard. The impedance of a planar load is, in general, complex with the load resistance $R_{LOAD} \neq 50\ \Omega$ [43, 44]. In conclusion, the calibration reference impedance is complex, frequency-dependent and difficult to define. Interpretation of the S-parameters measured with respect to such calibration conditions is a very hard task (see Chapter 2).

The concept of TMR was presented by Eul and Schiek in 1988 [45]. Later, they showed how TMR can be derived from the general Txx algorithm [46]. The match standard of TMR is a double-one port that is perfectly matched to the system reference impedance. Thus, its reflection coefficient r_M should satisfy $r_M = 0$ in the calibration reference impedance and its pseudo-S-parameters should vanish (the situation that was discussed in Chapter 2). Such conditions usually hold for coaxial applications.

Obviously, another solution for TMR is needed to relax requirements to the match standard regarding two aspects:

1. Enable the use of complex impedance elements;
2. Abolition of $S_{X2,11} = S_{X2,22}$ ([47], Table 3.1).

TMR that enables such conditions is often called "transfer TMR" [48]. The impedance of the match standard can be defined at each measurement port and can be arbitrary. This is the most common scenario for a custom load standard implemented on a silicon test chip.

Table 3.1 Conditions of transfer TMR calibration

Standards	Requirements	Unknown	Terms	Product
Thru	Known: $S_{11}, S_{21}, S_{12}, S_{22}$	n/a	4	n/a
Match	Known: S_{11}, S_{22}	n/a	2	n/a
Reflect	$S_{11} = S_{22}$ known within ± 90 degree	S_{11} (S_{22})	1	$r_X, S_{11} = S_{22}$

To the best of my knowledge, the complete mathematical solution for the transfer TMR has never been published. That is why it was re-derived and is presented in the next section. This derivation is the key to the solution when the calibration reference impedance, the characteristic impedance of the thru standard as well as the impedances of the paired match are all different.

3.3 Transfer TMR Solution

Figure 3.1 shows the seven-term systematic error model of a two-port double reflectometer VNA. Here, m are the ideal measurement receivers, a and b are the incident and reflected/transmitted waves, respectively, $[A]$ and $[B]$ are the 2×2 matrices of error terms and $[T_x]$ is the DUT matrix of transmission T-parameters.

As it was demonstrated in [46], the relationship between the measured wave quantities m and the wave quantities at the DUT terminal a and b can be fully described by the matrix equation:

$$\begin{bmatrix} m_1 \\ m_2 \end{bmatrix} = \begin{bmatrix} A_{11} & A_{12} \\ A_{21} & A_{22} \end{bmatrix} \begin{bmatrix} b_1 \\ a_1 \end{bmatrix}. \tag{3.1}$$

It is important to note that matrix $[A]$ represents the so-called error box and does not correspond to any physical two-port element. As stated in [48], such a nomenclature is chosen for calculation convenience as well as for accordance with previously published materials.

For the case of a one-port measurement of a DUT with the reflection coefficient:

$$r_X = \frac{b_1}{a_1} \tag{3.2}$$

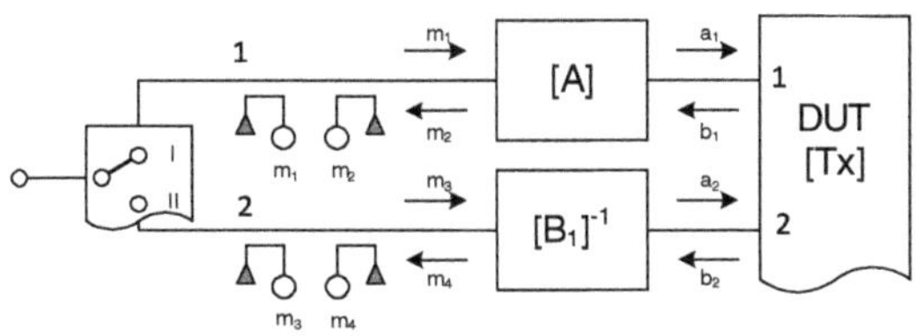

Figure 3.1 The seven-term systematic error model of a two-port double reflectometer VNA.

Equation (3.1) can be transformed to

$$\frac{m_1}{m_2} = \frac{A_{11} r_X + A_{12}}{A_{21} r_X + A_{22}}. \tag{3.3}$$

Equation (3.3) is the fundament of the one-port calibration and error correction.

It is easy to show that for the error correction step, elements of matrix $[A]$ can be defined up to one common constant factor α. The last allows normalization of the terms of $[A]$ to one arbitrary parameter. It is convenient to set the normalization factor to A_{22}. For simplification, the notation A_{ij} is further used for the normalized elements too. With this assumption, (3.3) can be re-written as

$$\frac{m_1}{m_2} = \frac{A_{11} r_X + A_{12}}{A_{21} r_X + 1}. \tag{3.4}$$

Similar to the first reflectometer, the second reflectometer can be described by

$$\begin{bmatrix} m_3 \\ m_4 \end{bmatrix} = \begin{bmatrix} B_{11} & B_{12} \\ B_{21} & B_{22} \end{bmatrix} \begin{bmatrix} a_2 \\ b_2 \end{bmatrix}. \tag{3.5}$$

The same assumption as for $[A]$ holds for the matrix $[B]$ too: it is a virtual two-port error box element. Addressing the way how it is connected, Equation (3.5) is linearly transformed to

$$\begin{bmatrix} a_2 \\ b_2 \end{bmatrix} = \begin{bmatrix} B_{11} & B_{12} \\ B_{21} & B_{22} \end{bmatrix}^{-1} \begin{bmatrix} m_3 \\ m_4 \end{bmatrix}. \tag{3.6}$$

Taking into account that the transmission matrix of a two-port DUT $[T_x]$ is given by

$$\begin{bmatrix} b_1 \\ a_1 \end{bmatrix} = \begin{bmatrix} T_{X11} & T_{X12} \\ T_{X21} & T_{X22} \end{bmatrix} \begin{bmatrix} a_2 \\ b_2 \end{bmatrix} \tag{3.7}$$

and with respect to (3.1) and (3.6), we get the mathematical description of a two-port measurement system:

$$\begin{bmatrix} m_1 \\ m_2 \end{bmatrix} = [A][T_X][B]^{-1} \begin{bmatrix} m_3 \\ m_4 \end{bmatrix}. \tag{3.8}$$

Equation (3.8) establishes the relationship between the actual DUT parameters, represented by the transmission matrix $[T_x]$ and the wave quantities m_i measured by the ideal VNA receivers over the error matrixes $[A]$ and $[B]^{-1}$ (Figure 3.1).

Modern VNAs automatically measure the DUT in both directions (Figure 2.5). Thus, (3.8) can be extended to

$$\begin{bmatrix} m_1' & m_1'' \\ m_2' & m_2'' \end{bmatrix} = [A][T_X][B]^{-1} \begin{bmatrix} m_3' & m_3'' \\ m_4' & m_4'' \end{bmatrix} \tag{3.9}$$

or, in short form,

$$[M_X] = [A][T_X][B]^{-1}, \tag{3.10}$$

where $[M_x]$ is the measurement matrix, given by

$$[M_X] = \begin{bmatrix} m_1' & m_1'' \\ m_2' & m_2'' \end{bmatrix} \begin{bmatrix} m_3' & m_3'' \\ m_4' & m_4'' \end{bmatrix} = [m_a][m_b]^{-1}. \tag{3.11}$$

The primed m_i' correspond to the forward, while double-primed m_i'' correspond to the reverse measurement directions. Equation (3.11) also proves that the bidirectional switch which leads the incident measurement signal is excluded from the measurement matrix reflectometer VNA architecture: it does not require a high-quality switch with high repeatability and isolation.

As it was shown in [46], the error-term matrixes $[A]$ and $[B]$ can be derived from three different standards:

$$[M_1] = [A][T_{X1}][B]^{-1} \tag{3.12}$$

$$[M_2] = [A][T_{X2}][B]^{-1} \tag{3.13}$$

$$[M_3] = [A][T_{X3}][B]^{-1}. \tag{3.14}$$

The first standard, the thru, is a perfectly matched, zero-length transmission element. Its transmission matrix is given by

$$[T_1] = \begin{bmatrix} 1 & 0 \\ 0 & 1 \end{bmatrix}. \tag{3.15}$$

Now, the measurement matrix for this standard according to (3.11) is

$$[M_1] = \begin{bmatrix} m'_{11} & m''_{11} \\ m'_{21} & m''_{21} \end{bmatrix} \begin{bmatrix} m'_{31} & m''_{31} \\ m'_{41} & m''_{41} \end{bmatrix} = [m_{a1}][m_{b1}]^{-1}. \tag{3.16}$$

From the measurements of the first (r_M) and second (r_R) reflection standards,

$$\frac{m'_{12}}{m'_{22}} = \frac{A_{11}r_{M1} + A_{12}}{A_{21}r_{M1} + A_{22}} = \Gamma_{A1} \tag{3.17}$$

$$\frac{m''_{32}}{m''_{42}} = \frac{B_{11} + B_{12}r_{M2}}{B_{21} + B_{22}r_{M2}} = \Gamma_{B1} \tag{3.18}$$

$$\frac{m'_{13}}{m'_{23}} = \frac{A_{11}r_{R1} + A_{12}}{A_{21}r_{R1} + A_{22}} = \Gamma_{A2} \tag{3.19}$$

$$\frac{m''_{33}}{m''_{43}} = \frac{B_{11} + B_{12}r_{R2}}{B_{21} + B_{22}r_{R2}} = \Gamma_{B2}. \tag{3.20}$$

Equations (3.17) and (3.19) can be re-written as

$$A_{11}r_{M1} + A_{12} - A_{21}r_{M1}\Gamma_{A1} - A_{22}\Gamma_{A1} = 0 \tag{3.21}$$

$$A_{11}r_{R1} + A_{12} - A_{21}r_{R1}\Gamma_{A2} - A_{22}\Gamma_{A2} = 0 \tag{3.22}$$

and (3.18) and (3.20) as

$$B_{11} + B_{12}r_{M2} - B_{21}\Gamma_{B1} - B_{22}\Gamma_{B1}r_{M2} = 0 \tag{3.23}$$

$$B_{11} + B_{12}r_{R2} - B_{21}\Gamma_{B2} - B_{22}\Gamma_{B2}r_{R2} = 0. \tag{3.24}$$

Using (3.15), we can replace elements of matrix $[B]$ in (3.23) and (3.24):

$$[B] = [M_1]^{-1}[A] \tag{3.25}$$

and obtain

$$A_{11}c_1 + A_{12}r_{M2}c_1 - A_{21}c_2 - A_{22}r_{M2}c_2 = 0 \tag{3.26}$$

$$A_{11}c_3 + A_{12}r_{R2}c_3 - A_{21}c_4 - A_{22}r_{R2}c_4 = 0 \tag{3.27}$$

where

$$c_1 = M_{1,22} + M_{1,21}\Gamma_{B1}$$

$$c_2 = M_{1,12} + M_{1,11}\Gamma_{B1}$$

$$c_3 = M_{1,22} + M_{1,21}\Gamma_{B2}$$

$$c_4 = M_{1,12} + M_{1,11}\Gamma_{B2}. \tag{3.28}$$

Equations (3.21), (3.22), (3.26) and (3.27) are the basis for the self-calibration step that calculates the originally unknown reflection coefficient r_R of the reflect standard. In a short form, they can be rewritten as

$$[C_R]\underline{[A]} = \underline{0} \tag{3.29}$$

where

$$\underline{[A]} = \begin{bmatrix} A_{11} & A_{12} & A_{21} & A_{22} \end{bmatrix}^T.$$

Except the trivial solution, when $\underline{[A]} = \underline{[0]}$ (what is out of our scope), Equation (3.29) holds if

$$det[C_R] = 0. \tag{3.30}$$

The last condition is used to find the missing parameter r_R. It is important to stress that

$$r_{R1} = r_{R2} = r_R \tag{3.31}$$

must be forced, while $r_{M1} = r_{M2}$ is not required. In fact, (3.30) can also be solved when

$$r_{M1} \neq r_{M2} \neq 0. \tag{3.32}$$

This is an essential condition that satisfies the initial requirements to the *in-situ* transfer TMR. The condition $r_M = 0$ does not need to be forced for all VNA ports (as discussed in Chapter 2).

Using (3.31) and (3.32), (3.30) yields

$$r_R = \frac{-k_1 \pm \sqrt{k_1^2 - 4k_0k_2}}{2k_2} \tag{3.33}$$

where

$$k_0 = r_{M1}r_{M_2}(a_2 - a_1) + (a_3 - a_2)$$
$$k_1 = (r_{M1} + r_{M2})(a_2 - a_1) + (a_3 - a_2)$$
$$k_2 = r_{M1}r_{M2}(a_3 - a_2) + (a_2 - a_1)$$

with

$$a_1 = c_2c_4 + c_1c_3\Gamma_{A1}\Gamma_{A2}$$
$$a_2 = c_1c_4\Gamma_{A2} + c_2c_3\Gamma_{A1}$$
$$a_3 = c_1c_4\Gamma_{A1} + c_2c_3\Gamma_{A2}.$$

The right solution for the sign in (3.33) can easily be found from the estimated phase of the reflection coefficient of the reflect standard. It is sufficient to know it within ± 90 degree for each frequency point.

Now, all seven parameters of standards are known and the calibration procedure can be executed as follows. Normalizing (3.17) to A_{22}, Equations (3.17) and (3.18) can be re-written as

$$-\tilde{A}_{11}r_{M1} - \tilde{A}_{21} + \tilde{A}_{21}\Gamma_{A1}r_{M1} = -\Gamma_{A1} \tag{3.34}$$

$$-\tilde{B}_{11} - \tilde{B}_{12}r_{M2} + \tilde{B}_{21}\Gamma_{B1} + \tilde{B}_{22}\Gamma_{B1}r_{M1} = 0 \tag{3.35}$$

where

$$\widetilde{[A]} = \begin{bmatrix} A_{11}/A_{22} & A_{12}/A_{22} \\ A_{21}/A_{22} & A_{22}/A_{22} \end{bmatrix} = \begin{bmatrix} \tilde{A}_{11} & \tilde{A}_{12} \\ \tilde{A}_{21} & 1 \end{bmatrix}$$

$$\widetilde{[B]} = \begin{bmatrix} B_{11}/A_{22} & B_{12}/A_{22} \\ B_{21}/A_{22} & B_{22}/A_{22} \end{bmatrix}.$$

From (3.19) and (3.20), two further equations can be derived:

$$-\tilde{A_{11}}r_R - \tilde{A_{12}} + \tilde{A_{21}}\Gamma_{A2}r_R = -\Gamma_{A2} \tag{3.36}$$

$$-\tilde{B_{11}} - \tilde{B_{12}}r_R + \tilde{B_{21}}\Gamma_{B2} + \tilde{B_{22}}\Gamma_{B2}r_R = 0. \tag{3.37}$$

The measurements of the thru give additional four equations:

$$\tilde{A_{11}} - M_{1,11} - M_{1,12}\tilde{B_{21}} = 0 \tag{3.38}$$

$$\tilde{A_{12}} - M_{1,11}\tilde{B_{12}} - M_{12}\tilde{B_{22}} = 0 \tag{3.39}$$

$$\tilde{A_{21}} - M_{21}\tilde{B_{11}} - M_{22}\tilde{B_{12}} = 0 \tag{3.40}$$

$$M_{21}\tilde{B_{12}} + M_{22}\tilde{B_{22}} = 1. \tag{3.41}$$

Equations (3.34) and (3.35), the combination of (3.36) and (3.37) and Equations (3.38–3.41) can be assembled in matrix form:

$$[C]\underline{[AB]} = \underline{[V]} \tag{3.42}$$

And finally,

$$\underline{[AB]} = [C]^{-1}\underline{[V]} \tag{3.43}$$

where

$$C = \begin{bmatrix} -r_{M1} & 1 & \Gamma_{A1}r_{M1} & 0 & 0 & 0 & 0 \\ 0 & 0 & 0 & -1 & -r_{m2} & \Gamma_{B1} & \Gamma_{B1}r_{M2} \\ -r & -1 & \Gamma_{A2}r_R & -1 & -r_R & \Gamma_{B2} & \Gamma_{B2}r_R \\ 1 & 0 & 0 & -M_{1,11} & 0 & -M_{1,12} & 0 \\ 0 & 1 & 0 & 0 & -M_{1,11} & 0 & -M_{1,12} \\ 0 & 0 & 1 & -M_{1,21} & 0 & -M_{1,22} & 0 \\ 0 & 0 & 0 & 0 & M_{1,21} & 0 & M_{1,22} \end{bmatrix} \tag{3.44}$$

$$\underline{[V]} = \begin{bmatrix} -\Gamma_{A1} & 0 & -\Gamma_{A2} & 0 & 0 & 0 & 1 \end{bmatrix}^T \tag{3.45}$$

and

$$\underline{[AB]} = \begin{bmatrix} \tilde{A_{11}} & \tilde{A_{12}} & \tilde{A_{21}} & \tilde{B_{11}} & \tilde{B_{12}} & \tilde{B_{21}} & \tilde{B_{22}} \end{bmatrix}^T. \tag{3.46}$$

3.3.1 Notes on Calibration Reference Impedance

Calibration condition (3.16) requires that the characteristic impedance of the thru Z_{LINE} must be fully matched with the calibration reference impedance $Z_{REF} = 50\,\Omega$, i.e. $S_{1,11} = S_{1,22} = 0$. While it is natural for coaxial applications, satisfying this requirement remains very challenging at the wafer level [21].

Adopting the TMR for a more general situation, such as $S_{1,11} = S_{1,22} = 0$, significantly complicates the derivation of the solution. An alternative way was proposed in [43] and then discussed in detail in [20]. This approach is used at NIST for a lumped-standard-based calibration, widely known as "NIST LRM" or "Translate LRM". The calibration calculations are proceeded as usual, resulting in matrices [A] and [B]. At this step, the calibration solution is defined for the reference impedance Z_{REF} equal to the characteristic impedance Z_{LINE} of the thru. Then, [A] and [B] are transformed to any target reference impedance Z_{REF}, using Equation (2.51). However, the characteristic impedance of the thru (i.e. the source reference impedance) must be known.

3.3.2 Calibration with a Thru of a Finite Length

Due to the TMR requirement (3.16), the thru is often called the "postulated standard" of a zero length. Its measurement is nothing else than looping back the measurement system to itself. RF probes cannot be connected together the same way as coaxial cables. Instead, the wafer-level thru is realized as a short line of a given length l, the propagation constant γ and the characteristic impedance Z_{LINE} (Figure 2.13). Hence, and under the previously discussed condition $Z_{REF} = Z_{LINE}$, its transmission matrix $[T_1]$ is

$$[T_1] = \begin{bmatrix} e^{-\gamma l} & 0 \\ 0 & e^{+\gamma l} \end{bmatrix}. \tag{3.47}$$

Again, replacing (3.15) by (3.47) would make all further calculations very bulky. An alternative way was proposed by [46]: the calibration steps can be performed in the same way as for the TMR procedure, yielding matrixes [A] and [B]. In this case, the calibration reference plane is set to the center of the thru standard. If required, it can be further moved to any arbitrary position. For instance, the following transformation sets the calibration reference plane to the end of the thru standard:

$$[\hat{A}] = [A][L]^{-1}, [\hat{B}] = [B][L] \tag{3.48}$$

where

$$[L] = \begin{bmatrix} e^{-\frac{1}{2}\gamma l} & 0 \\ 0 & e^{+\frac{1}{2}\gamma l} \end{bmatrix}. \tag{3.49}$$

3.3.3 Notes on the Probe-Tip Reflection Standards

Often, the impedance of the match standard r_{M1} and r_{M2} is given for the measurement reference plane at the end of the thru (e.g. at the probe-tip end). This is a common situation for commercial alumina calibration substrates. As I will show later, this is the most appropriate approach for the *in-situ* standards.

For this case, both r_{M1} and r_{M2} need to be recalculated as

$$r_0 = r_M e^{\gamma l}, \tag{3.50}$$

where r_M is the reflection coefficient of the match defined at the end of the thru. l and γ are, respectively, the length and the propagation constant of the thru (Figure 3.2). Furthermore, the impedance of the match r_0 should be transferred from the source impedance (typically $Z_{REF} = 50\ \Omega$) to the reference impedance $Z_{REF} = Z_{LINE}$ using (2.55).

A similar correction is required for the reflect standard. Its reflection coefficient r_R is given by (3.33) at the calibration reference plane, i.e. at the middle of the thru. This should be taken into account when determining the right sign of the root solution. Often, r_R is used for a quick qualitative

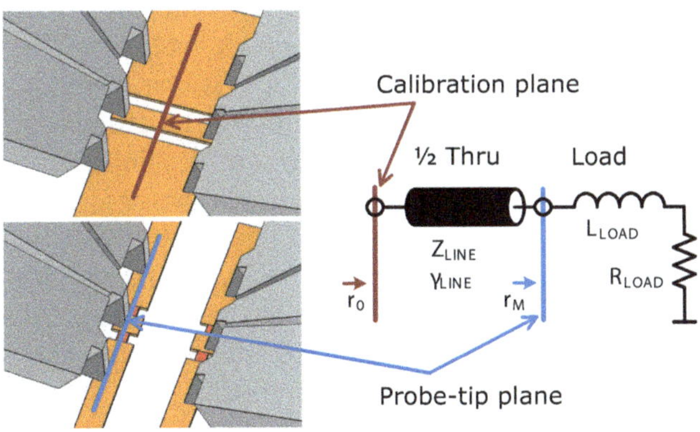

Figure 3.2 Definition of the match reflection coefficient at the probe-tip end (r_M) and at the middle of the thru standard r_0 for the transfer TMR.

verification of the TMR calibration accuracy (some examples of it are given in Chapter 6). In this case, r_R needs to be transferred to the more common reference position, e.g. at the end of the thru, using

$$r'_R = r_R e^{-\gamma l}. \tag{3.51}$$

3.4 Verification of the Transfer TMR

While the concept of the transfer TMR has been known for years, the quantitative accuracy verification of this method has never been investigated. I developed a verification methodology and presented the results in [49] together with Ralf Doerner from FBH (Berlin, Germany) who made all required measurements. We shared the calculation and the data analysis.

In this work, we proved the capability of the transfer TMR algorithm to accurately set the calibration reference impedance to the value of 50 Ω using two highly asymmetrical impedance elements which are used as match standards. Our experiments showed a very good agreement of the transfer TMR with the benchmark NIST multiline TRL up to 110 GHz (Attachment 13.1).

3.5 Multiport Calibration

Some special device characterization applications require multiport calibration and measurement techniques (e.g. [50]). Also, multiport calibration is used for measurements of differential passives and IC's.

I reviewed multiport calibration theory in [31]. For the *in-situ* calibration, the most promising approach, the RRMT method, integrates the 10-term-based algorithm with self-calibration techniques. It was presented for the first time in [51] (Attachment 13.2). My co-author Prof. Holger Heuermann (FH Aachen, Germany) developed the mathematical solution for the generalized RRMT procedure. In this work, I was responsible for collecting the measurement data and data analysis. Thanks to Steffen Schott (SUSS MicroTec, Germany) I had access to a 4-port VNA Ballmann S208 built on double-reflectometer architecture. It is important to note that in 2004, it was the only multiport VNA realizing this advanced architecture.

The RRMT procedure was developed specifically for the probe-tip calibration. The algorithm tolerates the reflection coefficients of the open and short standards. The partly unknown characteristics are calculated from the self-calibration step, based on the two-port transfer TMR (3.17–3.33). It is repeated twice: for the short and for the open. The match and thru standards must be fully known.

RRMT implemented two TMR self-calibration steps between measurement ports 1–2 and 3–4. The algorithm required all thru standards to be fully known, including the loop-back and cross-over elements (Figure 3.3). All thru standards were assumed to have a characteristic impedance of 50 Ω and to be lossless. As it was demonstrated in the paper, such a simplification was valid for the standards implemented in the multiport CSR-family of calibration substrates. Obviously, this might not be the case for on-wafer scenario or for higher frequencies.

The RRMT calibration was extended to integrate the unknown-thru USMO self-calibration steps from [52]. Now, the USMO self-calibration step could be used between measurement ports that require the loop-back and the cross-over thru elements (Table 3.2). The concept was proved on the 40 GHz multiport measurement system [53] (Attachment 13.3). For this paper, I planned the entire work, derived the mathematical solution, acquired measurement results, did the calculations and wrote the paper.

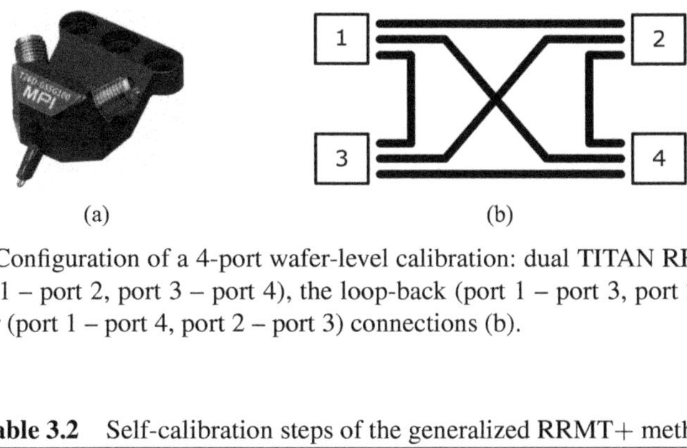

(a) (b)

Figure 3.3 Configuration of a 4-port wafer-level calibration: dual TITAN RF Probe (a), the straight (port 1 – port 2, port 3 – port 4), the loop-back (port 1 – port 3, port 2 – port 4) and the cross-over (port 1 – port 4, port 2 – port 3) connections (b).

Table 3.2 Self-calibration steps of the generalized RRMT+ method

Ports	1–2	3–4	1–4	2–3	2–4	1–3
Self-calibration steps	TMR	TMR	UMSO	UMSO	UMSO	UMSO
Self-calibration product (S-parameters)	Open, Short	Open, Short	Thru	Thru	Thru	Thru

My two co-authors, Prof. Holger Heuermann (FH Aachen, Germany) and Steffen Schott (SUSS MicroTec, Germany) provided scientific and management supervision, respectively.

The new calibration method received the commercial name "generalized RRMT+". It was further implemented in the commercial calibration software SussCal Professional from SUSS MicroTec and is patented under US7768271 [54]. The transfer TMR method was also implemented in SussCal Professional under the commercial name LRM+.

3.6 Conclusion

In this chapter, the key requirements to *in-situ* calibration methods were determined taking into account capabilities and limitations of modern silicon-based processes for fabrication of custom standards. It was also shown that, in this context, the transfer TMR is an appropriate method.

Mathematical solution for the transfer TMR was derived. The capability of this method to use arbitrary and asymmetrical impedance elements was proven using the NIST methodology. For the first time, the accuracy of the TMR was quantitatively verified for the case of application of highly asymmetrical loads. It was proved that the method is comparable with the reference NIST multiline TRL.

Also, the calibration solution RRMT+ was developed as the extension of the transfer TMR for multiport applications. It was demonstrated that the RRMT+ outperforms the reference multiport method and is also suited for *in-situ* implementation. During this work, I was in charge of implementing both calibration methods in the commercial calibration software SussCal Professional.

For further improvement, the error model shall be extended including crosstalk terms into it. It is of particular interest at sub-THz frequencies. While such extension goes beyond the scope of this work, it is important to note that the implementation steps of such calibration methods on a custom test chip will be similar to those discussed in the next chapters.

4

Design of On-Wafer Calibration Standards

For implementing the transfer TMR and the multiline TRL methods on the high-performance silicon processes, I initiated a close cooperation with engineering groups from IHP Microelectronics (Germany), IBM Microelectronics (Essex Junction, VT, USA) and STMicroelectronics (Crolles, France). We developed and proved concepts for on-wafer calibration for the bipolar SiGe:C process from IHP (published in [48]), the CMOS 8SF process from IBM (published in [55]) and the BiCMOS9MMW process from STMicroelectronics (published in [56–58]).

For these projects, I framed the design requirements and worked with every engineering group to adopt them to specifics of the selected process. Standards were laid out by in-house designers. IBM and IHP performed in-house measurements. I measured test chips from STMicroelectronics at facilities of Cascade Microtech GmbH (Germany), University of Dresden (TUD CEDIC, Germany) and Ferdinand Braun Institute for High Frequencies (FBH, Berlin, Germany). I performed calibration calculations, analysis of the results, planned all papers and wrote the related sections.

4.1 Design Tradeoff

RF probes and commercial calibration standards are always provided by the same vendor. Therefore, it is the responsibility of the probe vendor to optimize the design of commercial standards and to guarantee accurate probe-tip calibration. Electrical characteristics of planar commercial standards ensure that the calibration reference plane is accurately defined and that the calibration reference impedance $Z_{REF} = 50\ \Omega$.

In contrast, customized calibration standards should satisfy significantly more requirements balancing the design tradeoff:

1. The measurement reference place should be located as close as possible to the terminals of the intrinsic DUT. This will allow excluding all backend parasitics from the DUT measurement results;
2. Elementary devices (for example, transistors, spiral inductors, etc.) significantly vary in their geometries: short, long, single-finger, multi-finger, etc. There is not only one common reference plane, but one reference plane per geometry;
3. There are three common definitions of the device terminals (location of the measurement reference plane):
    (a) Input and output of an intrinsic device. This reference plane is requested by modeling engineers;
    (b) A certain location at the top metal level. This reference plane is requested by circuit designers;
    (c) At device contact pads (probe-tip end). It is a common view of characterization engineers on every device;
4. For both MOS and BiCMOS processes, only the top metal layers are suitable for broadband RF passives, e.g. calibration standards;
5. The lumped load element requires a resistor. Typically, resistors are made from poly-silicon and are located directly on the substrate.

Balancing these requirements, the following design rules for custom calibration standards for silicon processes were stated:

1. Design distributed standards (thru and lines) as well as reflect (both open and short) at the top metal layer;
2. Use poly-silicon for the load resistors. In this case, the equivalent impedance of the load is given by the load resistance and the impedance of via stack (load reactance);
3. Set the calibration reference plane at the top metal level and at a port distance corresponding to the widest intrinsic DUT;
4. When required, move the measurement reference plane to the DUT terminals (metal 1) using an appropriate de-embedding method. The S-parameters of the de-embedding elements should be corrected with respect to the on-wafer calibration, e.g. to the measurement reference plane that is located at the top metal level;
5. Implement additional elements for quantitative verification of the calibration accuracy. As it will be shown later, a serial attenuator can serve

as a simple verification element for the transfer TMR. A symmetrical pad open de-embedding element can also be chosen for the qualitative accuracy verification of both TMR and TRL methods (when short element was used as the calibration reflect).

The TMR standards were added on IHP's BiCMOS process test chip. IBM and STMicroelectronics processes allowed implementing both multi-line TRL and transfer TMR test chips. The test chip of every process included de-embedding elements. The type of these elements as well as their design were defined according to in-house rules and device characterization methodology applied by each partner.

Next, the most extensive set of calibration and de-embedding elements implemented on STMicroelectronic's BiCMOS9MMW process will be presented. Then, the specifics of custom standards from other processes will be discussed.

4.2 Design of RF Contact Pad and Device Interface

Design rules for RF DUT contact pads have been well established from the very beginning of the wafer-level RF device characterization (e.g. [59–61]). The basic requirement is that the pad configuration should support the Ground-Signal-Ground (GSG) RF probes if the measurement frequency exceeds 10 GHz. The size of an individual pad is subject to optimization under the following criteria:

1. Provide minimal pad parasitic capacitance;
2. Reduce the pad to the minimal possible size to save the on-wafer "real-estate" for devices;
3. Support mechanical capabilities of preferred RF probe technology. The pads should be long enough to keep the probe in contact when it skates forward due to overtravel;
4. Address specifics of contact pad metallization material. The required probe overtravel on Al pads is greater than that on soft materials, such as Cu and Au. In conclusion, the probe forward skate is longer; thus, the pads should be larger;
5. Allow multiple re-probing of DUT, e.g. for characterization across multiple temperatures.

Considering these five criteria, the RF contact pads are laid out as shown in Figure 4.1(b). They support probes with 100 μm pitch and have a probe-to-probe distance of 150 μm. It is the same spacing value as it is used

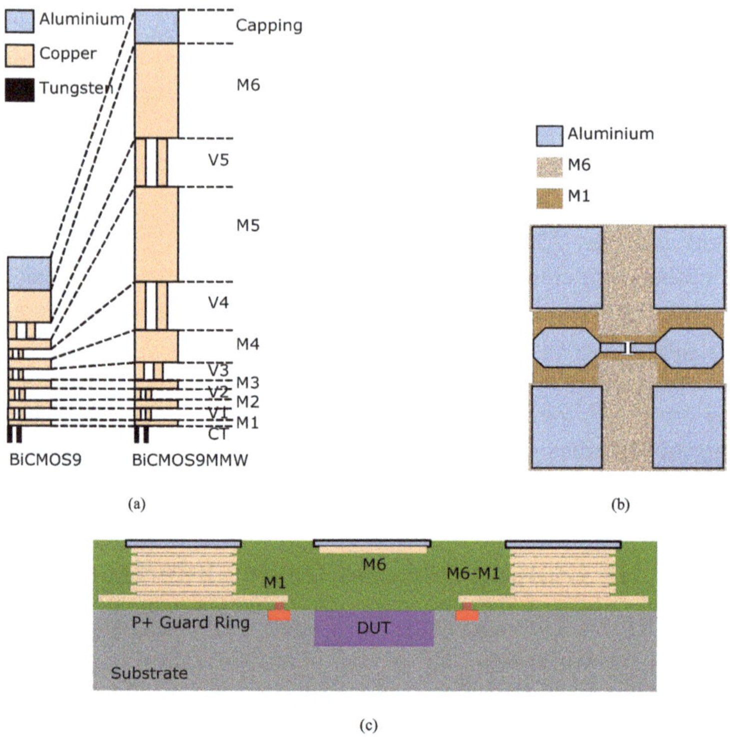

Figure 4.1 Schematic view of the BEOL cross section of the BiCMOS9 and BiCMOS9MW technologies (a), the RF contact pad design (b) and the cross section of the DUT interface (c).

on conventional commercial alumina calibration substrates. It enables an automated measurement and allows calibrating to the probe tip without probe repositioning.

The BiCMOS9MMW process uses aluminum for pad metallization. Aluminum oxidizes quickly under atmospheric conditions. It builds up an Al_2O_3 layer that is hard to break through. As a result, the probing on Al pads requires the probe overtravel from 25 to 75 μm (e.g. [62]) ensuring reliable and repeatable contact. Taking into account positioning accuracy and repeatability of conventional automated probe systems as well as mechanical capabilities of the Infinity Probe technology from Cascade Microtech (the in-house preferred probe design), the pad dimensions are laid out as 100 × 85 μm and 40 × 85 μm for the ground and the signal pads, respectively (see Table 4.1). Additionally, the signal pads are shaped out in an octagon-like form to further reduce the pad parasitic capacitance.

Table 4.1 Contact pad design of the BiCMOS9MMW

	Signal, Ground Length	Ground Width	Signal Width	Probe Pitch	Signal-to-Ground Gap	Probe Spacing
Parameter, μm	85	100	40	100	25	150

The interface to the device realizes the grounded coplanar design with a metal 1 ground shield and a meshed ground continuously spread over the via stack from the top metal 6 layer (M6) down to metal 1 (M1, Figure 4.1). The ground plane eliminates the losses caused by the substrate resistivity. Also, it makes the parasitic of the signal pads purely capacitive and facilitates the de-embedding up to high frequencies.

Transistors are laid out in the common collector configuration, i.e. the emitter and the substrate are shorted to the ground preventing high-frequeny resonances. In order to reduce substrate effects, all transistors have a P+ guard-ring at a minimum distance of the N-epitaxial well that is connected to the ground shield [57].

4.3 Location of the Reference Plane

As it was already briefly noted above, characterization, modeling and design engineering communities have different points of view on the "device". The location of the characterization reference plane is still a subject of debate: at the probe tip, at the lateral device terminals or at one of the levels of the BEOL. Up to now, the following procedures have prevailed as the industry's standard:

1. The system is calibrated to the probe-tip end by the SOLT method;
2. The BEOL parasitics are de-embedded using the open and short contact pad dummy elements [8];
3. The device RF Figures of Merits (FoM) are extracted at relatively low frequencies (typically at 20 GHz).

Device characterization engineers favor calibration of the measurement system to the probe-tip end using commercially available alumina calibration substrate (Figure 4.2). Dealing with optimized and characterized vendor calibration standards simplifies the task of system calibration. It also allows quick verification of the calibration accuracy with the help of additional reference elements available from the calibration substrate. Still, this approach may lead to a serious over- or under-estimation of FoM of sub-THz f_T/f_{MAX} HBTs [63].

Design engineers prefer SPICE models that include a minimum of BEOL parasitics and allow realistic pre-layout simulations (schematic mode). Advanced transistors showing very high f_T/f_{MAX} have junction capacitances of merely several femto Farads. Here, any small parasitics may strongly affect the transistor's FoM. Therefore, the top metal location (e.g. M6, Figure 4.2) is preferred. For designers, it is important that the process design kit (PDK) includes such SPICE libraries.

In contrast, modeling engineers aim to extract "physics-based" parameters of the device compact models. Therefore, the reference plane has to be shifted as close as possible to the intrinsic device terminals (e.g. M1, Figure 4.2). The requirements for device modeling are more challenging.

Designing *in-situ* broadband transmission lines (a crucial standard for the TRL and the TMR calibrations) close to the DUT terminals (e.g. M1 or M2 levels) is problematic for silicon processes. The first metal layers are not suitable for mm-wave passives. A more reasonable approach is using the top metal level (M6 for the BiCMOS9MMW, Figure 4.1) that facilitates good RF performance of the line [64]. Therefore, the optimal position of the *in-situ* calibration plane is at the top metal level M6.

If required, the on-wafer calibration reference plane can be shifted from the M6 to the M1 level by a conventional de-embedding approach. Because the parasitic impedance of the contact pads and interconnect lines are already calibrated out (included in the systematic measurement error model), the equivalent impedance of the de-embedding elements should represent the parasitic impedance of the M5-M1 via stack that is mostly of pure lumped nature. Measurement results later proved this statement (see Chapter 7).

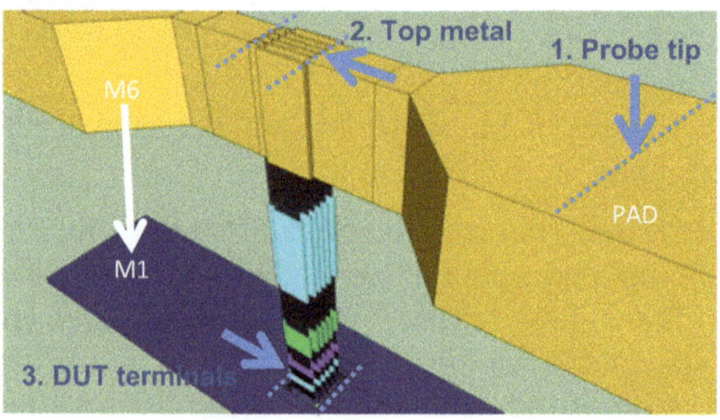

Figure 4.2 Three different locations of the reference plane: probe tip, top metal, DUT terminals for a complete open dummy. Example for the CBEBC HBT.

Table 4.2 Location of the reference plane

Calibration	Probe Tip	Top Metal	DUT Terminals
Probe tip	Straight	De-embedding of contact pad interconnect line parasitics	De-embedding of the entire DUT infrastructure
On-wafer	N/A	Straight	De-embedding of M6-M1 parasitics

For the probe-tip calibration, the de-embedding elements should capture the equivalent impedances of the contact pads and the entire DUT BEOL parasitics. Thus, the reference plane can be moved from the probe tip to the top metal M6 level or to the intrinsic DUT terminals at the M1 level (Table 4.2). This is challenging especially for small device geometries and at frequencies above 40 GHz [38].

4.4 Calibration Standards

A full set of standards was laid out addressing requirements for the multiline TRL and the transfer TMR calibration methods (Figure 4.3). Additional elements were placed on the test chip as well (e.g. a symmetrical attenuator).

Figure 4.3 Test chip of the STMicroelectronics' Si/SiGe:C BiCMOS9MMW process technology. The test chip includes the transfer TMR and the multiline TRL calibration standards. Reproduced with permission of STMicroelectronics.

4.4.1 Thru and Line

The thru and the line standards are designed as $Z_{LINE} = 50\ \Omega$ M1-shielded grounded coplanar waveguides with a signal width of about 10 μm and a gap to the grounds of about 9.5 μm (Figure 4.4). The implementation of the M1 shield minimizes the field penetration to the substrate and reduces the line conductance per unit length.

The effective length of standards is: 51, 549, 1545 and 3039 μm (Table 4.3). It is optimized to cover the frequency range from 1 to 110 GHz. The effective length of standards is found with the help of the approach presented in [65] and realized in Findlen.[1]

The ideal TRL calibration conditions are provided when the effective phase delay for a selected pair of lines is 90°. Consequently, the multiline

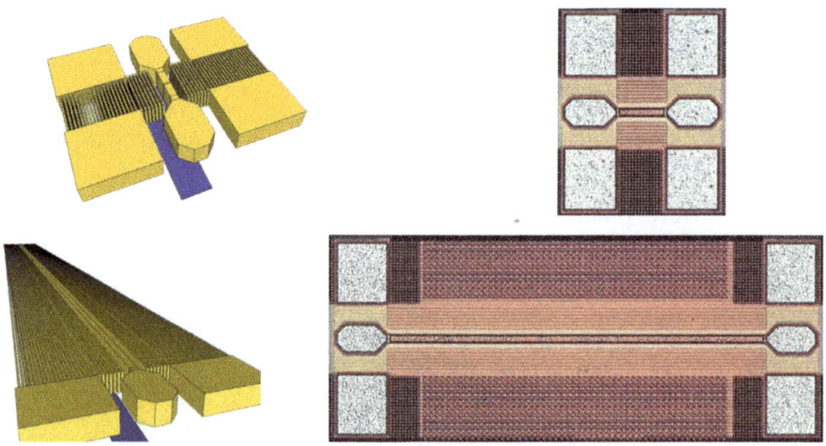

Figure 4.4 3D schematic view and a photograph of the thru (top) and line (bottom) calibration standards.

Table 4.3 The length of the line standards embedded on the BiCMOS9MMW chip

Standard	Design Length, μm	Equivalent Length, μm
Thru	51	0
Line 1	549	499
Line 2	1545	1494
Line 3	3039	2988

[1] A program available from NIST, Bolder CO, USA.

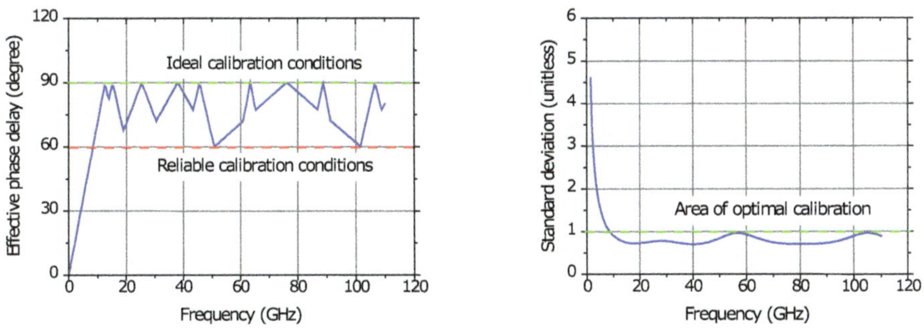

Figure 4.5 The effective phase delay (left) and the multiline TRL standard deviation (right) calculated for the designed calibration kit.

TRL is ill-conditioned if the phase delay is 0°. The effective phase delay for the designed calibration kit is in a range from 60° to 90° for frequencies from 10 to 110 GHz and thus ensures reliable calibration results (Figure 4.5, left).

For optimized conditions, the standard deviation parameter calculated by NIST Mutical program should not exceed the value of 1 across the frequency range of interest. Figure 4.5 (right) shows the standard deviation calculated for the designed calibration kit and proves that it meets this requirement from 10 to 110 GHz.

Taking into account the results for the effective phase delay and standard deviations, the multiline TRL can be further considered as accurate reference calibration for the frequency range from 10 to 110 GHz.

After analysis of device geometry variations (short one-finger and long multi-finger transistor from the BiCMOS9MMW design kit), we decided to set *in-situ* calibration reference plane to 7 μm away from the center of the thru. This requirement was considered when designing the reflect (short) and the match (load) standards.

4.4.2 Short

The high-reflective symmetrical reflect standard is designed as two shorting bars located 7 μm away from the center of the thru standard (Figure 4.6). This is the only difference from the thru standard.

The S_{11}, S_{21} and parameters are shown in Figure 4.7. The results are measured with respect to the on-wafer (M6) multiline TRL and compared against the probe-tip (ISS) short standard. The custom short is less reactive. As expected, the port-to-port crosstalk is about 15 dB higher compared to the short from the ISS. The *in-situ* shorts are much tighter to each other with only

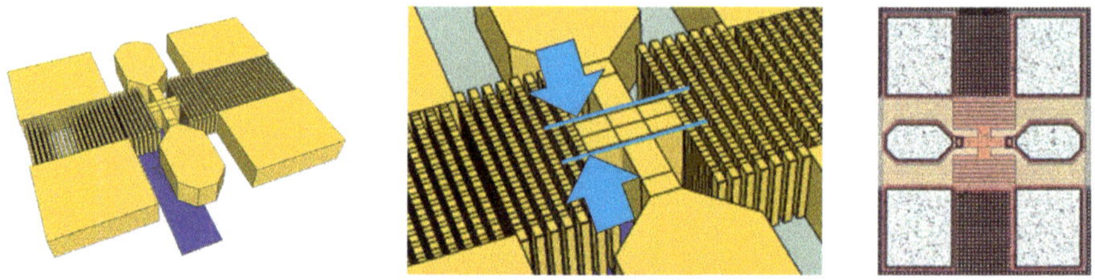

Figure 4.6 3D schematic view, a zoomed view and a photograph of the short standard.

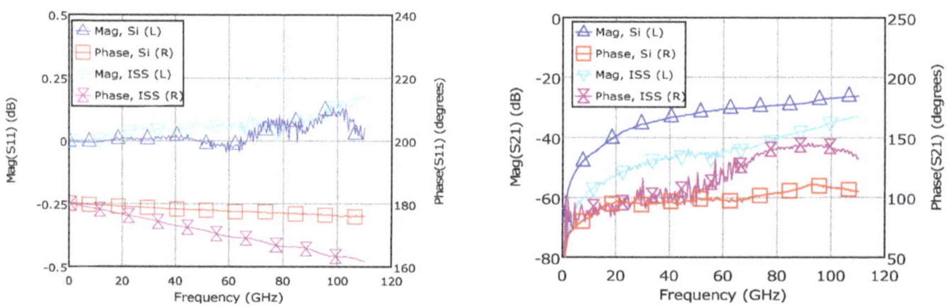

Figure 4.7 S_{11} and S_{21} parameters of commercial and custom short (reflect) standards measured with respect to the multiline TRL for both probe tip and on-wafer calibration.

14 μm distance (vs. 150 μm distance for the ISS short). The crosstalk for the *in-situ* short is less than –20 dB within the entire frequency range of interest. The element provides sufficient port isolation and is suitable for a reliable calibration.

It is interesting to note that the crosstalk for the custom short has almost purely capacitive behavior, whereas the ISS short shows a complex crosstalk nature above 50 GHz.

The magnitude of the S_{11} shows a similar pattern for both *in-situ* and ISS elements: it continuously increases with the frequency up to 0.2 dB at 110 GHz for the ISS short. The custom short repeats this effect from 60 GHz. I attributed this effect to the calibration residual error caused by the probe-tip design. The maximum error of 0.2 dB at 110 GHz can be neglected for now. I will come back to the probe-dependent calibration residual error in Chapter 9.

4.4.3 Load

The load is a single (per port) 50 Ω N+ salicided poly resistor located close to the M1 level (Figure 4.8). The resistors are laid out symmetrically for both

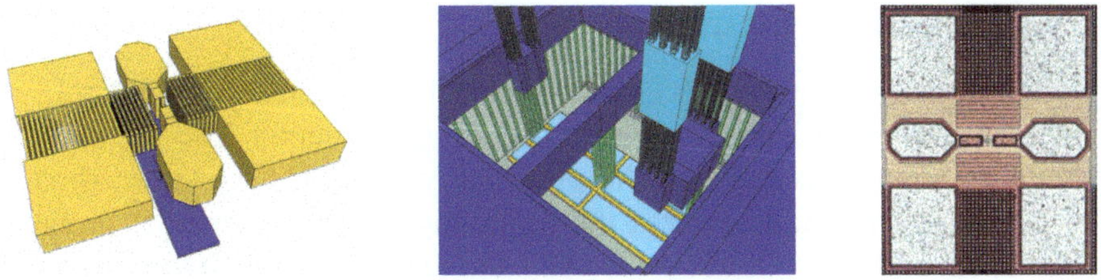

Figure 4.8 3D schematic view (left), a zoomed view (middle) and a photograph (right) of the load standard.

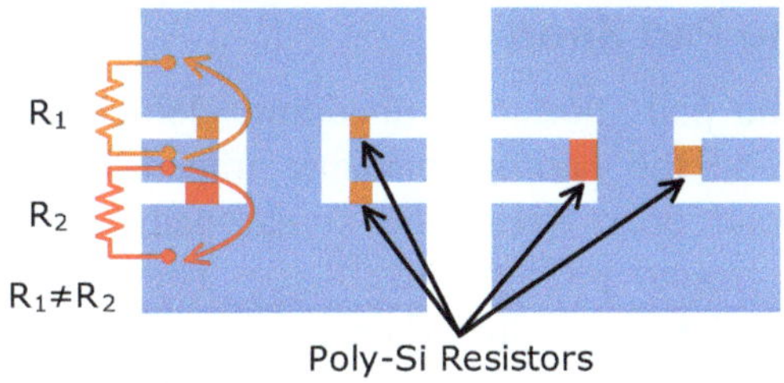

Figure 4.9 Schematic of the load: not-recommended (left) and recommended (right) design of the 50 Ω resistor for a custom match standard. The port 1 resistor of the first load is unbalanced due to fabrication inaccuracy.

ports. Often engineers repeat the layout of a probe-tip load with two 100 Ω resistors connected in parallel for the custom load standard; see Figure 4.9, [66,67]. The only advantage of such a layout is that it reduces the space taken by the standard on the substrate. So, the commercial vendors can put more loads on the substrate, increasing its lifespan and reducing the cost of RF calibration.

The fabrication process inaccuracy may lead to a high in-port asymmetry between the two 100 Ω resistors. The equivalent resistance of the load will be misbalanced. To solve this problem for commercial alumina standards, the resistors are trimmed to the specified values during the manufacturing process. In contrast, trimming *in-situ* resistors is very difficult, impractical and often not possible at all. That is why it is highly recommended to use one resistor per port, as shown in Figure 4.9 (right).

The via stack repeats the shortest single-finger transistor (Figure 4.8, middle). Therefore, the load reactance is slightly asymmetrical between ports. However, it can be accurately handled by the transfer TMR calibration capability, as it was shown in Chapter 3 and in [47,49].

4.5 De-Embedding Elements

The modeling test chip of the STMicroelectronics' BiCMOSMMW9 process includes an extensive set of de-embedding elements. The in-house de-embedding strategy is built on the proprietary scalable six-step method (Figure 2.11). The method requires the following dummy elements: probe short, pad open, pad short, thru, device open and device short. In addition, the complete open and the complete short elements are included. These elements are specific per device geometry, as we discussed in [58].

4.5.1 Pad Open and Pad Short

The pad open and the pad short de-embedding elements are laid out to represent the serial and the parallel parasitic impedances of the pad at the M6 level. The pad open keeps the signal pads (drawn at the top metal only) and the ground pads (Figure 4.10). The signal pad includes a few micrometers of the signal line (launch stab). It takes into account the pad-to-line discontinuity and provides a well-defined signal launching from the pad into the transmission line of the thru de-embedding element. The length of the launch stab is kept the same across all de-embedding elements. All interconnect metal levels except M6 and the via stack are removed.

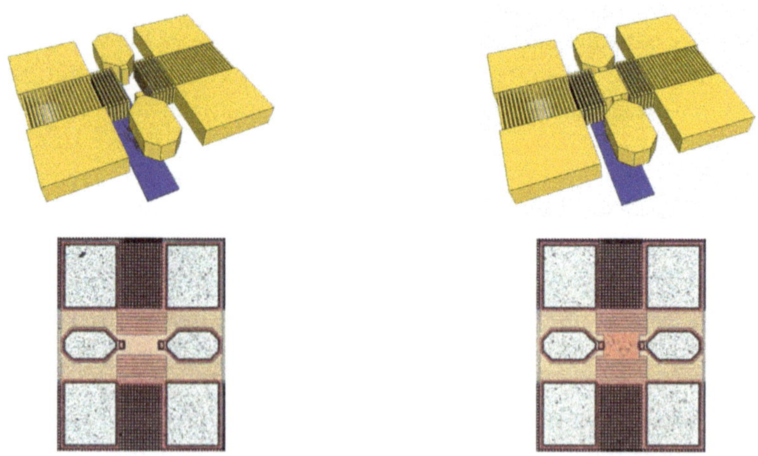

Figure 4.10 3D schematic view (top) and a photograph (bottom) of the pad open (left) and the pad short (right) elements.

The pad short has a top metal shorting area between the extremity of the pads and the closest ground metal stack. The via stack between M6 and M1 is kept and all M1 terminals are shortened over the DUT, while the DUT itself is removed.

4.5.2 Probe Short

The design of the probe short resembles the pad open, but the signal pads are shorted to the ground by the large M6 ground plate (Figure 4.11). As mentioned in [38], there are two purposes of this element. First, it evaluates any discrepancies of the probe-tip calibration. Second, it ensures the coherence of the RF/DC resistance measurements.

4.5.3 Complete Open and Complete Short

To move the reference plane to the intrinsic DUT terminals, the dummy elements must repeat the geometry of each transistor. Therefore, the complete open should replicate each DUT element. While excluding the contact vias, it includes all metals and the via stack between the end of the top layer line and the first metal above the DUT (Figure 4.12).

Figure 4.11 3D schematic view and a photograph of the probe short elements.

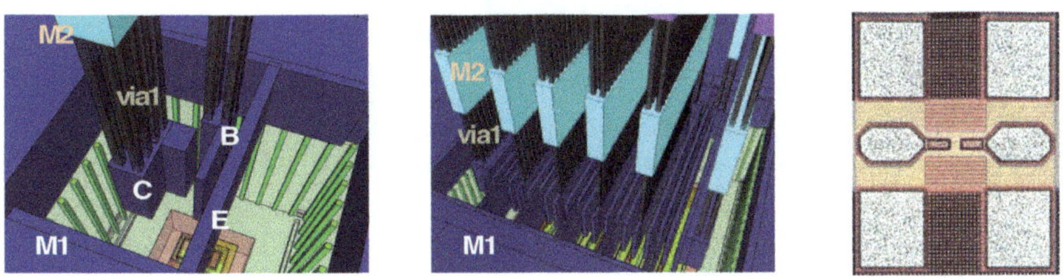

Figure 4.12 3D zoomed-in view of the M1 and M2 area, 3D schematic of the complete open element designed for short single-finger (left) and long multi-finger (middle) DUTs as well as its photograph (right).

Obviously, the equivalent capacitance of the complete open element depends on the DUT type. However, it is always greater than the equivalent capacitance of the pad open as it also includes a parasitic capacitance between M6 and M1 layers. A comparison for S_{11} and the port 1 equivalent capacitance C_1 of the complete open element is given in Figure 4.13 for the DUT in CBEBC configuration with $W_E = 0.23$ μm and $L_E = 0.15$ μm vs. the pad open.

Similar to the complete open, the complete short design depends on the DUT type. It is drawn from the DUT test structure where the device is removed and the access to the transistor pins is shortened over the M1 plate (Figure 4.14).

Again, the equivalent inductance of the complete short depends on its layout. It is always greater than the equivalent inductance of the probe short, pad short and reflect elements, as it includes the M6-M1 via stack parasitic

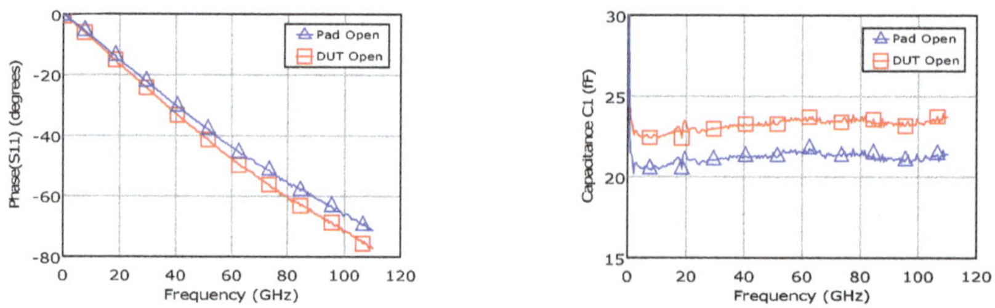

Figure 4.13 Comparison for the phase of S_{11} (left) and the equivalent capacitance C_1 for the complete open and the pad open elements measured with respect to the probe-tip multiline TRL calibration.

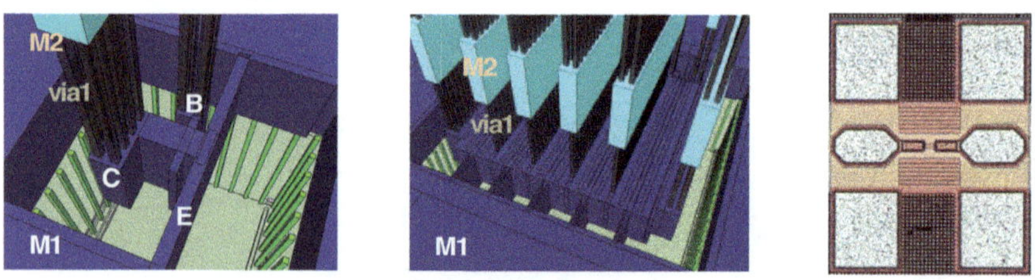

Figure 4.14 3D zoomed-in view of the M1 and M2 areas of a complete short element designed for short single-finger (left) and long multi-finger (middle) DUTs as well as a photograph (right).

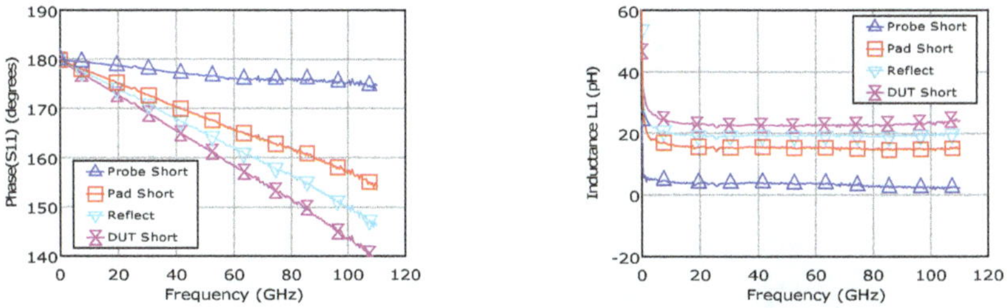

Figure 4.15 Comparison for the phase of S_{11} (left) and the equivalent inductance L_1 for complete short, probe short, pad short and reflect elements measured with respect to the probe-tip multiline TRL calibration.

inductance. For example, Figure 4.15 compares the S_{11} parameter and the port 1 equivalent inductance L_1 of the complete short for the DUT in CBEBC configuration with $W_E = 0.23$ µm and $L_E = 0.15$ µm vs. the probe short, pad short and the reflect measured and extracted with the respect to the probe-tip multiline TRL.

Both Figures 4.13 and 4.15 impressively show that the BEOL parasitics increase the closer we get to the terminals of the intrinsic device.

4.6 Calibration Standards and De-Embedding Elements Implemented in the IHP BiCMOS SiGe:C Process

The first implementation of the transfer TMR method was done on the IHP BiCMOS SiGe:C process (Figure 4.16). The modeling chip included thru, load and short calibration standards (Figure 4.17).

At the time of the experiment, the IHP in-house de-embedding strategy was based on the three-step method. The method uses open, open-short, short-open and thru dummy elements (Figure 4.18). The calibration standards and the dummies were designed such that the thru standard could also serve for the benchmark de-embedding step. It saved one slot on the test chip where we put the symmetrical attenuator for the purpose of calibration and de-embedding accuracy verification.

Other technical details of the standard design are proprietary information of IHP Microelectronics and cannot be published here.

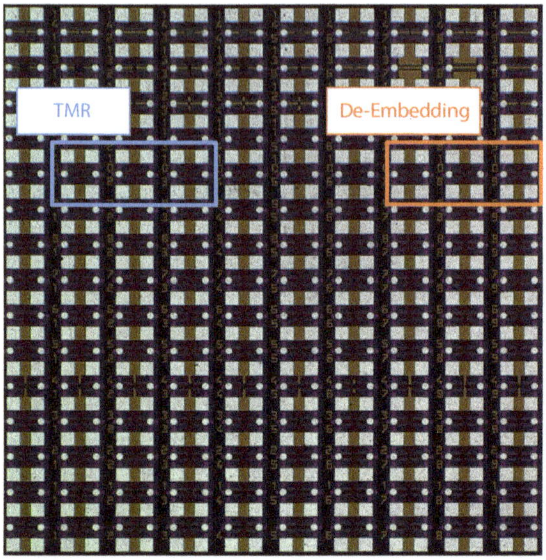

Figure 4.16 Test chip of the IHP SiGe:C BiCMOS process with transfer TMR standards and the three-step de-embedding dummy elements. Reproduced with permission of IHP Microelectronics.

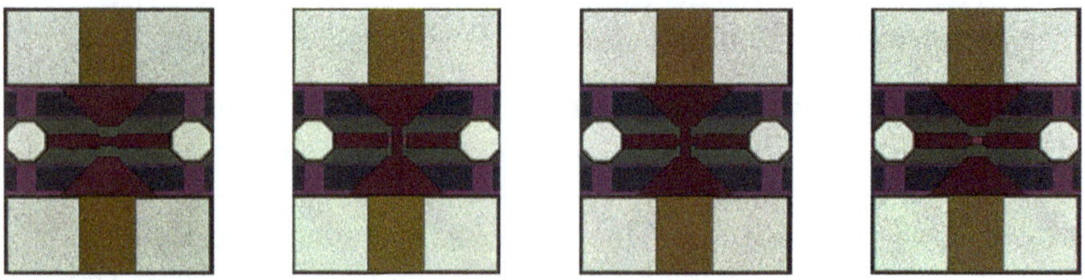

Figure 4.17 Transfer TMR calibration standards from the IHP SiGe:C process, from left to right: thru, load, short and the verification element attenuator.

Figure 4.18 Three-step de-embedding elements from the IHP SiGe:C process test chip. From left to right: open, open-short and short-open.

4.7 Calibration Standards Implemented in IBM RF CMOS 8SF Process

We implemented the multiline TRL as well as the transfer TMR standards on the IBM advanced RF CMOS 8SF 0.13 μm process. The calibration set included thru, three lines, open, short and load.

Standards were fabricated within a typical 2-port GSG on-wafer probe pad set with aluminum metallization. The ground pads, signal pad and signal launch lines were placed on an upper metal level layer, while a meshed ground plane was placed under the signal geometries on the first metal level layer. The transmission lines were laid out as extensions of the signal launch lines. The thru and line signal connector was placed one level below the pads and was made of copper. The thru length was 100 μm and the width was 18 μm. The length of lines was optimized for the frequency range of 1–110 GHz. The ground plane was placed on the first metal level. The space between the bottom of signal line and the top of ground was 8.4 μm. The CPW grounds are 45 μm wide. The gap between the edge of signal and the edge of ground was 68.5 μm. Such a design ensures that the main microstrip mode is the dominant line propagation mode in the frequency band of interest.

The signal contact pad measured 45 × 75 μm, while we used 95 × 75 μm ground pads. A photograph of the thru standard is shown in Figure 4.19.

The de-embedding elements were designed for the general 4-port method as described in [38]. Besides conventional open and shot elements, the test chip also included open-resistor, resistor-open, short-resistor and resistor-short. We used the same thru element for both de-embedding and calibration.

Other technical details of the standard and de-embedding element design are proprietary information of IBM and cannot be disclosed here.

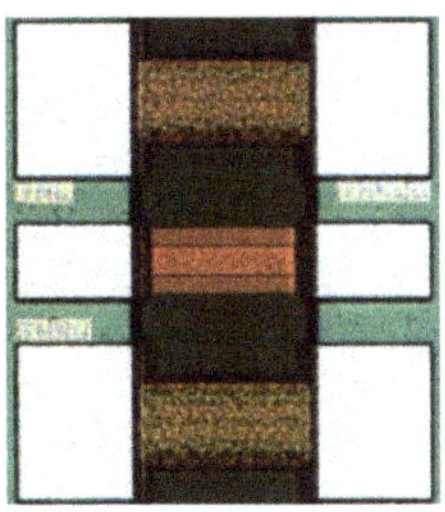

Figure 4.19 The thru standard implemented in IBM RF CMOS 8SF process test chip. Reproduced with permission of IBM Microelectronics.

5

Electrical Properties of Custom Standards

5.1 Multiline TRL and Transfer TMR Requirements

The key difference between calibration and de-embedding is the "view" on calibration standards and de-embedding elements. In a conventional de-embedding approach, the electrical characteristics of de-embedding elements remain unknown. The only assumption is that a single element represents a part of the pad and interconnects parasitic impedance. So, the parasitic impedance can be measured straightforward and further subtracted from the DUT results.

The calibration approach assumes that the measurement errors (e.g. the parasitic impedances) are captured by a specific error model. The complexity of parasitic impedance equivalent circuit is irrelevant. However, calibration requires a set of standards with known electrical characteristics. Manufacturers of commercial calibration substrates meet these requirements as follows [68, 69]:

1. The characteristic impedance of the line is kept within a defined specification and has a nominal value of $Z_{LINE} = 50\ \Omega$;
2. The conductive loss of the line is negligible;
3. The thin film resistor of the load standards is accurately trimmed to $R_{LOAD} = 50\ \Omega$ with an accuracy better than 0.3%;
4. The reflection standards (open, short and load) are optimized for minimal parasitic reactance for a given probe pitch. The equivalent parasitic reactance of a standard is specified (calculated or/and measured) for each probe pitch.

Obviously, these requirements are hard to address for custom standards. As a result, the first attempts to implement custom calibration process for silicon process failed [70].

As it was already discussed in Chapters 2 and 3, the multiline TRL and the transfer TMR require to know the following:

1. The characteristic impedance Z_{LINE} of the thru (line) for both TRL and TMR calibration;
2. Impedance of the load Z_{LOAD} for TMR calibration.

5.2 Measurement of the Characteristic Impedance of the Line

Following an extensive research, three techniques were selected to measure the characteristic impedance Z_{LINE} of the line for the IBM CMOS and STMicroelectronics BiCMOS processes:

1. Calibration comparison method from Arz [71];
2. Lumped load method from Williams [72];
3. Electromagnetic simulation.

The objective of my experiments was to achieve reliable results up to 110 GHz. Other methods required an initial probe-tip calibration followed by various pad de-embedding techniques [73–75]. They demonstrated insufficient extraction accuracy at frequencies above 20 GHz or they are cumbersome and less practical. Therefore, I left them out of the scope of this investigation.

The key work on this topic was done together with Phil Corson and Susan Sweeney from IBM Microelectronics (Essex Junction, VT, USA) and Uwe Arz from PTB (Braunschweig, Germany) and verified on the IBM advanced RF CMOS 8SF process. The results were presented in [76] (Attachment 14.1). I developed the entire verification methodology, did all calculations (except extraction of the DUT parameters) and wrote the major part of the conference paper. The design of the custom calibration standards, the electromagnetic simulations and the acquisition of measurement data were done by IBM engineers. The analysis of the results was done together with all co-authors.

For the first time, it was demonstrated that the calibration comparison method from [71, 77] can successfully be used for extraction of the characteristic impedance Z_{LINE} of the *in-situ* calibration lines. However, a well-defined reference calibration is required. Both NIST GaAs reference material RM8130 and the commercial calibration substrate CS-5 (from GGB Industries) were verified for establishing the reference calibration up to 110 GHz.

A strong impact of the probe-tip design and non-optimized reference calibration boundary conditions on the experimental results was found at mm-wave frequencies. This topic will be discussed later in Chapter 9.

It was proven that the lumped load method from [72] can be applied to measure Z_{LINE} of the tested lines from IBM's RF CMOS 8SF process test chip. Later, I verified the lumped load method for BiCMOSMMW9 test chip from STMicroelectronics, [56]. The lumped load method showed results comparable with the benchmark calibration comparison method. The lumped load method is the favorite choice for practical applications. However, it requires that the conductive loss of the line is minimized by special measures of the line design such that it can be neglected.

It remained challenging to achieve a sufficient level of simulation accuracy for custom standards realized in advanced multi-layer technologies.

On-the-fly measurement of some electrical parameters of standards may be needed because they vary from chip to chip, wafer to wafer and lot to lot due to instability of the fabrication process. This problem is analyzed in Chapter 8.

5.3 Establishing the Reference for the Calibration Comparison Method

One important topic remained behind the scenes in [76]: how to measure and how to verify the characteristic impedance of CPW lines for a commercial calibration substrate at mm-wave frequencies? Previously, the well-defined NIST GaAs reference material RM8130 was used to establish a reference calibration for the calibration comparison method (see e.g. [78, 79]). However, the RM8130 is specified for use up to 40 GHz [80]. To the best of my knowledge, a solid metrological reference for wafer-level measurements above 40 GHz is still missing.

Though RF probe vendors recommend commercial calibration substrates for the probe-tip calibration up to sub-THz frequencies, the reference impedance Z_{REF} of such a calibration is questionable and, in a more general term, undefined because:

1. The characteristic impedance Z_{LINE} of the CPW line can significantly differ from its nominal value of $Z_{LINE} = 50\ \Omega$ due to inaccuracies and instability of the fabrication process. Nevertheless, none of the commercial suppliers specifies it for a particular substrate;
2. The interpretation of the reference impedance of lumped-standard-based calibration methods (such as SOLT, conventional LRM or LRRM)

becomes difficult at mm-wave frequency range. That is why these methods cannot be considered as good candidates for the reference calibration.

To overcome these problems, the following strategies were developed:

1. Measure Z_{LINE} of CPW line from several commercial alumina calibration substrates and verify the measurement method up to 110 GHz;
2. Use the multiline TRL for the reference calibration on alumina substrate;
3. Transfer reference impedance of the multiple TRL to $Z_{LINE} = 50\ \Omega$.

It was executed as following. First, the lumped load method was applied extracted the characteristic impedance Z_{LINE} was extracted for three commercial calibration substrates: ISS 101-190 from Cascade Microtech, CSR-3 and CSR-8 from SUSS MicroTec. Next, the accuracy of the multiline TRL performed with and without correction for the line Z_{LINE} was verified against the NIST RM8130. Results proved that the lumped load method delivers a sufficient level of accuracy for alumina substrates. Taking into account that the commercial CPW lines implemented on ceramic exhibit negligible conductive loss, it was assumed that the extracted Z_{LINE} is valid (at least) up to 110 GHz. Therefore, the reference multiline TRL calibration can be executed at every commercial alumina substrate.

The results of this work were published in [22] and are co-authored by Ralf Doerner (FBH, Germany) and Steffen Thies (Rosenberger, Germany) (Attachment 14.2). For this paper, I developed the verification methodology, planned the work, defined the measurement campaign and wrote the conference paper. The measurement results were acquired by Ralf Doerner at the FBH facility. I did calculations and the data analysis together with Ralf Doerner.

Later, when working on paper [76], an extensive analysis of two CS-5 calibration substrates from GGB Industries was performed proving the earlier developed approach for the measurement of the line Z_{LINE} of commercial alumina substrates.

5.3.1 NIST GaAs Reference Material RM8130

The RM8130[1] consists of a coplanar wave guide multiline TRL calibration set: a 550 μm long thru line, five lines with lengths of 2685, 3750, 7115, 20.245 and 40.550 μm and two offset shorts located at a distance of 225 μm

[1] Available from NIST, Boulder, CO, USA.

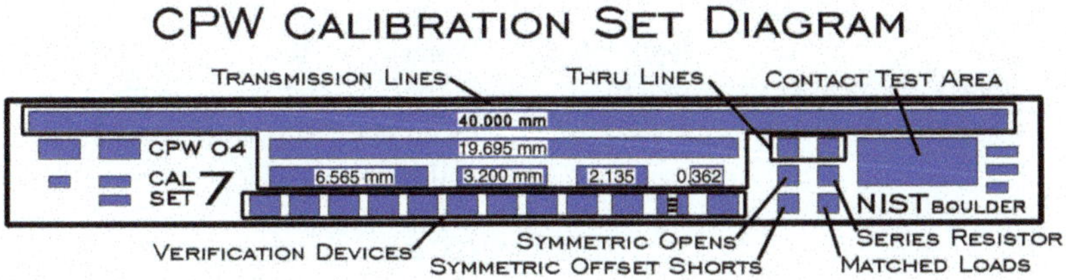

Figure 5.1 The layout of the NIST coplanar waveguide RM8130. Picture from [81].

(a) (b) (c) (d) (e)

Figure 5.2 Photographs of some calibration and verification element implementations on the RM8130 CPW, from left to right: thru, short, open, resistor and the series attenuator.

from the beginning of the line. Additionally, there are 12 verification reference elements (Figures 5.1 and 5.2). The calibration set is fabricated on a GaAs wafer and is fixed on a quartz holder for better handling.

The line capacitance per unit length C' of RM8130 CPW is specified according to the individual tests provided by NIST for each reference material. This value is used by the benchmark multiline TRL calibration for the accurate definition of the characteristic impedance Z_{LINE} of the RM8130 lines by (2.25) and the transformation of the measurement system reference impedance to $Z_{REF} = 50\ \Omega$ using (2.51).

Table 5.1 shows the configuration of two setups that I used in experiments from [49, 76]. The setup A at IBM and the setup B at FBH used the same Agilent 8510XF 110 GHz VNA but were equipped with different RF probes. Also, each laboratory had its own RM8130. That gave the opportunity to compare the electrical characteristics of the CPW lines from two RM8130 and being measured by two probe types.

Figure 5.3 shows the measured attenuation and the relative phase constants of the line for both RM8130 substrates. There is a noticeable difference in the attenuation constant while the difference in the relative phase constant is negligible. It was attributed to the impact of the RF probe design and was taken into account for further analysis of the measured data. The characteristic impedance shows comparable behavior for both setups (Figure 5.4).

Table 5.1 Experimental conditions

	VNA	RM8130	RM8130 C', pF/cm	Wafer Probes	Test CPW	CPW C', pF/cm
Setup A, IBM	8510XF (Agilent)	#11/4	1.7877	Picoprobe 110 GHz, GSG, 100 μm pitch	CS-5 (a)	1.629
Setup B, FBH	8510XF (Agilent)	#7/3	1.7877	ACP 110 GHz, GSG, 100 μm pitch	CS-5 (b)	1.522

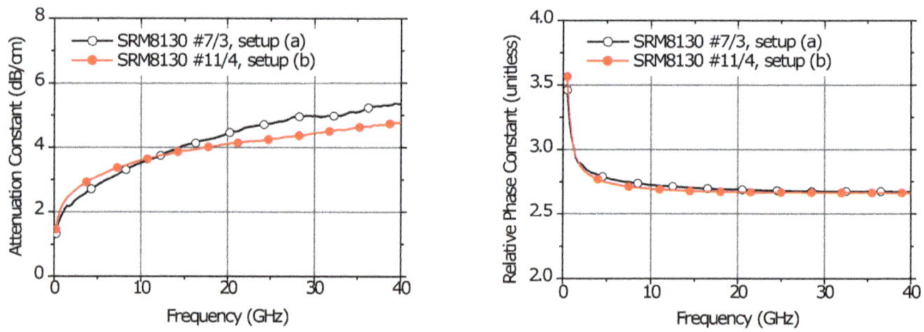

Figure 5.3 Attenuation constant (left) and the relative phase constant (right) of two RM8130 used in the experiments.

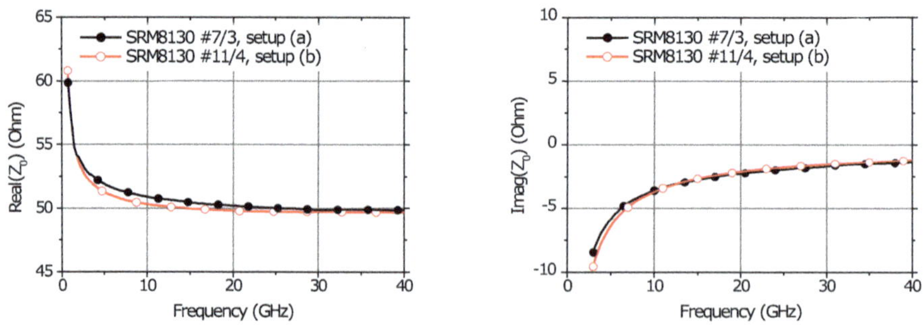

Figure 5.4 The real (left) and the imaginary (right) parts of the characteristic impedance of two RM8130 used in the experiments.

5.3.2 Commercial Alumina Substrate as the Reference

Next, the capability of the calibration comparison technique was verified for the measurement of the characteristic impedance of the commercial CPW lines up to 110 GHz on the CS-5 alumina calibration substrate. The CS-5 substrate consisted of lumped standards (open, short and load) as well as CPW line sets: a 150 μm long thru line and four lines with lengths of 6550, 1450, 950 and 500 μm. Its estimated effective dielectric constant is about 5.26. The substrate is 625 μm thick. In these experiments, the calibration substrate was placed directly on the metal chuck.

The verification strategy was built as follows: using the lumped load method, the capacitance per unit length C' of the CPW line was extracted from

$$C' \approx \Re\left(\frac{\gamma}{j\omega R_{LOAD}}\frac{1+\Gamma_{LOAD,Z_{LINE}}}{1-\Gamma_{LOAD,Z_{LINE}}}\right), \tag{5.1}$$

where $\Gamma_{LOAD,Z_{LINE}}$ is the load reflection coefficient corrected by the TRL calibration with $Z_{REF} = Z_{LINE}$; γ is the propagation constant of the line measured by the multiline TRL; R_{LOAD} is the resistance of the load. The component $G/\omega C$ from (2.25) was neglected.

To increase the extraction accuracy, R_{LOAD} was measured using an SMU with an accuracy better than 0.05 Ω. Therefore, the C' value was assumed to be the reference parameter. The characteristic impedance Z_{LINE} as well as other equivalent circuit parameters of the line: conductance G', resistance R' and inductance L' per unit length were calculated from the known C' and the propagation constant γ using the LINEPAR program (see Chapter 2 and Figure 2.3).

After that, the reference multiline TRL calibration was performed on RM8130, setting the calibration reference impedance to $Z_{REF} = 50$ Ω and shifting the calibration reference plane to the probe tips.

Finally, the second-tier multiline TRL was performed on the examined alumina substrate. The characteristic impedance Z_{LINE} of the alumina CPW line was extracted from the second-tier error boxes and the line propagation constant γ (the product of the second-tier calibration) with the help of the ExtractZ0[2] program. Same here, capacitance C', conductance G', resistance R' and inductance L' per unit length were calculated from the measured Z_{LINE} and (2.21) and (2.22).

[2]Both LINEPAR and EztractZ0 are available from NIST, Bolder CO, USA.

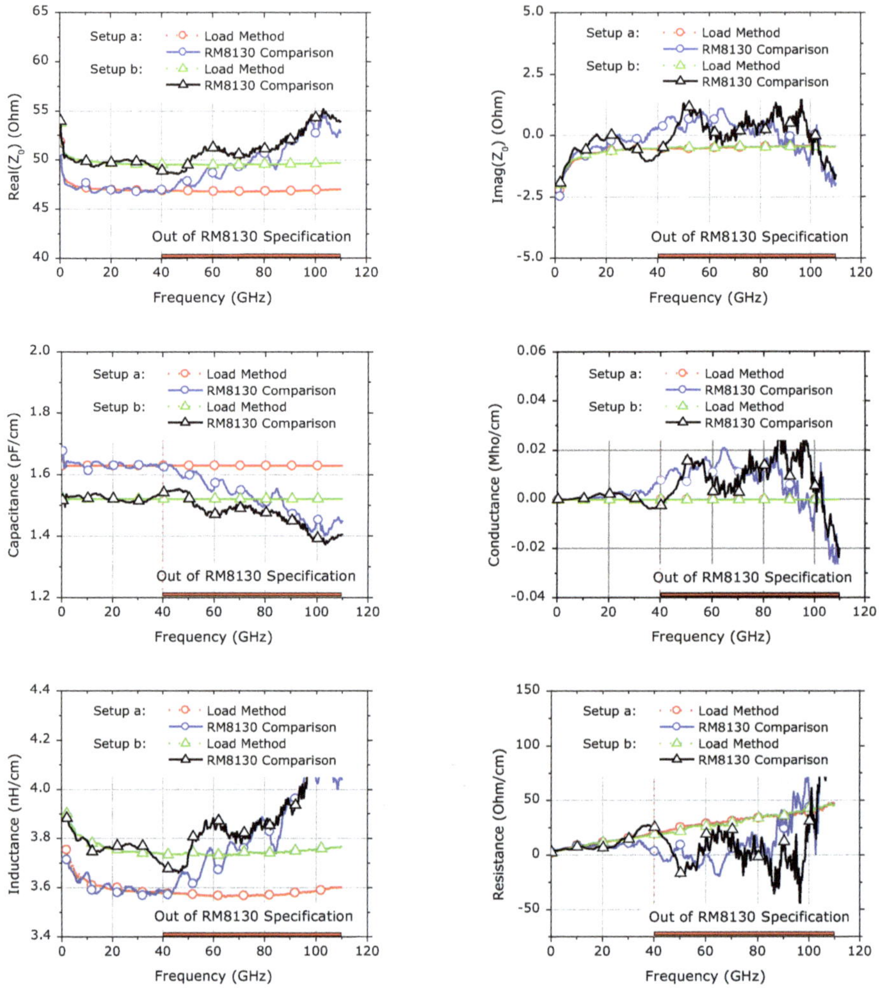

Figure 5.5 Parameters of the CS-5 alumina CPW line extracted from the calibration comparison to the reference RM8130 calibration and the load method for two setups.

To ensure the results, the same experiment was repeated for another CS-5 substrate that was measured on the second setup. Results are presented in Figure 5.5.

The results led to four important findings:

1. The calibration comparison method is in a good agreement with the lumped load method up to 40 GHz;

2. Parameters extracted from the calibration comparison method fall apart above 40 GHz;
3. Below 40 GHz, capacitance C' and inductance L' per unit length are different for tested CS-5 substrates;
4. All four parameters extracted from the calibration comparison method show the same trend but have noticeable deviation between setups.

As discussed in [76], the variation of about 7% for the capacitance C' and about 5% for the inductance L' per unit length for two CS-5 substrates roots in the instability of the thin film fabrication process.

Calibration comparison results strongly deviate above 40 GHz for both setups because the cross section of the RM8130 CPW lines is not optimized for those frequencies. Therefore, I considered the frequency band from 40 to 110 GHz to be out of the RM8130 specification for the calibration comparison.

The setup dependent "error patterns" that are noticeable for all parameters extracted from the calibration comparison can be attributed to the probe-dependent calibration residual errors. Here, I compared two calibrations performed on very different CPW designs and material. Thus, the probe-design-dependent parasitic effects such as crosstalk, coupling, radiation and propagation of the higher-order modes can be significantly emphasized. Residual errors are particularly well exhibited above the specification of RF8130.

This experiment proved the following hypotheses:

1. The alumina CPW lines have negligible conductance loss;
2. The extraction of the coefficient $G/\omega C$ is not needed. Thus, the lumped load method is valid for accurate measurements of the capacitance per unit length C' of commercial alumina CPW lines;
3. The RM8130 cannot be used as a reference above 40 GHz.

Summarizing, it was proven that the reference calibration can be established using the commercial calibration substrates. This enabled accurate extraction of the line characteristic impedance Z_{LINE} on conductive substrates from the calibration comparison method. However, results indicate that there are some fabrication variations and probe-type design dependencies. This issue highlights the need for a reference material that is valid at least up to 110 GHz. That is why the characterized alumina substrates will be used as the reference for further experiments. For a particular alumina substrate, it is mandatory to measure the capacitance per unit length C' of its lines. Also, it is highly recommended to use RF probes of the same design for all experiments.

5.4 Results of Measurement of the Characteristic Impedance of the Line

5.4.1 IBM Advanced RF CMOS 8SF Process

In paper [76], the propagation constant and the characteristic impedance of custom line standards were obtained from the simulated and measured results. Both the lumped load and the calibration comparison methods were used. For the calibration comparison, the reference multiline TRL calibration was established on both RM8130 (up to 40 GHz) and on the commercial alumina substrate CS-5 (up to 110 GHz). Before that, the characteristic impedance Z_{LINE} of the CS-5 CPW lines was accurately measured.

Also, the capability of the 3D simulation was evaluated. The line was simulated with a simplified model using the 3D full-wave electromagnetic field simulation package Ansoft HFSS®. The 3D structure included all passivation and metal layers with interconnects simplified to solid bars. Simulation results yielded the electrical parameters of the transmission line including the propagation constant and the characteristic impedance. The experiments were carried out on data measured at IBM laboratory (Table 5.1, setup A).

Figure 5.6 compares the simulated and the measured attenuation and phase constants. These parameters were measured directly through multiline TRL calibration, as shown in [21].

Figure 5.7 shows the characteristic impedance Z_{LINE} of the same line obtained from four different experiments: EM simulation, lumped load method, calibration comparison method with the reference CS-5 calibration and the calibration comparison method with the reference RM8130 calibration [55,76]. All results are in a good agreement in the entire frequency range.

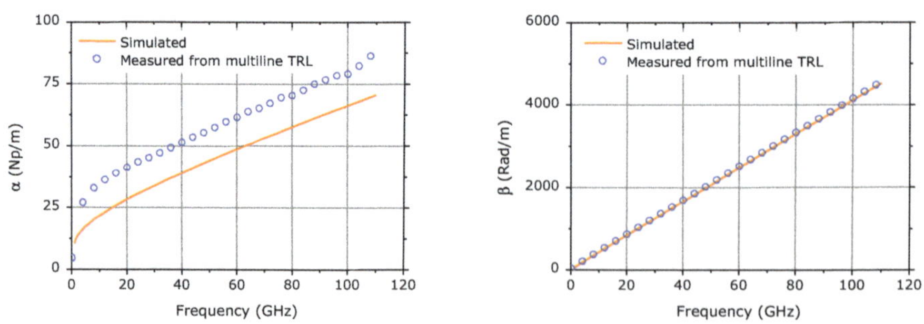

Figure 5.6 Simulated and measured attenuation (left) and phase (right) constants of silicon line from the advanced IBM CMOS process.

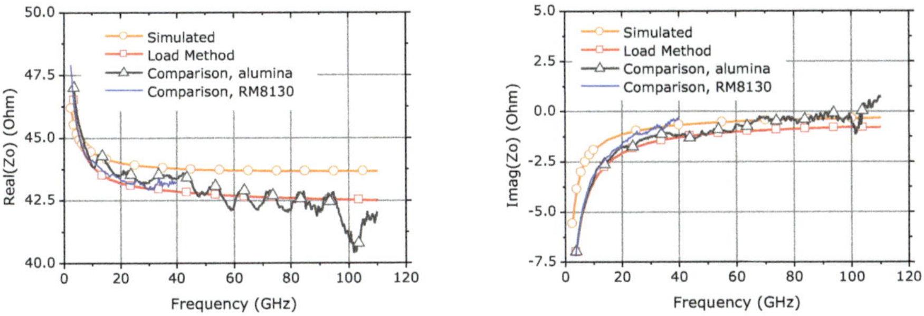

Figure 5.7 Simulated and extracted real (left) and imaginary (right) parts of the examined silicon line characteristic impedance Z_0 from the advanced IBM CMOS process. Extracted data were obtained from the load method and the calibration comparison method to reference material RM8130 and alumina substrate. Comparison to RM8130 is limited in a frequency of 40 GHz due to the specification of the used reference material.

The difference in the substrate material between the GaAs of the RM8130, the alumina and the silicon appears to be successfully compensated by the calibration comparison method.

Next, the equivalent circuit parameters of the line were extracted to evaluate the impact of each method. It is important to note that the lumped load method assumes that the capacitance per unit length C' is constant over frequency and that conductance losses are negligible, i.e. $G/\omega C = 0$. Again, calibration comparison and the lumped load method are in a good agreement (Figure 5.8).

5.4.2 STMicroelectronics' BiCMOS9MMW Process

Similar measurements of propagation constant γ and the characteristic impedance Z_{LINE} were performed for the line implemented in the STMicroelectronics' BiCMOS9MMW test ship. Both the lumped load and the calibration comparison methods were used. Figure 5.9 displays the propagation constant as well as the effective dielectric constant of the examined line. For comparison, the same characteristics of a commercial alumina substrate ISS 107-783A from Cascade Microtech are added to these graphs. The data for the ISS were obtained from a separate experiment.

The real part of ε_{EFF} is 3.5 vs. 5.0 for the ISS line. The imaginary part is comparably small. The attenuation constant α is smooth and the relative phase constant β/β_0 does not change over the frequency. Therefore, one can conclude that BiCMOS9MMW lines support a quasi-TEM propagation mode

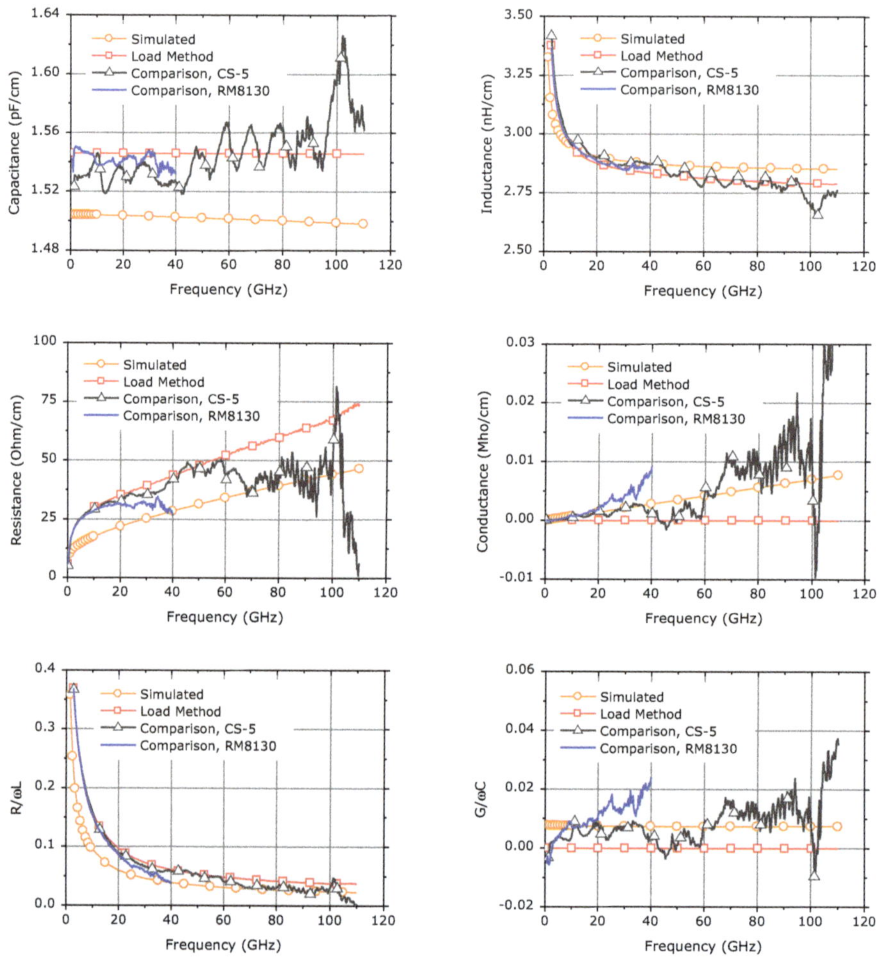

Figure 5.8 Simulated and extracted circuit parameters of the silicon line from the advanced CMOS IBM process. Extracted data were obtained from the load method and the calibration comparison method to reference material RM8130 and alumina substrate. Comparison to RM8130 is limited in frequency to 40 GHz due specification of the reference material used.

at least up to 110 GHz. These measurement results are in a good agreement with those published in [64].

Figure 5.10 presents the characteristic impedance Z_{LINE} extracted from the calibration comparison (probe-tip multiline TRL on CS-5 as reference) and the lumped load method [56]. The measured capacitance per unit length is $C' = 1.262$ pF/cm.

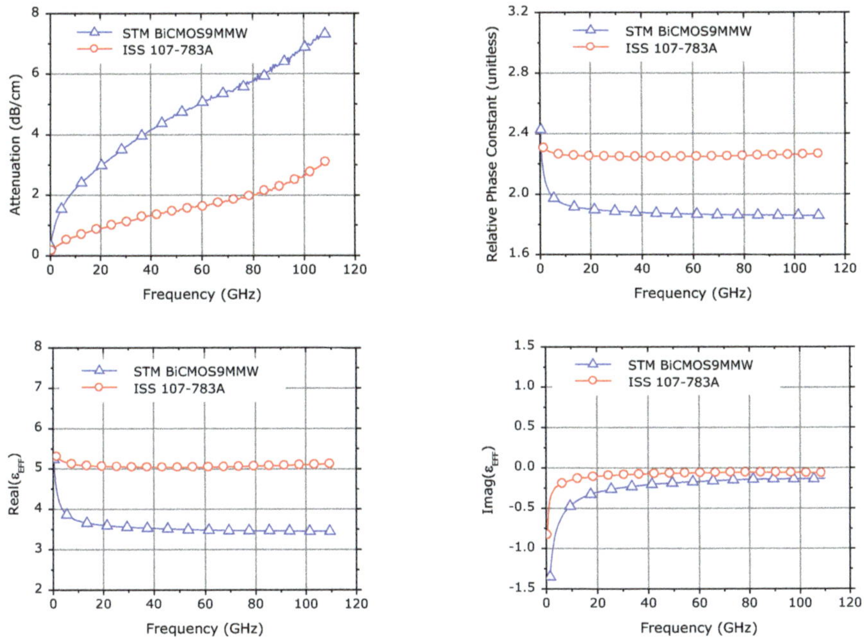

Figure 5.9 The attenuation, the relative phase constant, and the real and the imaginary parts of the effective dielectric constant of the calibration line standards measured by the multiline TRL. Comparison for the STMicroelectronic's BiCMOS9MMW process and the alumina calibration substrate ISS 107-783A from Cascade Microtech.

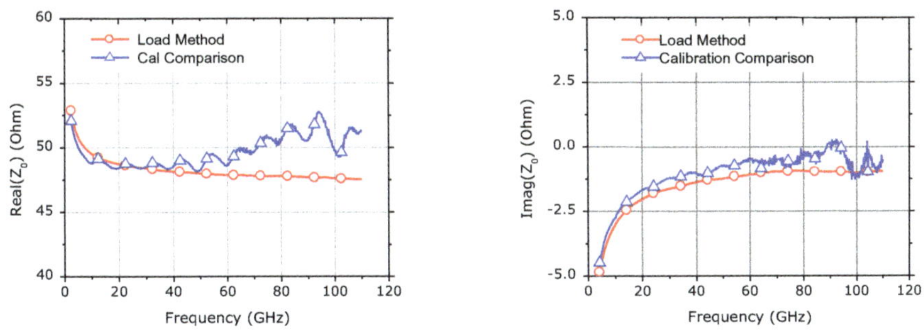

Figure 5.10 Characteristic impedance of line fabricated on ST Microelectronics' BiC-MOS9MMW process. Results obtained from the lumped load and calibration comparison method on CS-5 alumina substrate.

Finally, the line equivalent circuit parameters were extracted from the measured propagation constant and the characteristic impedance (based on the lumped load method): resistance R', inductance L' per unit length as well as $R/\omega L$ (Figure 5.11).

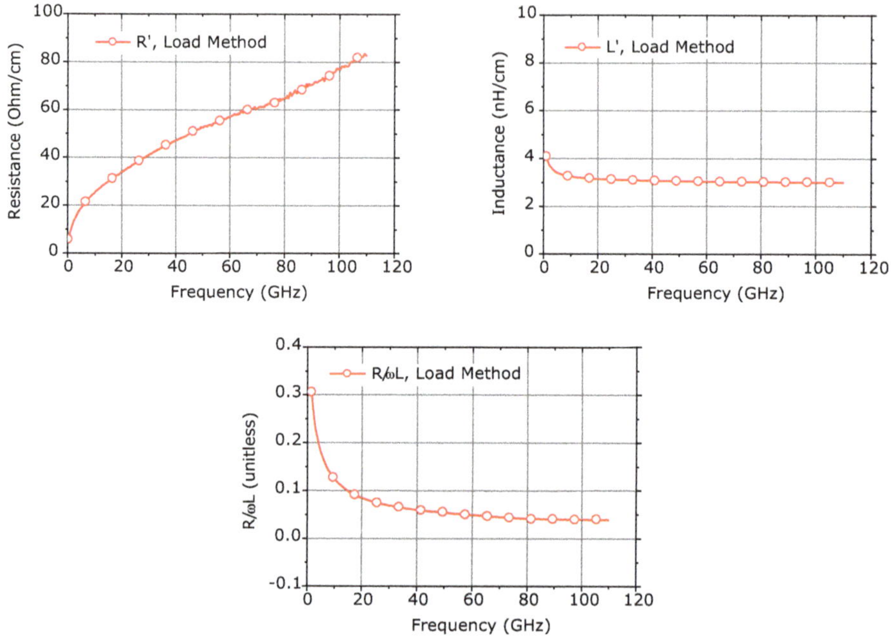

Figure 5.11 Resistance R', inductance L' per unit length as well as $R/\omega L$ for the calibration lines from the ST Microelectronics' B5T BiCMOS process.

A slight rippling was observed on the extracted parameters of the alumina CPW lines (Figure 5.10). The same effect occurs on silicon lines in both IBM RF CMOS and STMicroelectronics' BiCMOS processes measured with respect to the first-tier reference calibration on alumina (Figures 5.7 and 5.10). In both experiments, the CS-5 calibration substrate and 110 GHz Picoprobes from GGB Industries were used. Possible reasons for such effects might be the following:

1. The high-resistivity absorbing layer that is placed under the ground conductors and extends them (design specific of the CS-5 CPW lines);
2. Non-ideal calibration boundary conditions, because the CS-5 was placed on the grounded metal chuck in this experiment;
3. The design of wafer probes.

Most importantly, however, the lumped load method demonstrates an acceptable accuracy for determining the electrical characteristics of the transmission line standards on silicon up to 110 GHz. The success shown in this study is achieved by the design of the line standards on the silicon. The first metal level ground plane under the signal provides a dominant microstrip propagation mode in both technologies.

5.5 Definition of the Load Impedance

The EM simulation and the measurement methods were evaluated for characterization the load impedance Z_{LOAD} for various processes and layouts of the load standard [47, 55, 57]. As it was proven in [47], the load resistance is affected by fabrication process variability. Thus, it should be measured on every test chip.

The load reactance X_{LOAD} is defined for the reference plane located at the top metal. It is mainly represented by the equivalent reactance of the via stack (Figure 4.8). Typically, the via stack reactance is relatively small and is of lumped nature.

If the reference plane is shifted to the probe-tip end, the capacitance of the contact pads and the interconnect lines bring forth the largest contribution to the load reactance. This part is captured by the on-wafer calibration as a part of the systematic measurement errors and is calculated out.

5.5.1 Measurement Method

The on-wafer multiline TRL is the most accurate method to measure the load impedance and, thus, its reactance. This approach provides broadband results. If the size of the test chip is limited, the number of the required lines can be reduced. The extraction of the equivalent reactance of the load via stack should be done at frequencies of a well-conditioned TRL [55, 57].

The S_{11} and the S_{22} parameters of the load standard from the STMicroelectronics' BiCMOS9MMW process test chip are shown in Figure 5.12. The data are corrected by the on-wafer multiline TRL. The calibration reference impedance was accurately set to $Z_{REF} = 50\ \Omega$. The measurement reference plane was located at the top metal level (M6 from this process) and was shifted 7 μm away from the center of the thru in both directions towards the pads (as discussed in Chapter 4, and shown in Figure 4.8).

As shown in [47], the load reactance is about 10 times less sensitive to the process instability than its resistance. That is why it can be assumed to be stable enough for a given process. It significantly simplifies measurement of the load impedance. For instance, the impedance of a particular load element can be recalculated "on-the-fly" from the known (nominal) reactance and the measured resistance R_{LOAD}.

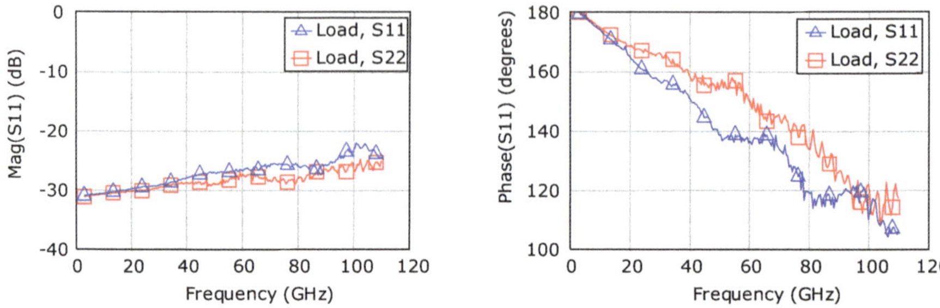

Figure 5.12 The S_{11} and S_{22} parameters of the STMicroelectronics' BiCMOS9MMW process test chip load standard measured with respect to the M6 multiline TRL.

It is not always required to extract the equivalent circuit parameters of a load model. In [47], we also proposed the following method: the real part of the load impedance can be normalized to the nominal value of its resistance R_{NOM}. For instance, the nominal resistance can correspond to a resistor laid out for $R_{LOAD} = 50\ \Omega$. Therefore,

$$R_{NORM}(f) = \frac{\Re(Z_{XX,NOM}(f) - Z_{YX,NOM}(f))}{Z_{XX,NOM}(f_0) - Z_{YX,NOM}(f_0))}, \quad X,Y = 1...2, \quad X \neq Y, \tag{5.2}$$

where Z_{NOM} are Z-parameters of the nominal two-port load and $Z_{NOM}(f_0) = R_{NOM}$.

The impedance $Z_{LOAD,ARB}$ of an arbitrary load element can be reconstructed from the measured resistance $R_{LOAD,ARB}$ (see Figure 5.13):

$$Z_{LOAD,ARB}(f) = R_{LOAD,ARB} \cdot R_{NORM}(f) + j\Im(Z_{NOM}) \tag{5.3}$$

5.5.2 EM Simulation of the Load Reactance

As it was demonstrated in [55], the simplified EM model of the load already provides an acceptable level of accuracy and is also a viable alternative for defining the load impedance. Indeed, the model does not require to include the whole element, instead only the via stack of the load resistor is needed. In other words, the simulation ports have to be located at the on-wafer calibration reference plane. It is much easier to simulate such a 3D structure accurately than the entire load element.

Figure 5.14 compares simulated vs. measured results for the S_{11} parameter for the load from the IBM advanced RF CMOS 8SF process. The variation

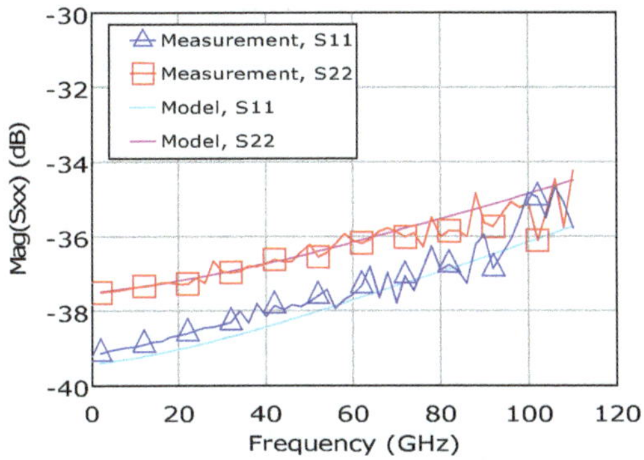

Figure 5.13 Modeled and measured results of the load standard S_{11} and S_{22} parameters implemented in SiGe:C BiCMOS process from IHP Technologies.

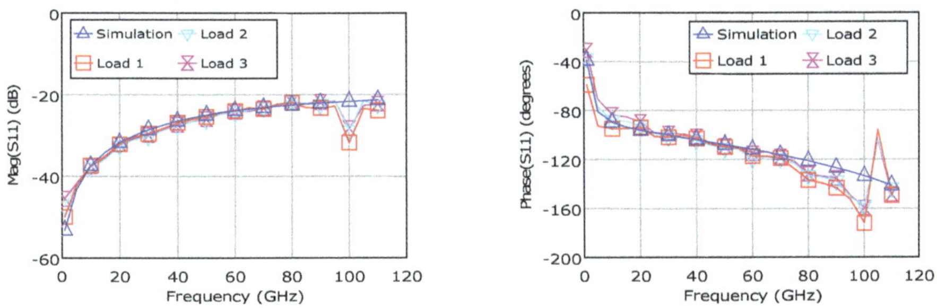

Figure 5.14 The S_{11} and S_{22} parameters of the STMicroelectronics' BiCMOS9MMW process test chip load standard measured with respect to the M6 multiline TRL.

across the measured three elements from the same chip is marginal and can be attributed to the contact repeatability error on aluminum pads. Also, as noted in [76], the measurement error at frequencies around 100 GHz is related to the specifics of the VNA used.

5.5.3 Two-Step De-Embedding for Measurement of Load Reactance

The possibility to use the probe-tip calibration followed by the two-step open-short de-embedding for measurement of the load reactance was evaluated in [57]. This method can be taken into consideration if it is not possible to implement the TRL standards on the test ship.

First of all, it proved that the tested load (from the BiCMOS9MMW test chip) can be assumed to be purely lumped at least up to 110 GHz. Thus, extraction of the equivalent load inductance can be successfully done

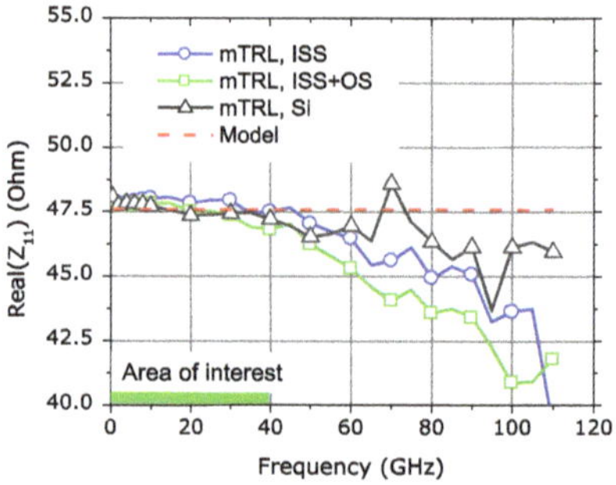

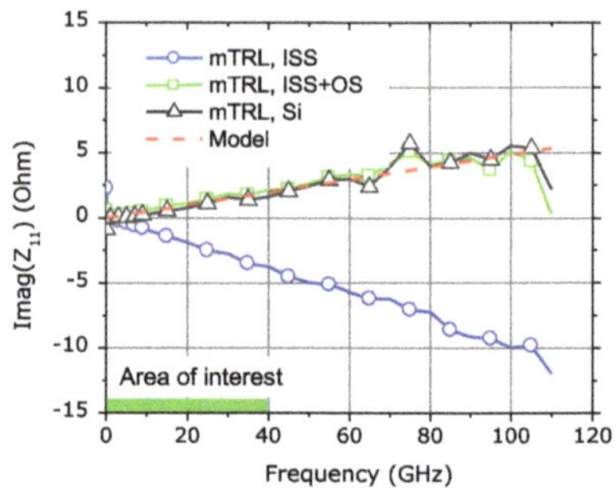

Figure 5.15 The real (left) and imaginary (right) parts of the Z_{11} parameter of the load from the STMicroelectronics' BiCMOS9MMW process.

at relatively low frequencies, i.e. below 40 GHz. At those frequencies, the two-step open-short de-embedding provides reliable results and makes the extraction procedure simple and practical. In our experiment, both the probe-tip calibration extended by the pad open, pad short (Figure 2.11.a, Figure 4.10) de-embedding and the on-wafer multiline TRL revealed good agreement for measured load reactance (Figure 5.15). For this particular design of the load, the equivalent circuit mode is given by a series resistor R_{LOAD} and an inductor L_{LOAD}. It is valid up to 110 GHz. For the example shown in Figure 5.15, the extracted parameters are $R_{LOAD} = 47.8\ \Omega$ and $L_{LOAD} = 7.8$ pH.

A slight-roll off of the measured real part of the load Z_{11} parameter might be caused by the extraction error. A simple one-port impedance element model was used for the load because the experiment was focused on the frequency range below 40 GHz (marked as "area of interest" in Figure 5.15). This simple model does not account for crosstalk between measurement ports that increase with the frequency. But, for the frequency range of interest, all methods demonstrated are in a good agreement.

5.6 Conclusion

Several methods were developed which are suitable for measurement of characteristic impedance Z_{LINE} of the line and the impedance of the load Z_{LOAD} standards implemented on a silicon test chip. Both parameters are crucial for the successful and accurate on-wafer multiline TRL and the transfer TMR calibrations.

For the first time, the calibration comparison method was verified for the measurement of the characteristic impedance Z_{LINE} of the line for frequencies up to 110 GHz. Thus, it became possible to establish a solid reference calibration on silicon.

Also, it was proved that a simple method based on the lumped load can be used for measurement of the line Z_{LINE} on silicon as well. If the calibration lines are designed properly (for instance, as on the IBM RF CMOS 8SF or STMicroelectronics' BiCMOS9MMW processes), the results of the lumped load method are valid over the entire frequency range (i.e. up to 110 GHz in my experiments).

Both methods were validated on the multiline TRL calibration kit implemented on the test chip from two different processes, RF CMOS and BiCMOS.

For the first time, the lumped load method was applied for the measurement of the characteristic impedance of CPW lines from commercial calibration substrates and verified it against metrological reference material RM8130 up to 40 GHz. This work enabled the use of the calibration comparison method for the measurement of the characteristic impedance Z_{LINE} of silicon lines above 40 GHz when conductive losses cannot be neglected.

The calibration residual errors caused by the RF probe design were identified. It was shown that they impact the results of the calibration comparison method. These effects call for future extensive and detailed investigation.

Several methods for measurement of the load impedance were developed and verified. It was proved that the load resistance can be measured "on-the-fly". Conventional SMUs that are connected to the RF probe by using the bias-T offer an acceptable accuracy for this kind of measurement. The load reactance depends on the load design and process specifics. It does not vary significantly with the process instability. Thus, it can be either simulated or measured with respect to either the probe tip or the reference on-wafer TRL calibration. Each method was validated and it was found that they yield comparable results. The final choice depends on individual preferences for a particular process, the design of the load as well as on the size of the test chip.

The methods, experiments and results demonstrated and discussed in this chapter formed a solid background for the implementation of the *in-situ* multiline TRL and the transfer TMR calibration methods for advanced RF silicon processes.

6

Verification Methods for On-Wafer Calibration

Several methods can be applied to verify the accuracy of the *in-situ* calibration. It is always recommended to implement some transmission lines into a test chip to enable the reference multiline TRL. With the help of the calibration comparison method from [77], other *in-situ* calibration methods can be quantitatively verified against the multiline TRL reference. Often, especially in the phase of technology development, it is mandatory to achieve the highest possible calibration and thus the measurement accuracy for devices. Also, technology development engineers should provide accurate PDK model parameters of the backend devices (including transmission lines, e.g. [64]). So, there are many benefits from implementing the multiline TRL structures into a test chip. However, in the mass production phase, the increase of the cost of test caused by additional space-intensive structures is not acceptable. Instead, cost-effective verification methods should be used for a qualitative verification of the accuracy of the *in-situ* calibration (e.g. [49, 51, 81]).

Both the multiline TRL and the transfer TMR were qualitatively verified and the results were published for IHP SiGe:C BiCMOS [47], IBM RF CMOS SF8 [55] and STMicroelectronics' BiCMOS9MMW [56–58] processes. For the first time, the quantitative verification of the transfer TMR was performed which was implemented on the IBM RF CMOS SF8 test chip up to 110 GHz. Results were reported in [55].

6.1 Verification of the Multiline TRL

In Chapter 5, it was proved that the transmission lines from both IBM RF CMOS and STMicroelectronics' BiCMOS processes can be considered as supporting a single propagation mode up to 110 GHz. Also, two methods

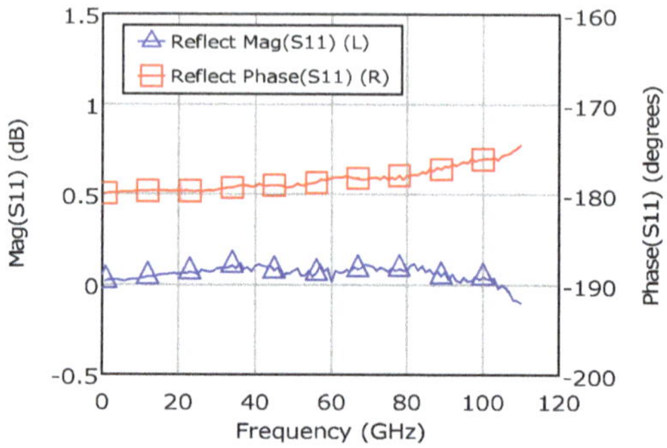

Figure 6.1 The magnitude and the phase of S_{11} of the reflect (short) standard implemented on STMicroelectronics' BiCMOS9MMW test chip measured with respect to the on-wafer M6-level multiline TRL.

were demonstrated for accurate measurement of the line characteristic impedance Z_{LINE} for both processes. Therefore, the *in-situ* multiline TRL is justified to be the benchmark reference for quantitative accuracy verification of other methods. The reference impedance of the benchmark TRL is transferred to $Z_{REF} = 50\ \Omega$ after accurate measurements of the line characteristic impedance Z_{LINE}. Here, some qualitative verification results are presented for the multiline TRL that were not reported in the referenced publications.

The common qualitative verification technique of a self-calibration method (such as TRL) is the evaluation of the reflection coefficient of the reflect standard r_X from (3.33). The verification criteria are described using the example of the TMR method later. The results for the multiline TRL are presented in Figure 6.1.

Also, the *S*-parameters of passive elements, such as calibration lines or de-embedding dummies, are good indicators of the calibration success.

Figure 6.2 exemplifies the return and the insertion loss of the thru standard implemented on the STMicroelectronics' BiCMOS9MMW test chip. The raw measurement results are corrected with probe tip as well as with the on-wafer multiline TRL. The reference plane of the on-wafer calibration is located at the top metal level and is shifted 7 μm away from the center of the thru towards the probe tips. The comparison proves that the contact pad parasitic impedance and the parasitic contribution of the interconnect section are accurately captured by the systematic measurement error model of the entire system and are calibrated out. The calibration reference impedance of both probe tip and on-wafer multiline TRL is set to $Z_{REF} = 50\ \Omega$.

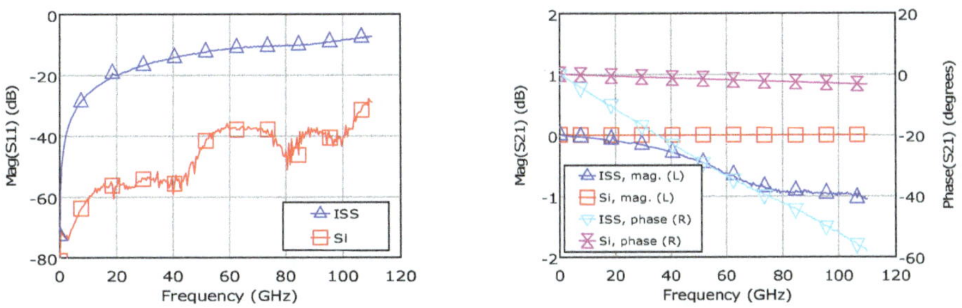

Figure 6.2 The magnitude of S_{11} and the magnitude and phase of S_{21} of the thru standard standards implemented on STMicroelectronics' BiCMOS9MMW test chip measured with respect to the probe tip (ISS) and on-wafer M6-level (Si) multiline TRL.

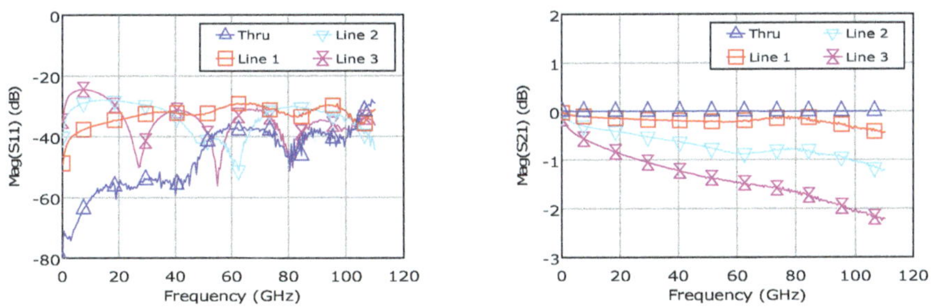

Figure 6.3 The magnitude of S_{11} and S_{21} of the calibration line standards implemented on STMicroelectronics' BiCMOS9MMW test chip measured with respect to the on-wafer multiline TRL.

The return and the insertion loss of all three lines from the test chip are given in Figure 6.3. Here, the parameters are corrected by the on-wafer TRL under the same conditions.

6.2 Quantitative Verification of the Transfer TMR

The quantitative accuracy verification of the transfer TMR was performed for the first time on Si for IBM 0.13 μm RF CMOS SF8 process [55]. The experimental setup for the 110 GHz wafer-level measurements of the RF CMOS test chip included an Agilent 8510XF VNA, a semi-automated wafer-probe station Summit 12k and 110 GHz wafer probes with a pitch of 100 μm from GGB Industries. The influence of contact repeatability on measurement data, calibration standards, verification elements and evaluation transistor S-parameters was reduced by acquiring raw data (with error correction turned off) in three measurement runs. Measurement data from the VNA were recorded with proprietary data acquisition software developed by IBM and with commercial software SussCal. Offline TMR and TRL calibration, error correction, and calibration comparison was performed using SussCal and MultiCal[1] software packages.

The following factors affect the results of the quantitative verification procedure:

1. Drift of the measurement system (basically, the drift of the VNA) within the experimental time;
2. Contact repeatability error on aluminum pads of the IBM RF CMOS SF8 process;
3. Accuracy of the reference calibration.

First, the system drift (instrument drift) was measured within the time interval of the DUT measurements using the calibration comparison technique and a commercially available alumina calibration substrate. The probe contact repeatability error is significantly less than the measurement instrument instability within the experimental time (about eight hours) due to fabrication of the standards with gold contact pads. Therefore, it was assumed that the error defined from the calibration comparison method represented merely the instability of the measurement instrument.

Also, the contact repeatability error was quantified for probing on aluminum pads from three identical measurement series A, B and C of the same wafer-embedded calibration set. Obtained results showed that the error between the series A–B, B–C and A–C are comparable and that it is larger than the instrument drift. It follows that the contact repeatability on aluminum pads affects experimental results more than the test instrument instability.

Next, the characteristic impedance Z_{LINE} of the RF CMOS line standards was extracted from both the lumped load as well as the calibration comparison methods. The commercially available calibration substrate CS-5 and the NIST RM8130 were used as the reference for the calibration comparison method of measurement for the reference multiline TRL on silicon (as shown in the previous chapter). Thus, the accuracy of the reference multiline TRL on silicon was ensured. The reference plane for the calibration comparison was set to the top metal level and the end of the thru standard. The reference impedance of the benchmark TRL was transferred to $Z_{REF} = 50\ \Omega$.

---

[1]Multical is available from NIST, Bolder CO, USA

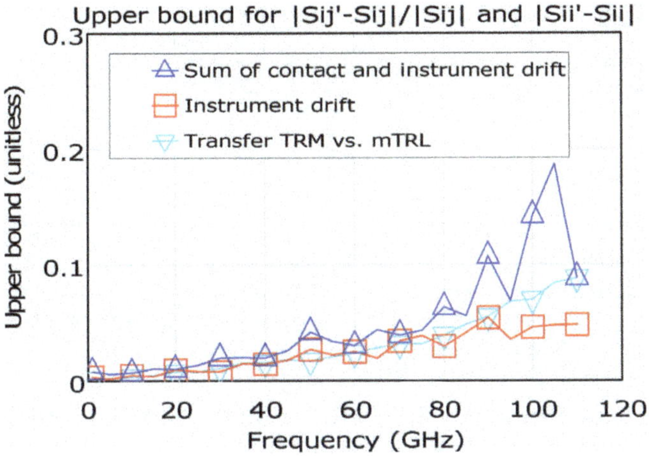

Figure 6.4 The instrument drift, the sum of the instrument drift and the contact repeatability error as well as the accuracy verification of the transfer TMR calibration on the IBM RF CMOS SF8 test chip.

Finally, the on-wafer TMR was verified against the benchmark multiline TRL. Similar to the previous experiment on the RM8130 (Chapter 5), the difference between the examined transfer TMR and the benchmark multiline TRL is comparable to the system instrument drift and the contact repeatability error for measurements on silicon up to 110 GHz (Figure 6.4) [55].

6.3 Qualitative Verification of the Transfer TMR

A qualitative verification of the *in-situ* TMR calibration can be applied when the reference calibration is not available.

6.3.1 Verification of the Reflect r_X

Similar to TRL, the TMR routine does not require the reflect standard to be fully known. Thus, the reflection coefficient r_X of the reflect standard calculated by the self-calibration step is a good indicator if the algorithm failed by any reasons. There are three evaluation criteria for r_X:

1. Its phase should be homogeneous within the entire frequency range;
2. Its magnitude should not show any gain, i.e. it must not exceed 0 dB;
3. The magnitude of S_{XY} (for paired elements) represented should be less than one (or negative in dB scale) across the entire frequency range.

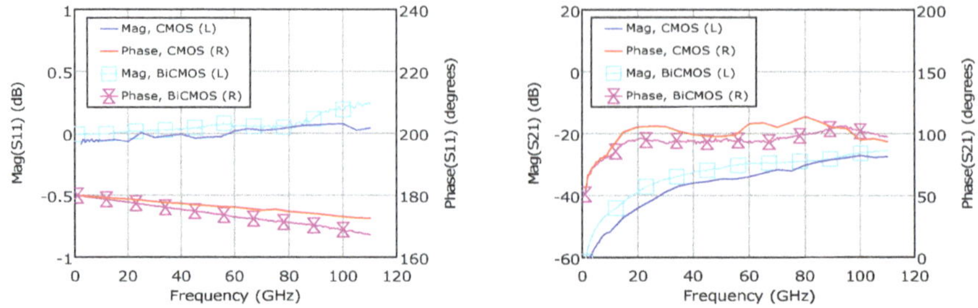

Figure 6.5 The magnitude and the phase of S_{11} (left) and S_{21} (right) of the reflect standard from the IBM RF CMOS SF8 and the STMicroelectronics' BiCMOS9MMW test chips corrected by the transfer TMR calibration.

Figure 6.5 compares the IBM RF CMOS SF8 and the STMicroelectronics' BiCMOS9MMW processes reflect standard S_{11} and S_{21} parameters corrected with respect to the *in-situ* transfer TMR calibration. Both measurements prove that the calibration is reliable within the entire frequency range. However, one observed a marginal increase of the magnitude S_{11} by 0.15 dB from 85 to 110 GHz in the BiCMOS experiment. This can be explained by the calibration residual error caused by the probe-tip design[2] (Chapter 9). Anyway, this error is relatively small and is rather comparable to the typical contact repeatability error on silicon pads (Figure 6.4).

6.3.2 Verification for Open and Attenuator

Accurate calibration aims at the exact definition of the reference impedance at both ports at $Z_{REF} = 50\ \Omega$). As it was shown in Chapter 2, the LRM-like procedures set the calibration reference impedance to the impedance of the load standard. If those loads are not equal at both VNA ports (e.g. due to the fabrication inaccuracy or the design specifics), the calibration reference impedance will be set inaccurately and will differ at the ports. Subsequently, the measurement of the same one-port device on different VNA ports gives results with either a positive or a negative offset from the expected value. The observed effect is often called "port asymmetry error". Figure 6.6 gives an example of calibration residual errors of the conventional LRM calibration. It was performed with an asymmetrical load resistance of $R_1 = 49\ \Omega$ and $R_2 = 51\ \Omega$ and the symmetrical short as reflect. All standards were implemented on a commercial calibration substrate. The magnitude of S_{11} and S_{22} parameters

[2] I used two different probe types for these experiments: Picoprobe from GGB for measurement of the IBM RF CMOS test chip and Infinity Probes from Cascade Microtech for measurement of the BiCMOS9MMW test chip from STMicroelectronics.

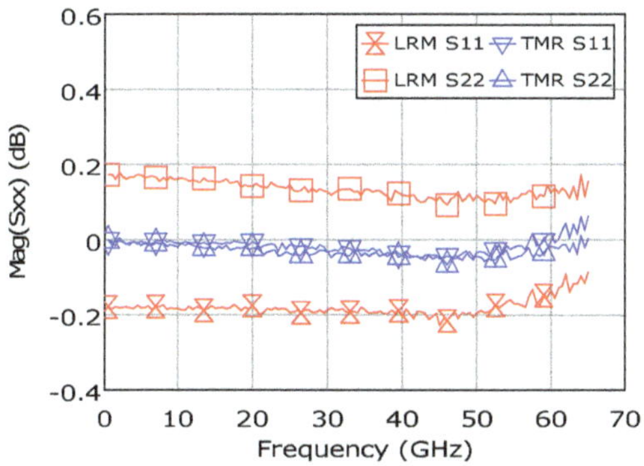

Figure 6.6 The magnitude of S_{11} and S_{22} of a symmetrical open element. Data are corrected with respect to the conventional LRM and transfer TMR calibration performed with 1 Ω asymmetry of the load resistance and short as reflect.

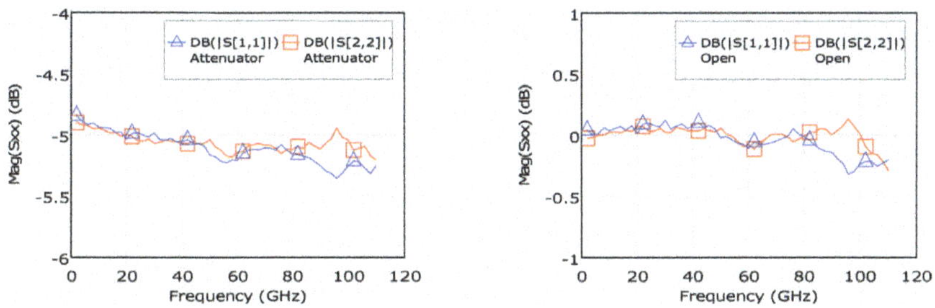

Figure 6.7 Results of port symmetry verification for the transfer TMR (short as reflect) using the series attenuator and the symmetrical open from the SiGe:C BiCMOS IHP test chip.

on a verification symmetrical open element shows an offset of ±1.85 dB from the expected value of 0 dB.

The method for the VNA ports symmetry verification was presented for the *in-situ* transfer TMR implemented on the SiGe:C BiCMOS process from IHP in [47]. Results from Figure 6.7 proved that S_{11} and S_{22} of the symmetrical open and attenuator (Figures 4.17 and 4.18) are in a good agreement. Hence, the transfer TMR algorithm has accurately compensated for the asymmetry of the load as discussed in Chapter 3 (see also Figure 5.13, Attachment 13.1, Figures 6.5 and 5.6).

However, there is an "unsymmetrical zone" between 80 and 105 GHz with the maximum error at 96 GHz of ±0.2 dB for attenuator and ±0.22 dB for open element. This is typically acceptable in practice for this frequency range. In [47], this effect was explained by the unsuspected simulation error of impedance of the load. This may also be caused by design errors of the load.

6.4 Conclusion

For the first time, the transfer TMR calibration was qualitatively and quantitatively verified on challenging high-performance RF silicon CMOS/BiCMOS technology up to 110 GHz. It was demonstrated that the difference of the benchmark multiline TRL and the TMR accuracy is comparable to the measurement system drift and the contact repeatability error.

It was shown that qualitative verification of calibration accuracy over the evaluation of the "port symmetry" is a good practical alternative to the calibration comparison method. Applying this method, the accuracy of the transfer TMR calibration can easily be verified in cases when the reference calibration is not available and/or when the use of the calibration comparison method is impractical or impossible. This may be the case during the in-line test or when implementing the calibration standards on the production test chip.

7

Advantages of *in-situ* Calibration

The benefits of the *in-situ* multiline TRL and the transfer TMR calibration methods were evaluated for every process: SeGe:C BiCMOS from IHP in [47], IBM RF CMOS SF8 in [55] and STMicroelectronics' BiCMOS9MMW in [56–58]. Several passive and active device parameters were used as test criteria. The advantages of the *in-situ* calibration over the conventional probe-tip calibration technique were demonstrated, which are the following:

1. Significant improvement of the measurement accuracy;
2. Simplification of the parasitic de-embedding.

This chapter gives an extensive comparison for devices of various geometries embedded in the STMicroelectronics' BiCMOS9MMW process modeling chip. Most of the results were not shown in the mentioned publications.

7.1 Choosing the Reference Probe-Tip Calibration Method for Comparison

Before verifying the *in-situ* calibration for characterization of active devices, it is necessary to have the benchmark probe-tip calibration available for the entire frequency range. The commonly used SOLT method cannot be considered as a solid reference at frequencies above 20 GHz. I evaluated various probe-tip calibration methods and proved that only the multiline TRL is capable to establish well-defined calibration reference impedance and to place accurately the measurement reference plane to the desired location. Accuracy of other methods is limited due to imperfection of planar calibration standards and probe misplacement errors. Results were presented in [63] (Attachment 16.1).

It was a joint work with STMicroelectroncs (Crolles, France) and TUD (Dresden, Germany). I developed the methodology of the analysis, planned the work, wrote the major part of the paper and presented the results at a conference. I acquired the measurement data together with Paulius Sakalas (TUD) at the measurement system of the TUD CEDIC laboratory. I did parameter extraction together with Nicolas Derrier and Didier Celi from STMicroelectronics. The data analysis was shared among all co-authors.

Based on the conclusions of this work, the probe-tip multiline TRL was chosen as the benchmark against the *in-situ* calibration. For comparison, this chapter will use the data corrected by the probe-tip SOLT as it is the industry-standard probe-tip calibration method [7].

7.2 Measurement Setup

The experiment on the BiCMOS9MMW chip was carried out on a broadband *S*-parameter measurement system from Cascade Microtech. It consisted of a semi-automated wafer-probe system Elite300^{TM}-AP, a pair of 110 GHz GSG Infinity Probes of 100 μm pitch and a fitting alumina calibration substrate (ISS), equipped with broadband 110 GHz PNA-X VNA and 4142B SMU from Agilent Technologies. The DUT input and output power levels (port 1 and port 2) were set to –35 and –15 dBm to protect the DUT from overload. The output power was calibration on the port 1 to ensure the desired measurement conditions over the frequency range.

The measured data of the ISS and *in-situ* standards, the de-embedding structures and the test transistors were acquired in raw format (with the *S*-parameter calibration turned off). The calibration and error correction were performed by WinCal XE^{TM}[1], and proprietary IC-CAP[2] script outside of the VNA on a computer and for the same raw data set. This approach ensured a minimal impact of contact repeatability and instrument drift on the accuracy of experimental results. Device parameters were extracted using IC-CAP.

Seven characterization strategies were applied to the measurement data of four HBT geometries (Table 7.1 and Table 7.2).

[1]WinCal XE is a product of Cascade Microtech, Inc.

[2]IC-CAP is a product of Agilent Technologies, Inc.

Table 7.1 Calibration and de-embedding experiments

Experiment	Standards	Calibration Algorithm	De-Embedding
1	ISS	SOLT	Two-step
2	ISS	LRRM	Two-step
3	ISS	LRM+	Two-step
4	ISS	mTRL	Two-step
5	*in-situ*	LRM+	Two-step
6	*in-situ*	mTRL	Two-step
7	ISS	SOLT	Six-step

Table 7.2 List of measured transistors

DUT	Geometry	Emitter Width, μm	Emitter Length, μm
DUT1	CBEBC	0.18	0.6
DUT2	CBEBC	0.18	5
DUT3	CBEBC	0.18	15
DUT4	C-5x(BEBC)	0.18	10
BUT5	CBE	0.18	5
BUT6	CBEB	0.18	5

The selected geometries represent a wide variety of DUT design, series and parallel parasitics. The comparison for DUT FoMs aimed at the following goals:

1. To show that *in-situ* calibration outperforms the conventional probe-tip calibration when the same two-step de-embedding is applied for shifting the measurement reference plane to the terminals of the intrinsic device (M6-level);
2. To compare the accuracy of two approaches: the most advanced six-step scalable de-embedding method augmented to the probe-tip calibration (Table 7.1, experiment 7) vs. the *in-situ* calibration augmented with the two-step de-embedding (Table 7.1, experiments 5 and 6).

The test included cold and biased measurement conditions of the DUTs.

7.3 Cold *S*-parameter Results

The Base/Emitter C_{BE}, Base/Collector C_{BC} and Collector/Emitter (Collector/Substrate) C_{CE} capacitances were extracted from the cold-mode *S*-parameters of all DUTs. Figure 7.1 compares the probe-tip SOLT and the reference multiline TRL calibration vs. the *in-situ* transfer TMR and the multiline TRL calibration methods (experiments 1, 4–6, Table 7.1)

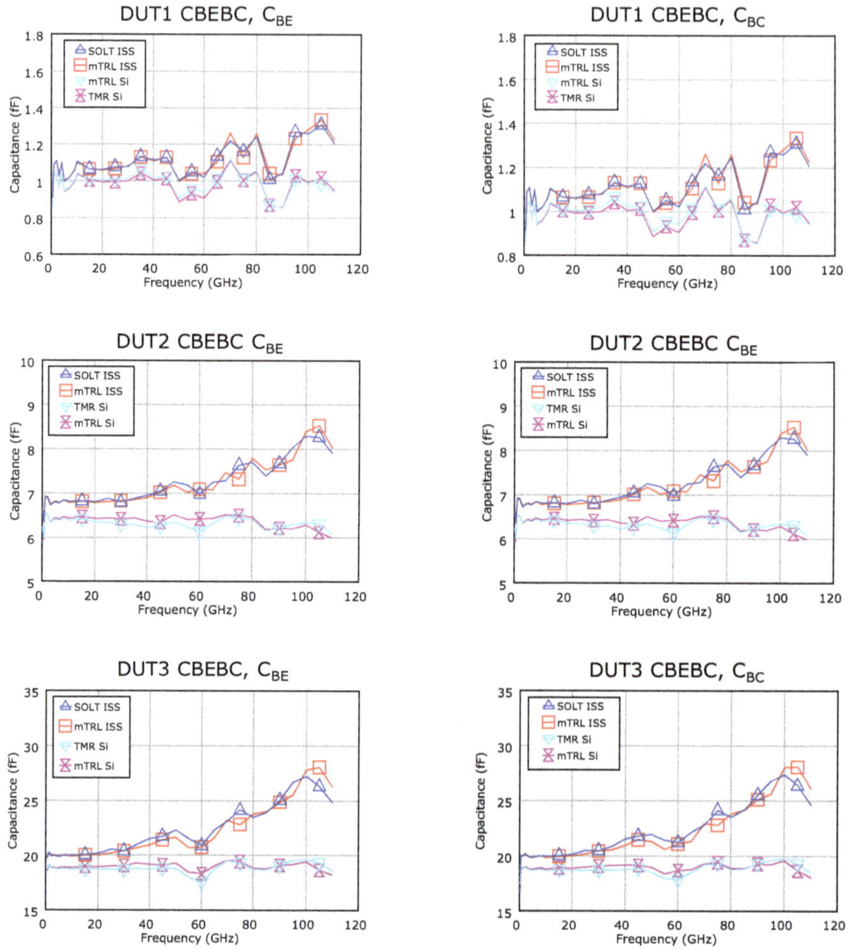

Figure 7.1 Comparison of the probe-tip SOLT and the multiline TRL vs. *in-situ* the transfer TMR and the multiline TRL for the Base/Emitter C_{BE} and the Base/Collector C_{BC} parameters measured for $V_{BE} = V_{BC} = 0$ V for $L_E = 0.6$ μm (top), $L_E = 5$ μm (middle) and $L_E = 15$ μm (bottom) CBEBC HBT with $W_E = 0.18$ μm.

for three DUTs of the CBCBE configuration with Emitter width of $W_E = 0.18$ μm and Emitter lengths L_E of 0.6, 5 and 15 μm (DUT1, DUT2 and DUT3, Table 7.2). The calibration data are de-embedded using the DUT-specific complete open and complete short elements. Similarly, Figure 7.2 gives the comparison results for the DUT5 and the DUT6 having the same W_E and L_E but different geometries (Table 7.2).

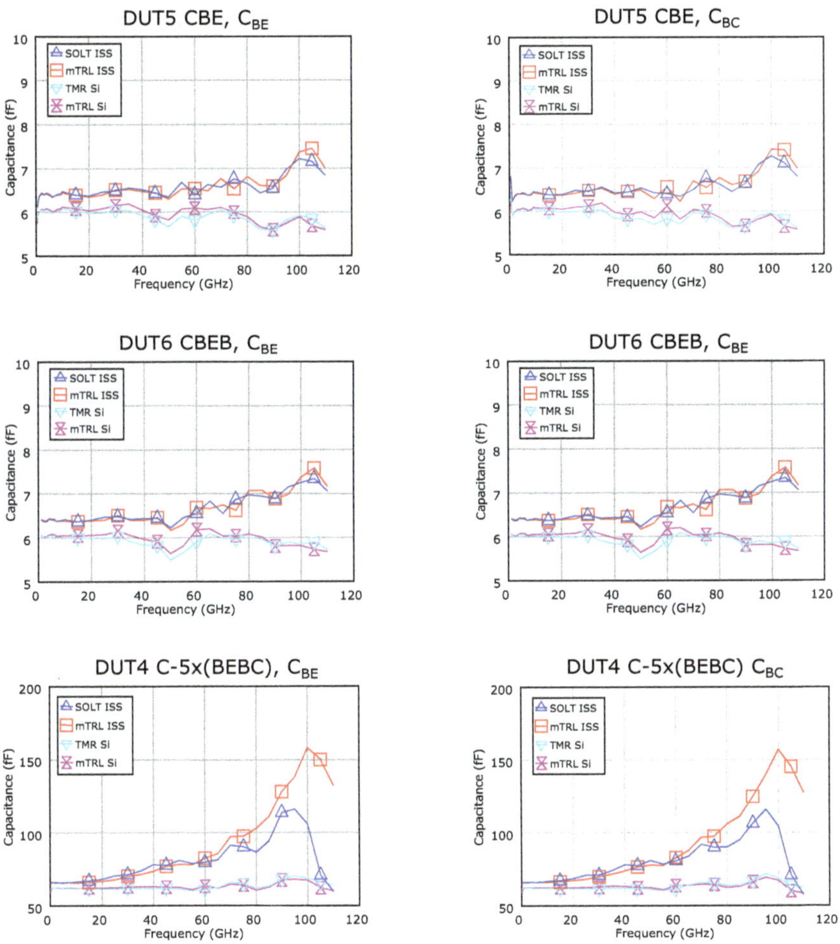

Figure 7.2 Comparison for the probe-tip SOLT and the multiline TRL vs. *in-situ* the transfer TMR and the multiline TRL for the Base/Emitter C_{BE} and the Base/Collector C_{BC} parameters of the CBE (top), the CBEB (middle) and HBT with $W_E = 0.18$ μm and $L_E = 5$ μm, and for the C-5x(BEBC) (bottom) HBT of $W_E = 0.18$ μm and $L_E = 10$ μm measured for $V_{BE} = V_{BC} = 0$ V.

Only the *in-situ* calibration (i.e. both the transfer TMR and the multiline TRL) gives the expected RF signature of the Π-equivalent circuit, which corresponds to the DUT in cold mode (0 V bias): the capacitances remain constant up to 110 GHz. The *in-situ* calibrations outperform probe-tip methods from 40 GHz for multi-finger and from 70 GHz for small transistors.

Obviously, the two-step de-embedding impedance model does not capture distributed effects of parasitics between the probe-tip and the M1 reference planes (Figure 4.2). The *in-situ* calibration includes these effects into the systematic measurement error model and thus calculates them out.

Figure 7.2: Comparison for the probe-tip SOLT and the multiline TRL vs. *in-situ* the transfer TMR and the multiline TRL for the Base/Emitter C_{BE} and the Base/Collector C_{BC} parameters of the CBE (top), the CBEB (middle) and HBT with $W_E = 0.18$ μm and $L_E = 5$ μm, and for the C-5x(BEBC) (bottom) HBT of $W_E = 0.18$ μm and $L_E = 10$ μm measured for $V_{BE} = V_{BC} = 0$ V.

Also, the probe-tip calibrations show a slightly greater value of the extracted capacitances across all DUT geometries and experiments. This systematic effect requires further investigation for a better understanding of its nature. Overall, this effect has a negligible impact on the results.

Comparison for other probe-tip calibration methods revealed a similar relationship. Figure 7.3 gives an example for the C-5x(BEBC) HBT (DUT4, Table 7.2) for experiments 1–6 (Table 7.1) [58].

In contrast to results presented in [63], the impact of the probe-tip calibration residual errors is not fully compensated by the two-step de-embedding in this experiment. It is explained as follows: the model of the VNA that was used for this experiment had a capability for accurate output power calibration at the mm-wave region (beyond 65 GHz) and for setting the power of the incident signal as low as minus 35 dBm. Thereby, the DUTs were not overloaded at mm-wave frequencies. It was possible to obtain more accurate results than in the experiment from [63].

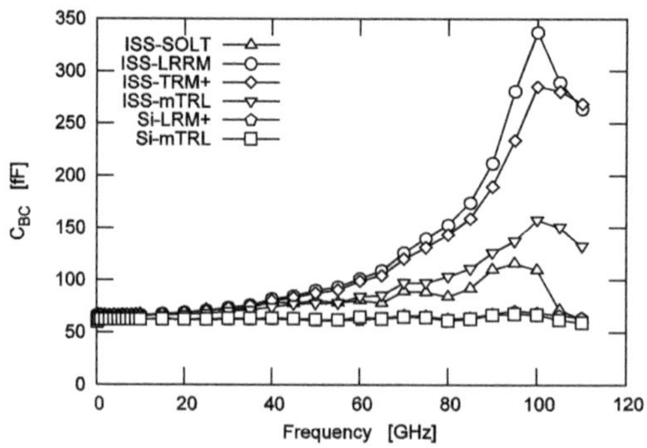

Figure 7.3 Base/Collector capacitance CBC of the C-5x(BEBC) HBT of $W_E = 0.18$ μm and $L_E = 10$ μm measured at $VBC = 0$ V.

7.4 Forward Mode *S*-parameter Results

7.4.1 *Y*-parameters

The DUT FoM Base/Emitter C_{BE}, Base/Collector C_{BC} and Collector/Emitter (Collector/Substrate) C_{CE} capacitances as well as maximum frequencies of the current gain f_T and the power gain f_{MAX} were extracted from the forward-mode *S*-parameters. Figure 7.4 compares extracted C_{BE} of the CBEBC HBT with $W_E = 0.18$ μm and $L_E = 5$ μm (DUT2, Table 7.2) for the probe-tip multiline TRL as well as the *in-situ* transfer TMR and the multiline TRL calibrations (experiments 4–6, Table 7.1) extracted from 20 and 100 GHz data. As before, the *in-situ* calibration provided more reliable results (i.e. minimal parameter variation over the frequency) for the same two-step de-embedding from the complete open and the complete short dummies.

7.4.2 f_T and f_{MAX}

Figure 7.5 shows peak f_T for a multi-finger device. Again, only the *in-situ* calibration methods give the expected results with constant f_T up to 110 GHz. While results for f_{MAX} demonstrated a comparable tendency, it is important to note that extraction of f_{MAX} is very challenging for advanced devices featuring small geometries. The port-to-port coupling that increases with

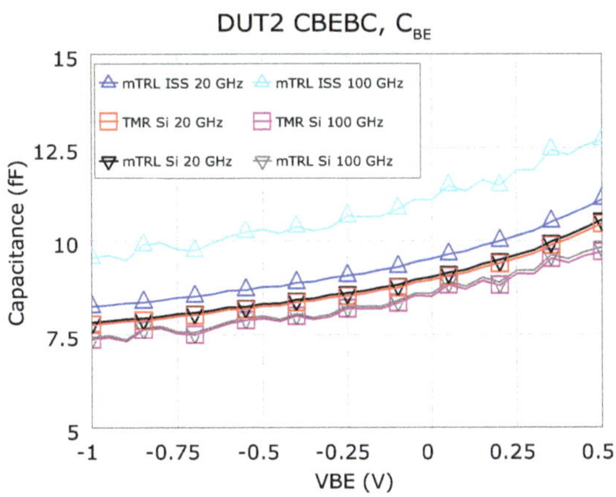

Figure 7.4 Base/Emitter capacitance CBE of the CBEBC HBT of $W_E = 0.18$ μm and $L_E = 5$ μm measured for forward-mode *S*-parameters extracted at 20 and 100 GHz.

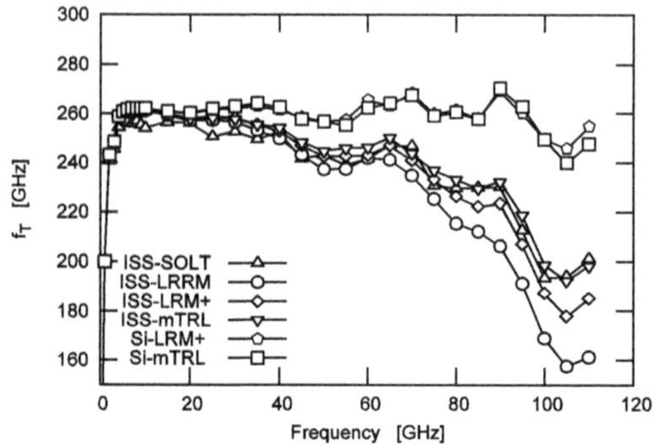

Figure 7.5 Peak f_T (for $JC = 15\ mA/\mu m^2$) vs. frequency [GHz] $@VBC = 0\ V$ for the C-5x(BEBC) HBT of $L_E = 10\ \mu m$.

the frequency affects the measurement accuracy of the DUT in the reverse mode. That is why crosstalk-corrected calibration methods (e.g. [82,83]) may increase the accuracy of f_{MAX} extraction of small devices.

Figure 7.6 compares f_T and f_{MAX} extracted from the *in-situ* transfer TMR calibration (experiment 5, Table 7.1) across various device geometries (DUT1 to DUT6, Table 7.2). f_T is presented for two locations of the reference plane: after the *in-situ* calibration (M6) and after the two-step de-embedding from the complete open and complete short dummies for each DUT (M1, Figure 4.2). Here, the crucial point is that f_T is comparable for DUT1, DUT2 and DUT3 (Figure 7.6, left). Often, engineers may observe a decrease of f_T for short transistors and attribute it to the so-called "3D effects" (e.g. [38]). Indeed, it is just a matter of how accurate and how close the measurement reference plane is shifted to the terminals of an intrinsic device. Results obtained here proved a very important statement from [58]: of the intrinsic DUT is constant regardless of its length.

So far, results of different calibration methods followed by the same two-step de-embedding step were compared (experiments 1–6, Table 7.1). As it was discussed in Chapter 2, the six-step scalable de-embedding method can accurately capture parasitics from the probe tip to M1. Consequently, the next experiment explores in how far this method can improve the accuracy of the two-step de-embedding for the probe-tip calibration. Figure 7.7 displays f_T of the same multi-finger transistor extracted from the experiments 1 and 5–7 (Table 7.1). Indeed, the six-step scalable de-embedding augmented to the less accurate probe-tip SOLT yielded results comparable to the *in-situ* calibration for the investigated frequency range. *Y*-parameters showed the same trend.

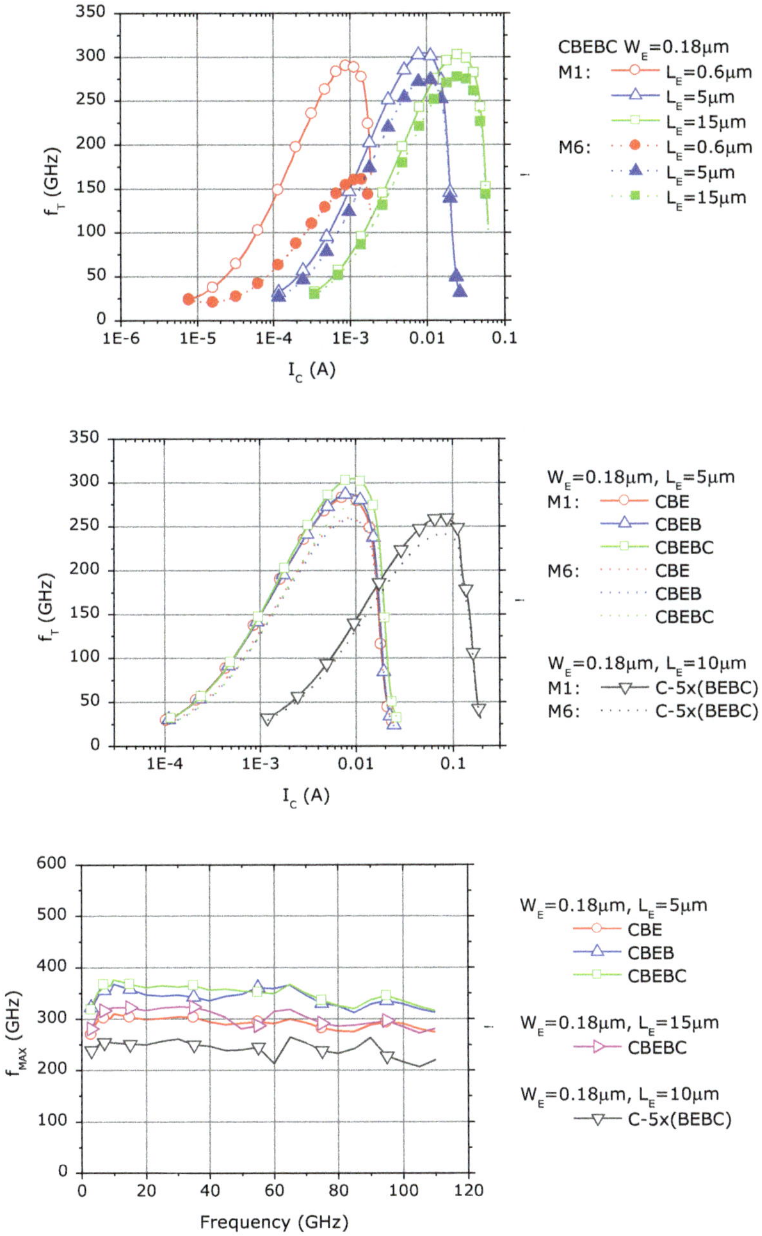

Figure 7.6 f_T over i_C (top, middle) and f_{MAX} over frequency (bottom) for various device geometries extracted from the *in-situ* transfer TMR calibration extracted from $f = 20\ GHz$ and $V_{BE} = 0.89\ V$, respectively.

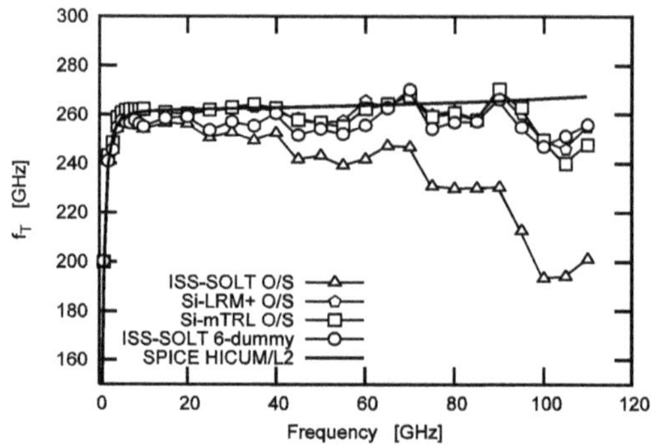

Figure 7.7 Peak f_T (for $J_C = 15mA/\mu m^2$) vs. frequency @$V_{BC} = 0\ V$, for a five-emitter finger HBT of $L_E = 10\ \mu m$, after the SOLT probe-tip calibration and complete open/complete short (white triangle) vs. six-step de-embedding (white circle), *in-situ* calibrations (white diamond and square) and the HICUM V2.30 compact model simulated with ADS software (solid line).

7.5 Comparison with a Compact Model

In [58], I also compared the extracted FoM with the HICUM Level2 compact model. I worked on this paper together with Nicolas Derrier from STMicroelectroncs (Crolles, France) (Attachment 16.2). He extracted parameters of the compact model HICUM Level2 (L2) v2.30 [84, 85] for a test transistor DUT4 (Table 7.2) from DC and S-parameter measurements data, calibrated by the probe-tip SOLT and de-embedded using the two-step method. RF FoM f_T and f_{MAX} were computed at 20 GHz and used inside the extraction sequence in order to avoid parasitic effects. ADS 2011.10 software was used to simulate the obtained model cards. The HICUM simulation results for f_T of the DUT4 are added to Figure 7.7. This comparison can be considered as an alternative proof because the HICUM compact model is based on the device's physics. As we noted in [58], the Y-parameters demonstrated the same trend.

7.6 Conclusion

The numerous experiments on HBTs of various geometries revealed a substantial accuracy improvement of parameter extraction when using the *in-situ* calibration (multiline TRL or transfer TMR) followed by the simple two-step de-embedding. The *in-situ* standards are optimized to calibrate out the major part of the backend parasitics and to move the measurement reference plane to the M6 level in one step. It was proved that the remaining M6-M1 parasitics can be removed by a simple two-step de-embedding. A comparison with the compact model (HICUM V2.30) confirmed the findings.

The probe-tip calibration augmented by the six-step scalable de-embedding and the *in-situ* calibration augmented by the two-step de-embedding yielded comparable results. Therefore, the choice of the method depends on the capability of the technology as well as on individual preferences.

The *in-situ* calibration requires some preparation work to characterize several electrical parameters of standards. In return, a simple two-step de-embedding can be applied. The probe-tip calibration demands an advanced six-step de-embedding with dummies easy to be realized. However, its application is cumbersome and needs additional silicon area.

8

Addressing Issues of On-Wafer Standards

8.1 How to Handle Standard Properties that Change over Temperature

Device characterization for model development and IC design debug requires measurement over a wide temperature range. For instance, a typical temperature test plan for silicon RF device characterization covers $-40C\ldots+125C$ with at least six temperature points. Advanced research requirements go far beyond these limits: down to 4°K and up to +500°C. This means that the *in-situ* calibration should be capable to deal with standards operating at those temperatures, e.g. standards with electrical characteristics that may change significantly with the temperature.

This is the opposite situation to a conventional probe-tip calibration. The auxiliary chucks that hold the calibration standards on a probe system are thermally isolated from the system thermal chuck (except special cryogenic probe systems). The calibration substrates are kept at relatively constant temperature conditions, ensuring that the electrical characteristics of planar standards remain almost the same during the temperature cycling (Figure 8.1).

Such a solution was developed in the early 1990s and since then it has remained the *de facto* industry standard [86]. There were no results or work previously published on the accuracy verification of the lumped-standard-based calibration over temperature. That is why an initial investigation of this problem was done as well.

The thermal sensitivity of standards depends on the design and on the specifics of a particular semiconductor process. Therefore, it was decided to verify the accuracy of over-temperature calibration using the following strategy:

1. Investigate the temperature impact on characteristics of commercial CPW standards;

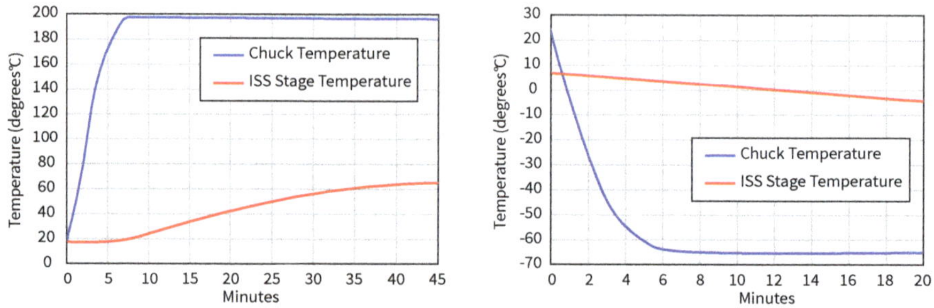

Figure 8.1 Temperature relationship between the main system and the auxiliary (ISS) chucks for a conventional RF probe system. The ISS stage is thermally isolated; therefore, the ISS standards stay at relatively constant temperature conditions. Picture from [86] (reconstructed).

2. Analyze the probe-tip calibration accuracy over multiple temperatures;
3. Identify the most and the least sensitive factors that influence the accuracy of different methods.

Here, the results of the probe-tip calibration investigation previously published in [87–89] are presented. The outcome of this approach can easily be adapted to any custom calibration.

8.1.1 Experiment with Commercial Standards at Cryogenic Temperatures

The first paper on this topic addressed the negative temperature range, from room temperature down to 4 K [87] (Attachment 17.1). The experiments were carried out on a cryogenic wafer-level measurement system from SUSS MicroTech [90]. It was a collaborative work with FBH (Berlin, Germany) and TUD (Dresden, Germany). My contribution to this work was the idea, planning of the work, development of the test and verification methodology, data acquisition, some calculations and data analysis as well as writing a paper. Ralf Doerner (FBH) greatly contributed to the calculation and data analysis of this work, while Paulis Sakalas (TUD) assisted the measurement session.

The test setup allowed measurements up to 40 GHz. That was enough to get a good understanding of important aspects of the multiline TRL and the transfer TMR methods related with temperature variation. We measured the propagation constant γ, the capacitance per unit length C' and the characteristic impedance Z_{LINE} of the line as well as the resistance R_{LOAD} of the load standards over multiple temperature points. We used the CSR-8 calibration substrate from SUSS MicroTec in this experiment.

It was found that the attenuation constant α, the relative phase constant β/β_0 as well as the capacitance per unit length of the tested substrate decrease with the temperature. The characteristic impedance Z_{LINE} increases negligibly and is about 1% greater at 4 K than at room temperature.

As expected, we observed a decrease of the attenuation constant of about 50% at 4 K from its room temperature value. It is a common practice to neglect the loss of the thru standard for lumped-standard-based calibration methods. This simplifies the calibration process but adds some errors. The decrease of the attenuation constant reduces this sort of errors.

A very stable and flat temperature coefficient of the CSR-8 load resistor was recognized. The resistance deviated from its room temperature value within $+/-3\%$. This is a typical trimming accuracy for commercial load resistors across the substrate [68]. Thus, the load resistance of the test substrate remains within the limits of common specification.

Finally, the accuracy of the transfer TMR calibration was verified for the edges of the temperature region of $+25°$C and 4°K and under the conditions of conventional room temperature models and fully characterized thru and load standards at the temperature. Results obtained from the calibration comparison method (with the benchmark multiline TRL) proved that no additional efforts in characterizing standards are required for accurate transfer TRM calibration on CSR-8 substrate at cryogenic temperatures. It is expected that the same will hold for *in-situ* standards, except the load resistance R_{LOAD}. It may be more sensitive to temperature and might require a correction.

8.1.2 Experiment with Commercial Standards at High Temperatures

The investigation was continued for a high temperature range. The results were presented at the 71st ARFTG Conference where the paper received the Best Interactive Forum Paper Award [88] (Attachment 17.2). For this work, I developed the idea, test and verification methodology, planned the work, and performed some calculations and data analysis, and was responsible for writing the paper. Ralf Doerner (FBH) did the experimental measurements and provided the calculation and data analysis.

The results for cryogenic temperatures reported in [87] were measured with the cryogenic model of the |Z|-Probe from SUSS MicroTec.

Because we did not have an alternative RF probe for measurement at cryogenic temperatures, the question whether the probe design had any impact on the extracted characteristics of standards remained unanswered.

So, first of all, two identical experiments were made. The attenuation constant α and the relative phase constant β/β_0 were measured at room temperature and at 150°C for the CSR-8 substrate, while using two different probes of the same configuration: |Z|-Probe from SUSS MicroTec and Picoprobe from GGB Industries (referred to as "probe P" in the article). It was demonstrated that the impact of the probe design is negligible for these experiments.

An ISS 101–190 substrate from Cascade Microtech was included in the experimental series (referred to as "substrate I" in the article). The setup had a measurement capability of 50 GHz and allowed testing from room temperature up to 150°C.

This work proved the previous statement that the characteristic impedance Z_{LINE} of the alumina CPW line decreases with increasing temperature. The temperature dependence of the attenuation constant α and the relative phase constant β/β_0 is in the same order of magnitude for both tested substrates. However, we identified a much higher temperature instability of the load resistance R_{LOAD} for the ISS 101–190. For the tested ISS, the extracted RTC was 92.9 ppm/°C, while the CSR-8 demonstrated RTC = 16.0 ppm/°C. This value is quite similar to the RTC = 10.2 ppm/°C extracted from the measurement data of another CSR-8 substrate, presented in [87]. Such a relatively high variation of the load RTC is caused by differences in the design of standards and manufacturing technologies used by both vendors. If it is not taken into consideration, it affects the accuracy of the lumped-standard-based calibration.

8.1.3 Sensitivity Analysis of Over-Temperature Probe-Tip Calibration for Multiline TRL, Transfer TMR and LRRM Methods

Changes in the electrical characteristics of standards over temperature may impact accuracy of various calibration procedures in different ways. For instance, the propagation constant γ of the line is automatically measured within the TRL algorithm. Therefore, TRL should be irrelevant to the variation in γ. Yet, it is sensitive to the line characteristic impedance Z_{LINE} as it leads to variation of the calibration reference impedance Z_{REF}.

The reference impedance of the lumped-standard-based methods (such as transfer TMR and LRRM) is defined by the impedance of the load standard Z_{LOAD}. Moreover, the characteristic impedance Z_{LINE} of the thru and its propagation constant γ also contribute to the calculation of the reference impedance Z_{REF} of both TMR and LRRM methods, as it was shown in Chapter 2 (see also Figure 3.2). Therefore, the accuracy of the over-temperature TMR and LRRM methods is affected by variations of three parameters: the impedance of the load Z_{LOAD}, the characteristic impedance Z_{LINE} and the propagation constant γ of the thru. The quantitative analysis of the contribution of certain standards to the overall accuracy of over-temperature S-parameter calibration methods has not yet been investigated. That is why I developed a practical method and concluded with recommendations on how to address the variation of electrical properties of standards across wide temperature ranges in the paper [90] (Attachment 17.3). I defined the scope of the work, developed the concept of the investigation, did several calculations, performed the data analysis and wrote the major part of the paper. Ralf Doerner (FBH, Berlin) acquired all measurement results, did the major part of calculation as well as data analysis. Gavin Fisher (Cascade Microtech) contributed to some calculation results as well as to the preparation of the final manuscript.

The multiline TRL, the transfer TMR and the LRRM methods were investigated up to 50 GHz and for the temperature range from +25°C to 150°C. The frequency and the temperature ranges were limited by the capability of the FBH measurement setup.

For the considered experimental setup, it was found that the lumped-standard-based calibration methods were less sensitive to temperature. In addition, it was shown that the load resistance R_{LOAD} is the main influencing factor for both LRRM and the transfer TMR methods. Therefore, a simple correction for resistance $R_{LOAD}(T)$ for a given temperature T can improve the calibration to the required level of accuracy:

$$R_{LOAD}(T) = R_{LOAD}(25) + \frac{RTC\, R_{LOAD}(25)(T-25)}{10^6}, \qquad (8.1)$$

where RTC is the resistor thermal coefficient of the load resistor, ppm/°C; R_{LOAD} is the initial resistance value measured at room temperature, $T = 25$°C. In this case, $R_{LOAD}(25) = 50\ \Omega$.

Addressing temperature variation of the R_{LOAD} reduced overall TMR calibration error from 2.2% down to negligible 0.6%. Correction for α and Z_{LINE} of the thru only had a marginal effect.

All wafer-level systems configured for the electro-thermal characterization of semiconductor devices include several SMUs. They are used for biasing the DUT to measure its output I/V characteristics as well as the hot (biased) *S*-parameters (Figure 2.6). As it was shown in paper [88], the resistance R_{LOAD} of the load standard can be measured at the current temperature over the RF probes and the measurement system's bias-Ts "on-the-fly". Such a configuration yields sufficient measurement accuracy. Thus, there is no need for the definition of the load resistor RTC prior to calibration.

The detailed description of the "on-the-fly" correction of electrical characteristics of a standard at a given temperature is also given in the patent application DE102009029906 [91]. This method can successfully be applied for other types of calibration substrates, temperature and frequency ranges as well as to the *in-situ* (on-wafer) calibration elements.

8.2 Impact of the Fabrication Inaccuracy of Custom Standards to the Accuracy of DUT Parameter Extraction

Instability of the fabrication process leads to variation of electrical parameters of custom calibration standards and reduces accuracy of calibrated measurement results. Here, one deals with the situation similar to the already discussed one in the previous section (the over-temperature calibration). The characteristic impedance Z_{LINE} of the line and the load resistance R_{LOAD} vary between different shots (test chips) on one wafer and also from wafer to wafer.

To evaluate this effect, the load resistance R_{LOAD} was measured and the capacitance per unit length C' of the line standards was extracted on a test wafer of the STMicroelectronics' BiCMOS9MMW process. Five test chips on different locations were selected to have the widest possible parameter distribution across the wafer (Figure 8.2). The measured and extracted results are presented in Figure 8.3.

Based on this test, the conclusion that the variations of C' and R_{LOAD} were not correlated to each other was made. It means that capacitance per unit length C' should be measured on every test chip for an accurate definition of the TRL reference impeadance Z_{REF}.

All known measurement methods for the characteristic impedance Z_{LINE} of the line are time consuming. They require multipe calculations which are difficult to automate. The intermediate results require evaluation and

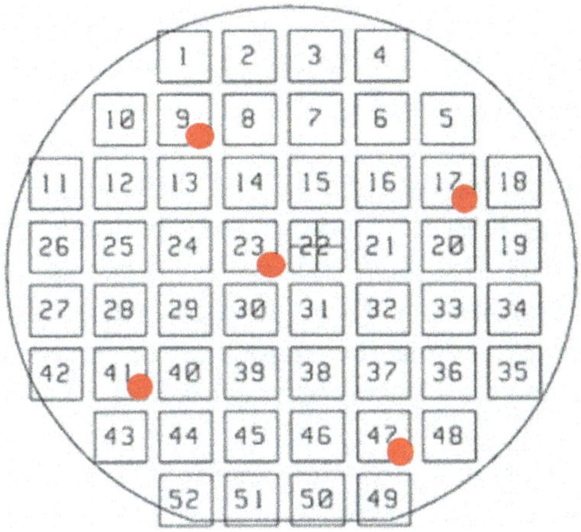

Figure 8.2 Test shots of the STMicroelectroncs' BiCMOS9MMW process.

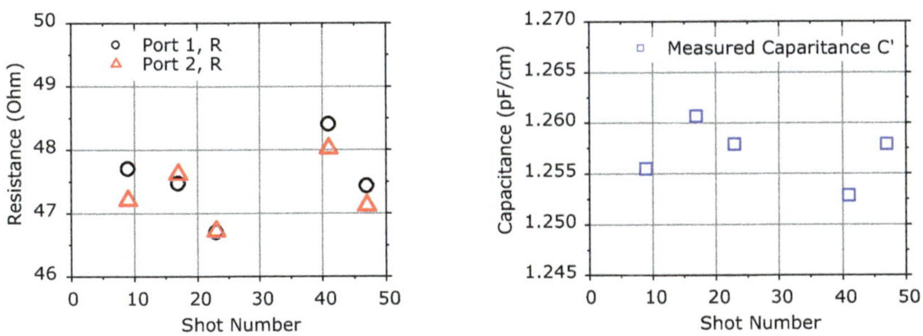

Figure 8.3 Variation of the load resistance (left) and the extracted capacitance per unit length (right) of the load and line standards over five shots for one wafer of the STMicroelectronics' BiCMOS9MMW process.

validation by the operator. Moreover, the capacitance per unit length C' measured at NIST and supplied with every reference material RM8130 is defined with an accuracy of up to the fifth digit after comma. It is very challenging to achieve the same level of accuracy at normal laboratory conditions and on the typical wafer-level RF device characterization system, especially for silicon processes. The contact repeatibility on the aluninum pads is significantly worse than on gold, as I showed in Figure 6.4. Therefore, it is cumbersome to implement the on-wafer multiline TRL into the in-line device characterization workflow.

Further, potential errors of transistor parameter extraction were investigated for the multiline TRL calibration when the calibration reference impedance Z_{REF} was not accurately measured. The analysis method and the results are given in [92] (Attachment 17.4). This work evaluated multiline TRL implemented in two processes: GaAs from FBH and BiCMOS9MMW from STMicroelectronics. I developed the analysis methodology, planned the work, acquired data for the BiCMOS process, did a part of calculations and parameter extraction, performed the data analysis with my co-author Ralf Doerner (FBH, Berlin) and wrote the major part of the conference paper. Ralf Doerner acquired the data for GaAs process and did the major part of calculations.

We investigated the accuracy requirements of the measurement of the characteristic impedance Z_{LINE} of the line using the lumped load method from [72] on GaAs and silicon processes for parameter extraction of passive and active devices. It was demonstrated that load resistance R_{LOAD} required for the extraction of the line capacitance per unit length C' can be measured with an accuracy of up to the second digit after comma. This corresponds to the measurement capability of a conventional wafer-level RF device characterization setup equipped with the SMUs and bias-Ts.

Also, it was shown that the transformation of the calibration reference impedance to $Z_{REF} = 50\ \Omega$ was not needed for extracting equivalent junction capacitances of a tested device for a given level of uncertainty of 10% and for the tested processes. Both findings proved that the implementation of the *in-situ* TRL into a characterization workflow in the investigated processes is significantly easier than we originally expected.

8.3 Conclusion

In this chapter, the analysis of the specific problems of the *in-situ* RF calibration was made. It included the instability of the electrical characteristics of standards due to temperature changes and inaccuracy of the fabrication process. These problems do not arise for the conventional probe-tip calibration. That is why they had never been investigated before.

It was demonstrated that it is necessary to know the actual resistance of the load R_{LOAD} on every calibration chip and at every calibration temperature because the load resistance has the main impact on accuracy of the transfer TMR calibration. Also, it was shown that the unknown R_{LOAD} can be measured over the bias-Ts using the RF probes and the SMU. The accuracy of this method is sufficient.

It was proved that the measurement of the line capacitance per unit length C' with the accuracy of up to the second digit after comma is acceptable for most practical cases. It is not necessary to transfer the multiline TRL calibration reference impedance to $Z_{REF} = 50\ \Omega$ for extraction of the small signal model parameters, such as junction capacitances at frequencies greater than 20 GHz. This holds for the evaluated BiCMOS9MMW test chip from STMicroelectronics.

All these findings significantly simplified the requirements that have to be satisfied for the *in-situ* transfer TMR and the multiline TRL.

9

Selected Aspects of Measurement and Calibration Assurance

As it was discussed in Chapter 5, commercial standards are currently the only available reference for measuring the characteristic impedance Z_{LINE} of custom line standards above 40 GHz. Establishing the measurement assurance at the wafer level is still not completely solved. Research at NIST and later at PTB (Germany) has substantially contributed to the definition of the tractability chain at the wafer level. Current investigations are focused on the development of robust calibration methods, uncertainty analysis as well as the design improvement for planar standards. This chapter discusses the impact of the boundary conditions and the RF probe design on the electrical characteristics of planar standards. As it is shown here, the impact of these effects is often underestimated and ignored. However, it must be taken into consideration when establishing the reference calibration at mm-wave frequencies. Furthermore, the modified ripple-test is presented as a simple practical first-order estimate of wafer-level calibration residual errors.

9.1 Optimization of Boundary Conditions for Coplanar Standards

With increasing frequency, parasitic effects caused by the environment become important, including the mechanical support on which the calibration set and the DUT are placed. They may give rise to unwanted higher-order modes and/or resonances [93,94]. Multi-mode wave propagation on a calibration substrate is a very critical situation since common calibration procedures assume only single-mode propagation. Any additional modes present in a system corrupt calibration and, finally, measurement data.

The line structure may not only support the desired CPW mode but also other modes. These parasitic modes like surface-wave modes and parallel-plate modes may couple to the CPW mode beyond certain frequencies or even over the whole frequency range [95]. The critical frequencies are related to the mode order and depend on the physical dimensions of the line structure, including substrate thickness and substrate boundary conditions.

The calibration substrate should be thin enough to assure that the critical frequencies for surface mode excitation occur above the band of interest. This common solution avoids losses in propagation [96]. Unfortunately, the fundamental parallel-plate mode still can be excited and, thus, has to be considered.

The further optimization of the standard boundary conditions for mm-wave frequencies was shown in [97]. In a joint work with Ralf Doerner (FBH, Berlin) and Ed Godshalk (Maxim Integrated, USA), I developed the method of analysis and the design of experiment, did some initial measurements and calculations, planned and wrote the major part of the paper. Ralf Doerner's main contribution was in data acquisition for all experiments, extensive calculations and data post-processing. Analysis of the results was a cooperative work with Ralf Doerner and Ed Godshalk. Ed Godshalk also greatly contributed to writing and presenting the final paper.

This work was focused on distinctive boundary conditions:

1. The calibration substrate resting on a metal plate. This is a "floating ground" condition;
2. The calibration substrate resting on a material similar to the substrate material. This is a "matched" condition;
3. The calibration substrate suspended in air. This is close to an "open" condition;
4. The calibration substrate resting on radiation-absorbing material (RAM). This is a "lossy" condition (Figure 9.1).

The work [98] studied influences of surface-wave modes in dielectric substrates that can affect electrical characteristics of coplanar standards at mm-wave frequencies. A couple of recommendations were presented, such as suspending the calibration substrate, decreasing its thickness and the use of radiation-absorbing materials (RAM). The practical results demonstrated noticeable improvement of the probe-tip calibration accuracy when the calibration substrate was suspended over the RAM.

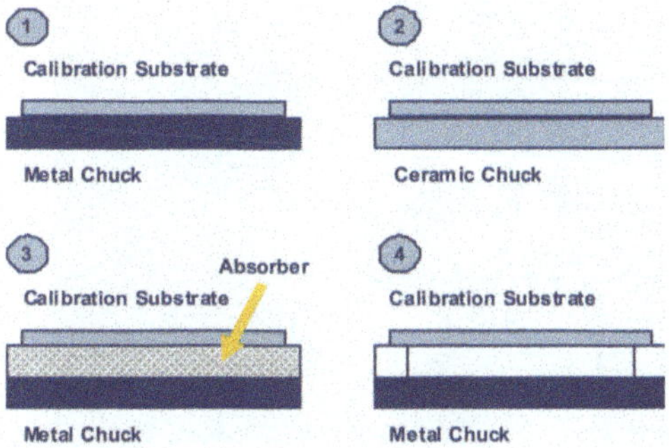

Figure 9.1 Experimental setups providing different calibration boundary conditions: (1) metal "grounded", (2) ceramic "matched", (3) absorber "lossy" and (4) suspended "open".

Later, [96] suggested placing the calibration substrate directly on the RAM, which is more practical and handy for daily use. Since then, it has become an industry-standard technique for accuracy improvement of the probe-tip calibration at mm-wave frequencies and beyond. Today, advanced RF probe systems include a calibration auxiliary chuck made from RAM [99, 100].

When measurement frequencies reached sub-millimeter wave bands, the substrate thickness could not be reduced further without the risk of damaging it during handling. For this reason, [97] investigated the approach of placing the calibration substrate on an extremely thick (compared to its own thickness, i.e. about 10 mm) ceramic support. It was shown that the coupling between the CPW and surface wave modes can be reduced when using small enough CPW dimensions such as those recommended in [98]. Experimental results invalidate the initial concern that CPW transmission lines on such a thick "substrate" would suffer from the coupling into the numerous possible modes. Nowadays, this solution is implemented in mm-wave and sub-THz probe systems (Figure 9.2) [101, 102].

Next, I show how boundary conditions influence electrical characteristics of calibration structures on a 250 μm thick calibration substrate ISS 104-783A. The results were obtained using Infinity I110-A-GSG-100 GSG probes with 100 μm pitch. Both calibration substrate and probe are from Cascade Microtech.

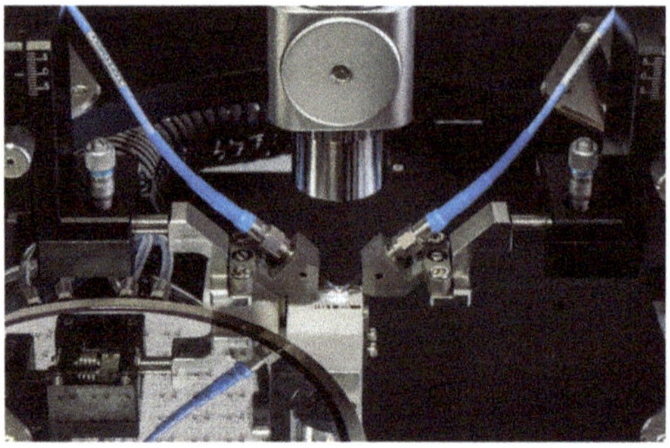

Figure 9.2 Auxiliary chuck made of ceramic on MPI TS150 manual probe station.

9.1.1 CPW Transmission Line Parameters

Propagation constant γ and characteristic impedance Z_{LINE} of a line are important figures of merit for distributed calibration standards. The propagation constant γ is extracted directly from the measurement set of lines by using the multiline TRL calibration method. The line capacitance per unit length C' is extracted from the lumped load method allowing the determination of Z_{LINE} from [21, 72].

The experiment was carried out for four boundary conditions (Figure 9.1). Additionally, the parameters of the tested line were calculated according to the analytical CPW model from [103].

Results showed a noticeable impact of boundary conditions on the propagation constant γ. It should be emphasized that the propagation constant γ was extracted under the assumption of a single-mode propagation. The non-ideal behavior of the propagation constant could be attributed to a possible multi-mode propagation. Consequently, the propagation constant γ and the deduced characteristic impedance Z_{LINE} of the line differ from the case of the pure CPW mode. The "grounded" case shows a periodic ripple above 20 GHz. In contrast to this, the attenuation constant for the "lossy" case is "smoother", but beyond 60 GHz, it rises faster than expected. Owing to its erratic behavior, the "open" boundary is not a viable alternative above 50 GHz. The CPW line on the "matched" ceramic chuck gives the lowest attenuation and a monotonic response up to 110 GHz.

Figures 9.3 and 9.4 show that the "lossy" boundary leads to a strong increase of the attenuation constant and roll-off of the imaginary part of Z_{LINE} at higher frequencies. The "matched" boundary leads to the least reactance of Z_{LINE} above 85 GHz [97]. In conclusion, the calibration and measurement accuracy is improved.

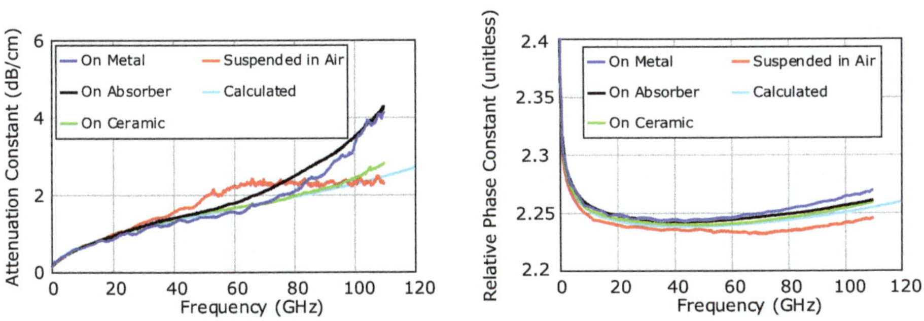

Figure 9.3 Extracted attenuation and relative phase constants of the same CPW line from a 250 μm alumina calibration substrate for four measurement boundary conditions: direct on the metal chuck (grounded), on metal chuck and absorber (lossy), suspended in air (open) and on ceramic chuck (matched) in comparison to a model.

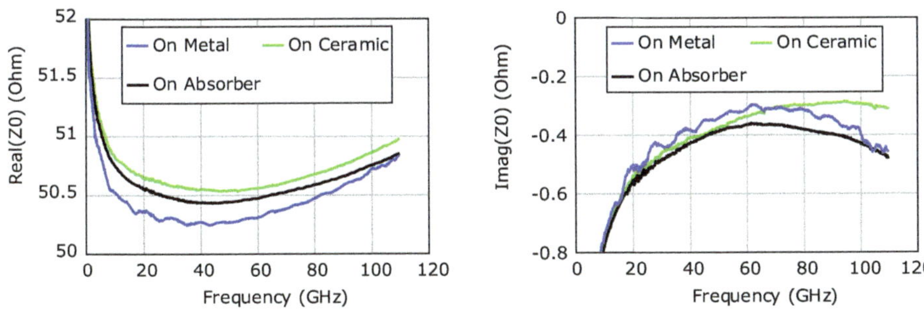

Figure 9.4 The extracted characteristic impedance of the same CPW line from a 250 μm alumina calibration substrate for three measurement boundary conditions: direct on the metal chuck (grounded), on metal chuck and absorber (lossy), and on ceramic chuck (matched).

9.1.2 Properties of Lumped Standards

The reflection coefficient of the short element implemented on the same calibration substrate was measured with respect to the probe-tip multiline TRL. A bare ceramic area of the substrate surface was used to simulate the open standard, which was originally missing on the test substrate. The reflection coefficients of both the short and the "open-on-ceramic" element are presented in Figures 9.5 and 9.6. Similar for lines, it was found that the boundary conditions influence characteristics of the lumped standards already above 30 GHz.

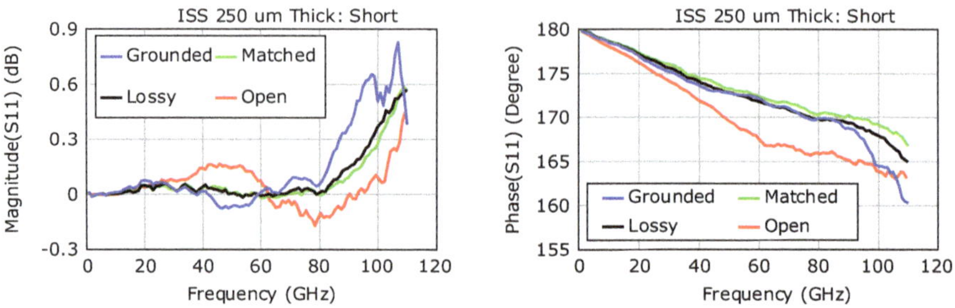

Figure 9.5 Magnitude (left) and phase (right) of the reflection coefficient of a short on a 250 μm alumina calibration substrate, measured with respect to the multiline TRL for different measurement boundary conditions: direct on the metal chuck (grounded), on metal chuck and absorber (lossy), on the ceramic chuck (matched) and elevated in air (open).

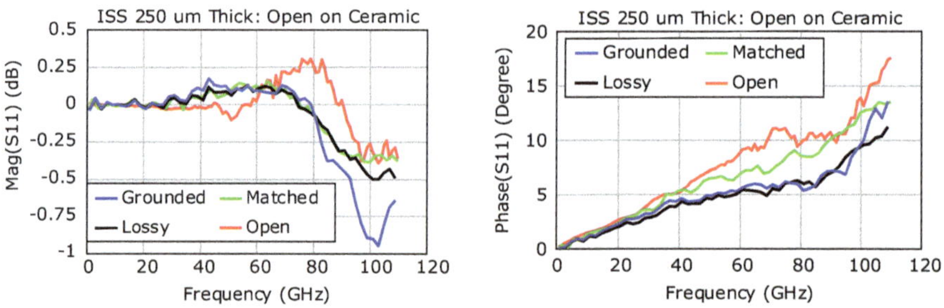

Figure 9.6 Magnitude (left) and phase (right) of the reflection coefficient of an "open-on-ceramic" on a 250 μm alumina calibration substrate, measured with respect to the multiline TRL for different measurement boundary conditions: direct on the metal chuck (grounded), on metal chuck and absorber (lossy), on the ceramic chuck (matched) and elevated in air (open).

9.2 Impact of the Probe-Tip Design

When experimenting with boundary conditions for the calibration substrate, a fairly unexpected phenomenon was observed. The reflection coefficients of the same short and open elements corrected by the same calibration method and under identical boundary conditions differed slightly on two measurement systems. The deviations noticeably increased with frequency.

It was ensured that the calibration reference plane was accurately set to the probe-tip end. Thus, applying the benchmarking multiline TRL algorithm from NIST should exclude any differences between setups, such as probes, cables, connectors, etc. After repeating the measurements several times, it was realized that the measurement errors were of systematic nature and caused by the wafer probes. In fact, one setup was equipped with

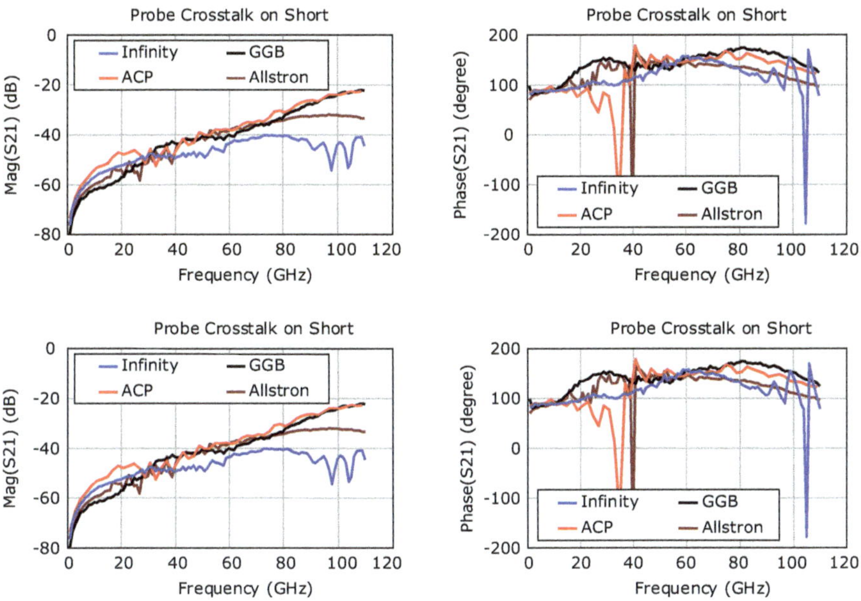

Figure 9.7 Magnitude (left) and phase of crosstalk measured for a pair of shorts on the same calibration substrate on the ceramic chuck (matched) corrected by the multiline TRL.

Infinity Probe, while another one had ACP probes. After changing probes, it became possible to verify the original hypothesis: the residual calibration errors are also dependent on the probe geometry.

The probe-to-probe crosstalk measured on a pair of coplanar shorts with a spacing of 150 μm is shown in Figure 9.7. For this experiment, the boundary conditions were "matched" by placing the substrate on an about 10 mm thick ceramic holder. Four different designs of 110 GHz 100 μm pitch GSG probes were evaluated. The system was calibrated every time by the multiline TRL.

Despite the relatively simple test structure and the wide separation between the elements, the crosstalk had a more complex nature than expected. Distinct discontinuities and resonances were found in the range from 18 to 50 GHz for coplanar probe types and above 80 GHz for the thin-film microstrip probe. A similar situation was observed for the "open-on-ceramic" reflection element, when probes touch a bare ceramic area on the calibration substrate.

The magnitude of the reflection coefficient of the short differs by more than 1 dB above 100 GHz (Figure 9.8). It is obvious that extracting the equivalent circuit parameters of a DUT becomes a very hard task under such

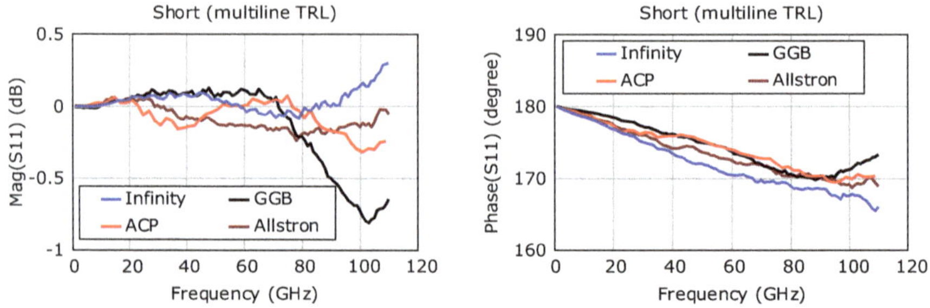

Figure 9.8 Magnitude (left) and phase (right) of the reflection coefficients of the short on the calibration substrate on the ceramic chuck (matched) corrected by multiline TRL.

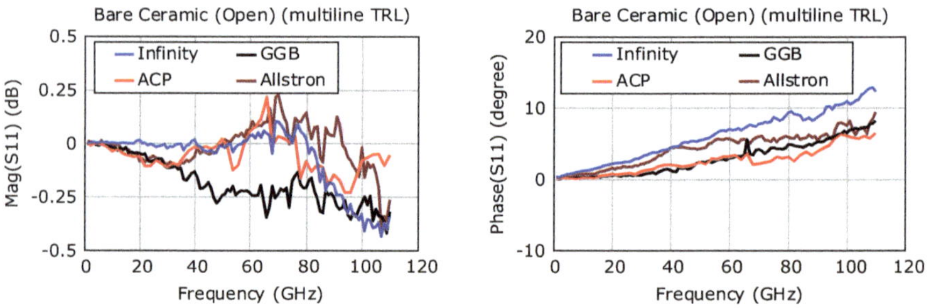

Figure 9.9 Magnitude (left) and phase (right) of the reflection coefficients of the probes contacting a bare ceramic (open) of the calibration substrate placed on the ceramic chuck (matched conditions) corrected by multiline TRL.

conditions. Similar to the magnitude, the phase of the reflection coefficient shows ripples and different trends. Results for the open element reveal a similar effect (Figure 9.9).

The nature of probe-dependent residual errors is complex and still not absolutely clear. The transition of the electro-magnetic fields via the probe to the DUT is not ideal and varies depending on the probe design. Conventional calibration methods cannot adequately address such parasitic phenomena as signal radiation, conversion to the higher-order modes and coupling (probe-to-probe, probe-to-DUT and probe-to-chuck).

To prove this hypothesis, a similar experiment was performed in different environment and with a custom designed test chip realized on a SeGe:C process from IHP Microelectronics. The calibration standards were laid out as microstrip elements of completely different to the commercial CPW elements geometries. However, the same kind of the probe-dependent calibration residual errors was identified (Figure 9.10) [104]. Thus, the probe design indeed

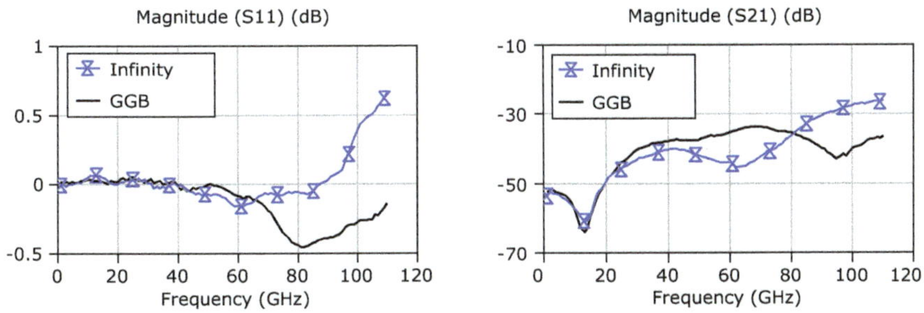

Figure 9.10 Magnitude of the reflection (left) and the transmission coefficients (right) of the short element measured by two different RF probes. The short is design as a microstrip element and is realized as a SiGe:C test chip. Results are corrected by the TRL performed on the microstrip lines on a custom SeGe:C test chip.

contributes to the calibration residual errors independent of the calibration standard design.

Obviously, the presented results are just phenomenological. A detailed investigation and more experiments (both practical and simulation) are required to develop a good understanding of the parasitic effects discussed here. This might be an important topic for future work.

Because the discussed phenomenon was seen at nearly every stage of this work on the *in-situ* calibration, the following rules were kept across all experiments:

1. Always use probes of the same design when comparing data from different setups;
2. Place the commercial calibration substrates on a thick ceramic holder.

These rules helped me to minimize possible sources of calibration residual errors in my experiments.

9.3 Estimation of Calibration Residual Errors

A simple practical method of estimation of the calibration residual error over "ripple test" is widely used for coaxial VNA measurements. The method is based on measurement of a precision air-line that is terminated with a highly reflective short standard at its end (e.g. [105, 106]). This configuration allows direct measurement of an effective source match that contributes the most to the residual errors according to [107].

When measuring the residual source match, the operator estimates the maximum magnitude of ripples that appear on the reflection coefficient curve

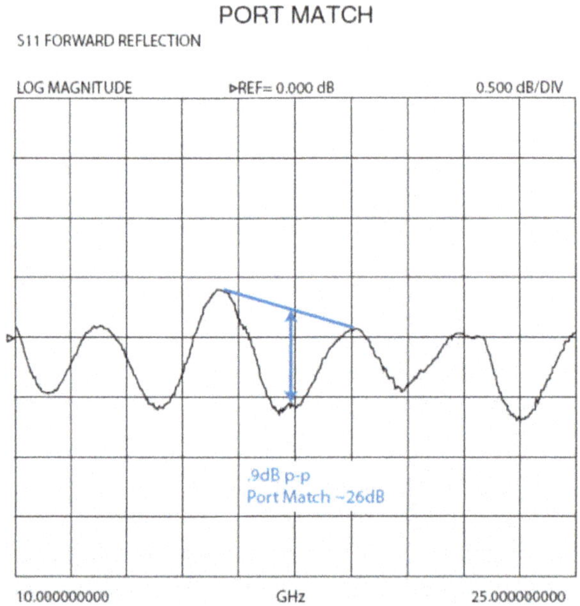

Figure 9.11 Measuring of a port match with a conventional ripple test. Picture from [106].

(Figure 9.11). The effective port match is then calculated from the ripple magnitude value (or can be found in a reference table) [106]. For the given example, the ripple magnitude of 0.9 dB corresponds to the minus 26 dB source match.

The conventional procedure for measurement of the port match is a manual process that requires the operator to interact with the measurement system. A more efficient way for an automated port match measurement was proposed in [107]. However, it is not applicable for wafer-level calibration: the automated calculation procedure requires the reference line to be perfectly matched with the system reference impedance.

Together with Prof. Holger Heuermann (FH Aachen, Germany), I developed a special algorithm of an automated ripple test for wafer-level applications. The algorithm was presented in [108] (Attachment 18). In this work, I provided a list of requirements, planned the work and made all measurements. I also calculated electrical characteristics of the reference coplanar line required for the algorithm. Prof. Heuermann implemented the impedance transformation step into the automated ripple test, did all calculations and wrote the major part of the paper. I presented this work at the ARFTG-69th conference. The new algorithm is also described in patent applications [109, 110]. I was in charge for implementing it into the commercial calibration software SussCal Professional. This method and the software tool allowed a simple and automated estimation of the S-parameter magnitude and the phase error bounds at the wafer-level.

9.4 Conclusion

In this chapter, the boundary conditions of CPW standards and the design of the wafer probes were presented as important aspects that influence calibration and measurement accuracy at the wafer level. Before this work, these two aspects were out of scope of the research work on establishing the wafer-level measurement assurance. It was demonstrated that they should be taken into consideration in the same manner as the variation of standard geometries and capabilities of calibration methods.

It was shown that using a thick ceramic support of the commercial calibration substrates is a preferable system configuration for measuring at mm-wave frequencies. For the novel concept probe system chuck optimized for mm-wave measurements, the US patent US7,999,563 was awarded [111]. The chuck is implemented in some advanced RF wafer-level measurement systems.

The probes of the same design were used for measurement experiments when possible. It minimized the potential impact of the probe-tip design on the measurement results obtained from different system setups. Developing an explicit solution for this problem was out of the scope of this work.

A modified ripple test was developed for simple and automated evaluation of the calibration and measurement errors at the wafer level. This new method was implemented in the commercial calibration software SussCal Professional and it is protected by pending international patent applications.

10

Conclusion and Further Research

10.1 Conclusion

Presently, calibrating the wafer-level S-parameter system for characterization of advanced silicon devices consists of two steps. First, the system is calibrated using the commercially available calibration standards and the measurement reference plane is set at the probe tips. Next, the pad de-embedding calculates out the parasitic impedances associated with the device contact pads, the access lines and the via stack. This approach has several drawbacks, such as limited accuracy and its complexity that increases with the frequency.

This work proposed an alternative way to calibrate the system using *in-situ* standards. The method presented here makes the intermediate probe-tip calibration step obsolete. The measurement reference plane can be set close to the terminals of a lateral device. Thus, the largest portion of the contact pads and the BEOL parasitics is included in the systematic measurement error model and is eliminated by the conventional S-parameter error correction procedure. In conclusion, the measurement accuracy is significantly improved while the system calibration procedure is simplified. The developed method enables accurate parameter extraction of advanced CMOS and BiCMOS devices at mm-wave frequencies. It appears to be a promising candidate for application at the sub-mm-wave frequency range.

The developed technique was implemented and verified on two BiCMOS and one RF CMOS processes from leading semiconductor manufacturers. The experiments were carried out up to 110 GHz for all processes. The measurement frequency was solely limited by the capability of the test equipment. The verification results demonstrated that the new technique significantly outperformed the conventional method in measurement accuracy already from 20 GHz. It was also demonstrated that the accuracy of the conventional method can be increased by implementing complex de-embedding methods, such as the scalable six-step procedure. However, the new *in-situ* calibration developed in this work wins over the advanced de-embedding methods thanks to its ease of implementation.

The key contribution of this work was to realize the lumped-standard-based transfer TMR calibration method on silicon. The main advantage of such a method over the distributed-standard-based calibration methods (such as TRL and the multiline TRL) is that all standards take relatively little space on the test chip and share the same geometry. Therefore, the transfer TMR has two significant advantages over TRL-like methods:

1. Saving expensive space of the test chip: for both CMOS and BiCMOS implementations, the transfer TMR standards required about 15 times less of the test chip size than the multiline TRL standards;
2. Repositioning of the wafer probes during the calibration process not required: calibration can be easily automated, increasing the repeatability while reducing the calibration time and, finally, the calibration costs.

Previously, there were several attempts to implement the *in-situ* lumped-standard calibration on silicon (e.g. [70]). Since then, it has become a common understanding in industry that it was generally difficult to replace the pad de-embedding step by the straightforward *in-situ* calibration for silicon processes. In particular, the application of conventional lumped-standard calibration methods (such as SOLT) on silicon is very challenging and, therefore, is considered to be worthless. By adopting the transfer TMR, this work opened new perspectives on how to improve the accuracy of the system calibration and the device characterization. It enabled accurate device parameter extraction and model verification at upper mm-wave frequency bands.

Based on a detailed analysis, this work proposed an optimal solution for each step of the development, implementation and verification of the *in-situ* calibration. That included the analysis of the capability of modern VNAs, the development and validation of calibration solutions suitable for silicon applications, the location of the *in-situ* calibration and the measurement reference plane, the creation of generalized calibration standard design rules for both CMOS and BiCMOS technologies, the characterization of calibration standards and, finally, the verification of calibration and measurement accuracy.

The transfer TMR (commercial name: LRM+) and the general RRMT+ developed in this work were implemented in the wafer-level calibration software SussCal Professional from SUSS MicroTec (later WinCal 4.5 from Cascade Microtech). Three patents have been granted and four patent applications were filed for methods and techniques presented here.

10.2 Further Research

Going beyond the original scope of this work, some aspects of the *in-situ* calibration technique could be subject to additional research. It particularly concerns such topics as the calibration over multiple temperatures, multiport calibration, solutions for systems affected by a strong port crosstalk and the multimode signal propagation as well as establishing *in-situ* calibration and measurement assurance.

10.2.1 Calibration over Multiple Temperatures

Modeling, parameter extraction and model verification require several measurement series of the same DUT at multiple temperatures. The DUT has to be heated up or cooled down in a wide temperature range (at least from minus 40°C to plus 125°C) to characterize the variation of its temperature-dependent parameters. A similar dependency is to expect for the electrical properties of the *in-situ* calibration standards and has to be taken into account. Chapter 8 presented a method to analyze and improve the over-temperature calibration accuracy. For reason of generalization, all investigations were carried out on commercially available ceramic calibration substrates. It was shown that the variation of the load resistance with the measurement temperature is the most influencing factor on the accuracy of the lumped-standard-based calibration. While a similar constellation is highly probable to expect for the *in-situ* calibration, an experimental proof of this hypothesis is still outstanding. Also, the characteristic impedance Z_{LINE} and the propagation constant γ of the thru standard may differ in their variation over temperature depending on the design of standards, process specifics and the frequency range.

10.2.2 Multiport Calibration

As discussed in Chapter 3, some advanced device characterization techniques as well as the IC design verification demand a multiport *in-situ* calibration solution. The general multiport RRMT+ method was developed and verified to address these challenges. It was proved that the RRMT+ has significant advantages over conventional methods: it enables the use of not-ideal reflect and asymmetrical load standards. In addition, RRMT+ relaxes requirements to the cross-over and the loop-pack thru standards that are very hard to characterize and to realize on a test chip. This work proved capabilities of the RRMT+ method up to 40 GHz on a commercial ceramic calibration substrate. As RRMT+ is a further development of the transfer TMR algorithm, their design requirements to the *in-situ* standards mostly resemble each other. The verification of the RRMT+ on a specific silicon process is the subject of future investigations.

10.2.3 Solution for Multimode Propagation Conditions

This work revealed some specific aspects of mm-wave on-wafer calibration and measurements that may require a special attention. It became obvious that the design of the RF probe tip contributes to calibration residual errors at mm-wave frequency range. This effect can be explained by unwanted probe coupling with the test structure, the excitation and propagation of higher-order modes and the signal radiation. These factors should be taken into consideration when expanding the proposed method to sub-THz frequencies. All efforts must be taken to prevent (or to suppress) propagating of the higher-order modes in calibration standards. However, this might become difficult to realize. Alternatively, the systematic error term model can be extended to address the multimode measurement conditions. This will require a corresponding multimode calibration solution, which is not present at the moment.

10.2.4 Crosstalk

The evaluated multiline TRL and the transfer TMR calibration methods are based on the commonly used seven-term model of systematic measurement errors. Therefore, the final error correction of the DUT S-parameters is provided by the test instrument. The device characterization software can download error-corrected DUT S-parameters from the VNA memory and perform further data processing (de-embedding, parameter extraction, etc.). This is a significant advantage because the management of big arrays of the error terms is under the responsibility of the VNA firmware. However, the seven-term model does not address the crosstalk between the device terminals. Further improvement of the proposed *in-situ* calibration method may be done by including the port crosstalk terms into the error model. This will lead to a more complex calibration solution but will support measurements close to THz frequencies. Most important, however, is that the optimization and the implementation steps of such advanced methods will remain similar to those presented in this work.

Bibliography

[1] P. Yue and M. Rodwell, "mm-Wave IC design: The transition from III–V to CMOS circuit techniques," in *IEEE Compound Semiconductor IC symposium (CSIC). Short course RF and high speed CMOS*, Nov. 12 2006.

[2] D. Harame, "RF CMOS and SiGe BiCMOS technologies for RF analog an mixed signal applications," in *Semiconductor Integrated Circuit Design and Applications Short Course*, (Stanford University, Palo Alto, CA (USA)), August 2–6 2004.

[3] "European DotFive Project," technical report, DotFive Consortium, 2011.

[4] "DotSeven. Towards 0.7 THz Silicon-Germanium heterojunction bipolar technology (0.7 THz SiGe HBT)," tech. rep., DotSeven Consortium, 2012.

[5] "CT209. From RF to MMW and THz silicon SoC technologies [RF2THZ SISOC]," tech. rep., CATRENE, 2012.

[6] A. Rumiantsev, "S-parameter calibration of two-port setup: How to choose the optimal calibration method," *IEEE International Microwave Symposium. WMD Workshop "Laboratory Class: Wafer-Level S-Parameter Calibration Techniques*", June 2011.

[7] "Radio frequency and analog/mixed-signal technologies for wireless communications," technical report, International Technology Roadmap for Semiconductors (ITRS), 2011.

[8] H. Ruecker, B. Heinemann, W. Winkler, R. Barth, J. Borngraber, J. Drews, G. G. Fischer, A. Fox, T. Grabolla, U. Haak, D. Knoll, F. Korndorfer, A. Mai, S. Marschmeyer, P. Schley, D. Schmidt, J. Schmidt, K. Schulz, B. Tillack, D. Wolansky, and Y. Yamamoto, "A 0.13 um SiGe BiCMOS technology featuring fT/fmaxof 240/330 GHz and gate delays below 3 ps," in *IEEE Bipolar/ BiCMOS Circuits and Technology Meeting (BCTM)*, pp. 166–169, 2009.

[9] J. Berkner, *Kompaktmodelle fuer Bipolartransistoren: Praxis der Modellirung, Messung und Parameterbestimmung – SGP, VBIC, HICUM und MEXTRAM*. Renningen: expert verlag, 2002.

[10] D. Celi, "Appendix B: Fréquence de Transition fT, Introduction à la Modélisation du Transistor Bipolaire," manuscript, ST Microelectronics, 2003.

[11] D. Celi, "Appendix C: Definition of the maximum oscillation frequency," in Introduction à la Modélisation du Transistor Bipolaire," manuscript, ST Microelectronics, 2003.

[12] R. Singh, M. Oprysko, and D. Harame, *Silicon Germanium. Technology, modeling, and design*. IEEE Press, Wiley-IEEE Press, 2004.

[13] M. Urteaga, R. Pierson, P. Rowell, V. Jain, E. Lobisser, and M. J. W. Rodwell, "130 nm InP DHBTs with ft $>$ 0.52 THz and fmax $>$ 1.1 THz," in *69th Annual Device Research Conference (DRC)*, pp. 281–282, 2011.

[14] D.-H. Kim, B. Brar, and J. A. del Alamo, "fT = 688 GHz and fmax = 800 GHz in Lg = 40 nm In 0.7Ga0.3As MHEMTs with $gm_{max} > 2.7mS/um$," in *IEEE International Electron Devices Meeting (IEDM)*, 5–7 Dec 2011.

[15] S. Munkyo, M. Urteaga, J. Hacker, A. Young, Z. Griffith, V. Jain, R. Pierson, P. Rowell, A. Skalare, A. Peralta, R. Lin, D. Pukala, and M. Rodwell, "InP HBT IC technology for terahertz frequencies: fundamental oscillators up to 0.57 THz," *IEEE Journal of Solid-State Circuits*, vol. 46, no. 10, pp. 2203–2214, 2011.

[16] M. Rodwell and M. Seo, "100–1000 GHz bipolar ICs: Device and circuit design principles," in *IEEE Bipolar BiCMOS Circuits and Technology Meeting (BCTM). Short Course.*, (Atlanta, GA), October, 9th 2011.

[17] M. Nishimoto and J. Laskar, "Study of millimeter-wave on-wafer calibration techniques and effect upon HEMT parameter extraction," in *ARFTG Microwave Measurements Conference-Spring, 43rd*, vol. 25, pp. 19–23, 1994.

[18] D. M. Pozar, *Microwave engineering*. John Weliy & Sons, Inc., 3rd edition ed., 2004.

[19] K. Kurokawa, "Power waves and the scattering matrix," *IEEE Transactions on Microwave Theory and Techniques*, vol. 13, no. 2, pp. 194–202, 1965.

[20] R. B. Marks and D. F. Williams, “A general waveguide circuit theory,” *Journal of research of the National Institute of Standards and Technology*, vol. 97, no. 5, pp. 533–562, 1992.
[21] R. B. Marks and D. F. Williams, “Characteristic impedance determination using propagation constant measurement,” *IEEE Microwave and Guided Wave Letters*, vol. 1, no. 6, pp. 141–143, 1991.
[22] A. Rumiantsev, R. Doerner, and S. Thies, “Calibration standards verification procedure using the calibration comparison technique,” in *European Microwave Conference, 36th*, pp. 489–491, 2006.
[23] “Implementing the Line-Reflect-Match (LRM) calibration techniue for on-wafer measurements,” The HP8510B Technical Feature, Cascade Microtech, Inc., Beaverton, OR, USA, 1990.
[24] A. Davidson, K. Jones, and E. Strid, “LRM and LRRM calibrations with automatic determination of load inductance,” in *ARFTG Microwave Measurements Conference-Fall, 36th*, vol. 18, pp. 57–63, 1990.
[25] S. Lautzenhiser, A. Davidson, and K. Jones, “Improve accuracy of on-wafer tests via LRM calibration,” *Microwaves & RF*, vol. 29, no. January, pp. 105–109, 1990.
[26] D. Rytting, “ARFTG 50 year network analyzer history,” in *ARFTG Microwave Measurements Conference-Spring, 71th*, pp. 11–18, 20 June 2008 2008.
[27] “Agilent PNA Microwave Network Analyzers,” Product Brochure 5989-7604EN, Agilent Technologies, Inc., Santa Clara, CA, USA, 2009.
[28] “R&S ZVA Vector Network Analyzer. High performance up to 110 GHz with up to four test ports,” Product Brochure, Rohde & Schwarz GmbH, Muenchen, Germany, 2011.
[29] “MS4640A Series. Family of RF to Microwave and Millimeter-wave Vector Network Analyzers,” Product Brochure 11410-00435, Anritsu, Morgan Hill, CA, USA, 2009.
[30] “Dual-source 4-port network analyzers true-mode stimulus application (TMSA),” Technical Overview, Agilent Technologies, Inc, Santa Clara, CA, USA, 2007.
[31] A. Rumiantsev and N. Ridler, “VNA calibration,” *IEEE Microwave Magazine*, vol. 9, no. 3, pp. 86–99, 2008.

[32] J. Martens, T. Roberts, K. Noujeim, and S. Reyes, "70 kHz–145 GHz broadband S-parameter measurements: calibration and verification possibilities," in *Asia-Pacific Microwave Conference (APMC)*, pp. 433–435, Dec. 4–7 2012.

[33] A. Lord, "Advanced waveguide probe technologies for mm-wave and terahertz probing," in *European Microwave Week. Agilent Technologies Workshop*, (Amsterdam, The Netherlands), 2012.

[34] T. J. Reck, C. Lihan, Z. Chunhu, C. Groppi, X. Haiyong, A. Arsenovic, N. S. Barker, A. Lichtenberger, and R. M. Weikle, "Micromachined on-wafer probes," in *IEEE MTT-S International Microwave Symposium Digest*, pp. 65–68, 2010.

[35] W. Wiatr, D. K. Walker, and D. F. Williams, "Coplanar-waveguide-to-microstrip transition model," in *IEEE MTT-S International Microwave Symposium Digest*, vol. 3, pp. 1797–1800, 2000.

[36] I. M. Kang, S.-J. Jung, T.-H. Choi, J.-H. Jung, C. Chung, H.-S. Kim, H. Oh, H. W. Lee, G. Jo, Y.-K. Kim, H.-G. Kim, and K.-M. Choi, "Five-step (Pad-Pad Short-Pad Open-Short-Open) de-embedding method and its verification," *IEEE Electron Device Letters*, vol. 30, no. 4, pp. 398–400, 2009.

[37] X. Wei, G. Niu, S. Sweeney, Q. Liang, X. Wang, and S. Taylor, "A general 4-port solution for 110 GHz on-wafer transistor measurements with or without impedance standard substrate (ISS) calibration," *IEEE Transactions on Electron Devices*, vol. 54, no. 10, pp. 2706–2714, 2007.

[38] F. Pourchon, C. Raya, N. Derrier, P. Chevalier, D. Gloria, S. Pruvost, and D. Celi, "From measurement to intrinsic device characteristics: Test structures and parasitic determination," in *IEEE Bipolar/BiCMOS Circuits and Technology Meeting (BCTM)*, pp. 232–239, 2008.

[39] K. H. K. Yau, I. Sarkas, A. Tomkins, P. Chevalier, and S. P. Voinigescu, "On-wafer S-parameter de-embedding of silicon active and passive devices up to 170 GHz," in *IEEE MTT-S International Microwave Symposium Digest*, pp. 600–603, 2010.

[40] M. Seo, B. Jagannathan, J. Pekarik, and M. J. W. Rodwell, "A 150 GHz amplifier with 8 dB gain and +6 dBm Psat in digital 65 nm CMOS using dummy-prefilled microstrip lines," *IEEE Journal of Solid-State Circuits*, vol. 44, no. 12, pp. 3410–3421, 2009.

[41] E. Strid, "Planar impedance standards and accuracy considerations in vector network analysis," in *ARFTG Microwave Measurements Conference-Spring, 27th*, vol. 9, pp. 159–166, 1986.

[42] R. B. Marks, "Multi-line calibration for MMIC measurement," in *ARFTG Conference Digest-Fall, 36th*, vol. 18, pp. 47–56, 1990.

[43] D. F. Williams, R. B. Marks, and K. R. Phillips, "Translate LRL and LRM calibrations," *Microwaves & RF*, vol. 30, no. February, pp. 78–84, 1991.

[44] A. Davidson, E. Strid, and K. Jones, "Achieving greater on-wafer S-parameter accuracy with the LRM calibration technique," in *ARFTG Conference Digest-Winter, 34th*, vol. 16, pp. 61–66, Nov. 30–Dec. 1 1989.

[45] H. J. Eul and B. Schiek, "Thru-Match-Reflect: one result of a rigorous theory for de-embedding and network analyzer calibration," in *European Microwave Conference, 18th*, pp. 909–914, 1988.

[46] H. J. Eul and B. Schiek, "A generalized theory and new calibration procedures for network analyzer self-calibration," *IEEE Transactions on Microwave Theory and Techniques*, vol. 39, no. 4, pp. 724–731, 1991.

[47] R. F. Scholz, F. Korndorfer, B. Senapati, and A. Rumiantsev, "Advanced technique for broadband on-wafer RF device characterization," in *ARFTG Microwave Measurements Conference-Spring, 63rd*, pp. 83–90, 2004.

[48] H. Heuermann, *Sichere Verfahren zur Kalibrierung von Netzwerkanalysatoren fuer koaxiale und planare Leitungssystemme.* Thesis, Ruhr-Universitaet Bochum, 1996.

[49] R. Doerner and A. Rumiantsev, "Verification of the wafer-level LRM+ calibration technique for GaAs applications up to 110 GHz," in *ARFTG Microwave Measurements Conference-Spring, 65th*, pp. 15–19, 2005.

[50] U. Mahalingam, S. C. Rustagi, and G. S. Samudra, "Three-port RF characterization of MOS transistors," in *ARFTG Conference Digest-Spring, 65th*, pp. 56–61, 17 June 2005.

[51] H. Heuermann, A. Rumiantsev, and S. Schott, "Advanced on-wafer multiport calibration methods for mono- and mixed-mode device characterization," in *ARFTG Microwave Measurements Conference-Spring, 63rd*, pp. 91–96, 2004.

[52] A. Ferrero and U. Pisani, "Two-port network analyzer calibration using an unknown 'thru'," *IEEE Microwave and Guided Wave Letters*, vol. 2, no. 12, pp. 505–507, 1992.

[53] A. Rumiantsev, H. Heuermann, and S. Schott, "A robust broadband calibration method for wafer-level characterization of multiport devices," in *ARFTG Microwave Measurements Conference-Spring, 69th*, pp. 56–60, June 2007.

[54] A. Rumiantsev, S. Schott, and S. Kanev, "Method for calibration of a vectorial network analyzer having more than two ports," US Patent US7, 768, 271, SUSS MicroTec Test Systems, GmbH, 2010.

[55] A. Rumiantsev, S. L. Sweeney, and P. L. Corson, "Comparison of on-wafer multiline TRL and LRM+ calibrations for RF CMOS applications," in *ARFTG Microwave Measurements Conference-Fall, 72nd*, pp. 132–135, 2008.

[56] A. Rumiantsev, P. Sakalas, F. Pourchon, P. Chevalier, N. Derrier, and M. Schroter, "Application of on-wafer calibration techniques for advanced high-speed BiCMOS technology," in *Bipolar/BiCMOS Circuits and Technology Meeting (BCTM)*, pp. 98–101, 2010.

[57] A. Rumiantsev, F. Pourchon, N. Derrier, P. Sakalas, and D. Celi, "Designing on-wafer calibration standards for advanced high-speed BiCMOS technology," in *European Microwave Integrated Circuits Conference (EuMIC)*, pp. 57–60, 2011.

[58] N. Derrier, A. Rumiantsev, and D. Celi, "State-of-the-art and future perspectives in calibration and de-embedding techniques for characterization of advanced SiGe HBTs featuring sub-THz fT/fMAX," in *Bipolar/BiCMOS Circuits and Technology Meeting (BCTM)*, pp. 92–99, 2012.

[59] "Layout rules for GHz-probing," Application Note, Cascade Microtech, Inc., Beaverton, OR, USA, 2005.

[60] "Introduction to bipolar device GHz measurement techniques," Application Note, Cascade Microtech, Inc., Beaverton, OR, USA, 2002.

[61] "Mechanical layout rules for Infinity Probes," Application Note, Cascade Microtech, Inc., Beaverton, OR, USA, 2006.

[62] "Probe selecation guide," Product Note, Cascade Microtech, Inc., Beaverton, OR, USA, 2012.

[63] A. Rumiantsev, P. Sakalas, N. Derrier, D. Celi, and M. Schroter, "Influence of probe tip calibration on measurement accuracy of small-signal parameters of advanced BiCMOS HBTs," in *IEEE Bipolar/BiCMOS Circuits and Technology Meeting (BCTM)*, (Atlanta, GA), 2011.

[64] G. Avenier, M. Diop, P. Chevalier, G. Troillard, N. Loubet, J. Bouvier, L. Depoyan, N. Derrier, M. Buczko, C. Leyris, S. Boret, S. Montusclat, A. Margain, S. Pruvost, S. T. Nicolson, K. H. K. Yau, N. Revil, D. Gloria, D. Dutartre, S. P. Voinigescu, and A. Chantre, "0.13 um SiGe BiCMOS technology fully dedicated to mm-wave applications," *IEEE Journal of Solid-State Circuits*, vol. 44, no. 9, pp. 2312–2321, 2009.

[65] R. Marks and K. Phillips, "Wafer-level ANA calibrations at NIST," in *ARFTG Microwave Measurements Conference-Fall, 34th*, vol. 16, pp. 11–25, 1989.

[66] R. Gillon, W. Tatinian, and B. Landat, "Application of the TRM self-calibration on standard silicon substrates," in *International Conference on Microelectronic Test Structures (ICMTS)*, pp. 109–112, 2003.

[67] N. Jain and D. Wells, "Design of a DC-to-90-GHz resistive load," *IEEE Microwave and Guided Wave Letters*, vol. 9, no. 2, pp. 69–70, 1999.

[68] "Impedance Standard Substrate to support all your high-frequency probing applications," Data Sheet ISS-DS-0108, Cascade Microtech, Inc., Beaverton, OR, USA, 2008.

[69] "Calibration substarte CS-8," Operation Manual, GGB Industries, Inc., Naples, FL, USA, 2004.

[70] A. Fraser, R. Gleason, and E. W. Strid, "Ghz on-silicon-wafer probing calibration methods," in *Proceedings of the Bipolar Circuits and Technology Meeting (BCTM)*, pp. 154–157, 1988.

[71] U. Arz, D. F. Williams, and H. Grabinski, "Characteristic impedance measurement of planar transmission lines," in *URSI-GA*, Aug. 17–24 2002.

[72] D. F. Williams and R. B. Marks, "Transmission line capacitance measurement," *IEEE Microwave and Guided Wave Letters*, vol. 1, no. 9, pp. 243–245, 1991.

[73] A. M. Mangan, S. P. Voinigescu, Y. Ming-Ta, and M. Tazlauanu, "De-embedding transmission line measurements for accurate modeling of IC designs," *IEEE Transactions on Electron Devices*, vol. 53, no. 2, pp. 235–241, 2006.

[74] T. Zwick, Y. Tretiakov, and D. Goren, "On-chip SiGe transmission line measurements and model verification up to 110 GHz," *IEEE Microwave and Wireless Components Letters*, vol. 15, no. 2, pp. 65–67, 2005.

[75] W. R. Eisenstadt and Y. Eo, "S-parameter-based IC interconnect transmission line characterization," *IEEE Transactions on Components, Hybrids, and Manufacturing Technology*, vol. 15, no. 4, pp. 483–490, 1992.

[76] A. Rumiantsev, P. L. Corson, S. L. Sweeney, and U. Arz, "Applying the calibration comparison technique for verification of transmission line standards on silicon up to 110 ghz," in *ARFTG Microwave Measurements Conference-Spring, 73rd*, (Boston, MA, USA), pp. 1–6, 2009.

[77] D. F. Williams, R. B. Marks, and A. Davidson, "Comparison of on-wafer calibrations," in *ARFTG Microwave Measurements Conference-Fall, 38th*, vol. 20, pp. 68–81, 1991.

[78] D. F. Williams, U. Arz, and H. Grabinski, "Accurate characteristic impedance measurement on silicon," in *ARFTG Microwave Measurements Conference-Spring, 51st*, vol. 33, pp. 155–158, 1998.

[79] D. F. Williams, U. Arz, and H. Grabinski, "Characteristic-impedance measurement error on lossy substrates," *IEEE Microwave and Wireless Components Letters*, vol. 11, no. 7, pp. 299–301, 2001.

[80] "Reference Material 8130. Coplanar waveguide calibration set," Report of Investigation, NIST, Gaithersburg, MD 20899, USA, 1998.

[81] I. Martinez, "In-situ calibration and verification techniques for the characterization of microwave circuits and devices," in *IEEE International Conference on Microelectronic Test Structures (ICMTS)*, pp. 161–165, 2012.

[82] H. Heuermann and B. Schiek, "Calibration of network analyser measurements with leakage errors," *Electronics Letters*, vol. 30, no. 1, pp. 52–53, 1994.

[83] J. V. Butler, D. K. Rytting, M. F. Iskander, R. D. Pollard, and M. A. Vanden Bossche, "16-term error model and calibration procedure for on-wafer network analysis measurements," *IEEE Transactions on Microwave Theory and Techniques*, vol. 39, no. 12, pp. 2211–2217, 1991.

[84] A. Pawlak, M. Schroter, J. Krause, D. Celi, N. Derrier, and A. Mukherjee, "A HICUM extension for medium current densities," in *9th European HICUM Workshop*, (Würzburg, Germany), 2009.

[85] F. Stein, Z. Huszhka, N. Derrier, C. Maneux, and D. Celi, "Extraction of the emitter related space charge weighting factor parameters of HICUM L2.30 using the Lambert W function." Unpublished Work, 2012.

[86] "High-frequency, over-temperature measurements and modeling," Application Note 10600AP-0893, Cascade Microtech, Inc., Beaverton, OR, USA, 1994.

[87] A. Rumiantsev, R. Doerner, and P. Sakalas, "Verification of wafer-level calibration accuracy at cryogenic temperatures," in *ARFTG Microwave Measurements Conference-Fall, 68th*, pp. 134–140, 2006.

[88] A. Rumiantsev and R. Doerner, "Verification of wafer-level calibration accuracy at high temperatures," in *ARFTG Microwave Measurements Conference-Spring, 71st*, pp. 103–106, 2008.

[89] A. Rumiantsev, G. Fisher, and R. Doerner, "Sensitivity analysis of wafer-level over-temperature RF calibration," in *ARFTG Microwave Measurement Symposium-Fall, 80th*, pp. 1–3, 2012.

[90] H. Geissler, A. Rumiantsev, S. Schott, P. Sakalas, and M. Schroter, "A novel probe station for helium temperature on-wafer measurements," in *ARFTG Microwave Measurements Conference-Fall, 68th*, pp. 67–73, 2006.

[91] A. Rumiantsev, V. Khutko, and S. Kanev, "Verfahren zur Ermittlung elektrischer Eigenschaften elektronischer Bauelemente und Verfahren zur Kalibrierung der Messeinheit," Patent DE102009029906, SUSS MicroTec Test Systems, GmbH, 2009.

[92] A. Rumiantsev, R. Doerner, and F. Lenk, "Impact of in-situ TRL reference impedance determination on parameter extraction," in *European Microwave Conference, 42nd*, (Amsterdam, NL), pp. 593–596, 2012.

[93] T. H. Miers, A. Cangellaris, D. Williams, and R. Marks, "Anomalies observed in wafer level microwave testing," in *IEEE MTT-S International Microwave Symposium Digest*, vol. 3, pp. 1121–1124, 1991.

[94] F. J. Schmueckle, R. Doerner, G. N. Phung, W. Heinrich, D. Williams, and U. Arz, "Radiation, multimode propagation, and substrate modes in w-band cpw calibrations," in *European Microwave Conference (EuMC)*, (Manchester, UK), 2011.

[95] E. M. Godshalk, "Generation and observation of surface waves on dielectric slabs and coplanar structures," in *IEEE MTT-S International Microwave Symposium Digest*, vol. 2, pp. 923–926, 1993.

[96] A. J. Lord, "Comparing the accuracy and repeatability of on-wafer calibration techniques to 110 GHz," in *European Microwave Conference (EuMC)*, vol. 3, pp. 28–31, 1999.

[97] A. Rumiantsev, R. Doerner, and E. M. Godshalk, "The influence of calibration substrate boundary conditions on CPW characteristics and calibration accuracy at mm-wave frequencies," in *ARFTG Microwave Measurements Conference-Fall, 72nd*, pp. 168–173, 2008.

[98] E. M. Godshalk, "Surface wave phenomenon in wafer probing environments," in *ARFTG Microwave Measurements Conference-Fall, 40th*, vol. 22, pp. 10–19, 1992.

[99] "Summit. 200 mm semi-automatic probe system," Data Sheet SUMMIT-DS-0613, Cascade Microtech, Inc., Beaverton, OR, USA, 2013.

[100] "Elite300. 300 mm semi-automatic probe station," Data Sheet Elite-DS-0612, Cascade Microtech, Inc., Beaverton, OR, USA, 2012.

[101] "EPS200RF. A complete 200mm manual probe system SMART package for RF applications up to 67 GHz," Technical Review EPS200RF-TR-0212, Cascade Microtech, Inc., Beaverton, OR, USA, 2012.

[102] "EPS150MMW. A dedicated 150 mm manual probing solution for mmW, THz and load pull application," Product Flyer EPS150MMW-FLYER-0512, Cascade Microtech, Inc., Beaverton, OR, USA, 2012.

[103] W. Heinrich, "Full-wave analysis of conductor losses on MMIC transmission lines," *IEEE Transactions on Microwave Theory and Techniques*, vol. 38, no. 10, pp. 1468–1472, 1990.

[104] A. Rumiantsev, "Calibration residual error sources in wafer-level S-Parameter measurement," in *Asia-Pacific Microwave Conference (APMC). Workshop "Establishing Confidence in Microwave Measurements"*, (Singapore), 2009.

[105] "Verifying the performance of Vector Network Analyzers," Application Note 5C-026, Maury Microwave, Inc., Ontario, CA, USA, 1999.

[106] "What is your measurement accuracy?," Application Note 11410-00270, Anritsu, Morgan Hill, CA, USA, 2001.

[107] H. Heuermann, "Old and new accuracy estimation of S-parameter measurements with the ripple-test," in *MTT-S Microwave International Symposium. Workshop TMB*, (San Francisco, CA, USA), Juni 2006.

[108] H. Heuermann and A. Rumiantsev, "The modified ripple test for on-wafer S-parameter measurements," in *ARFTG Microwave Measurements Conference-Spring, 69th*, pp. 1–5, June 2007.

[109] H. Heuermann and A. Rumiantsev, "Verfahren zur Messfehlerermittlung bei Streuparametermessungen," Patent DE102008009338, SUSS MicroTec Test Systems, GmbH, 2008.

[110] H. Heuermann and A. Rumiantsev, "Method for determining measurement errors in scattering parameter measurements," US Patent Application 12/030295, SUSS MicroTec Test Systems, GmbH, 2008.

[111] A. Rumiantsev, S. Kanev, S. Scott, and K. Stoll, "Chuck for supporting and retaining a test subsrate and a calibrataion substrate," US Patent US7, 999, 563, SUSS MicroTec Test Systems, GmbH, 2011.

PART II
Attachments

11

Attachment to Chapter 2

11.1 Paper [31]: "VNA Calibration"

A. Rumiantsev and N. Ridler, "VNA calibration," *Microwave Magazine, IEEE*, vol. 9, pp. 86–99, 2008.

In this paper, I presented a detailed investigation of the history and the state-of-the-art of VNA measurements and S-parameter calibration techniques. My co-author, Nick Ridler from National Physical Laboratory (NPL, UK) added metrological aspects to this paper, such as the definition of measurement traceability, measurement assurance as well as clarification of the term "calibration" that, in fact, has multiple meanings for vector network analysis. This research helped to form a strategy of how to approach the specific problem of the on-wafer calibration on silicon.

© CREATAS

VNA Calibration

Andrej Rumiantsev and Nick Ridler

It was during the late 1950s that the need for reliable measurement, and therefore reliable measurement standards, at RF and microwave frequencies began to emerge. This led to the introduction of precision coaxial air lines as primary reference standards of impedance [1], [2]; see Figure 1. These lines use conductors made from very-high-conductivity metals and air as the dielectric, due to the simple and predictable electromagnetic properties (i.e., permeability and permittivity) of air at RF and microwave frequencies [3]. This ensured that the properties of these lines were very close to those of ideal lines [4].

Also during the late 1950s and throughout the 1960s, much work was undertaken to develop precision coaxial connectors to ensure that very repeatable and reproducible measurements could be made at microwave frequencies [5], [6]. To help focus this effort, committees were established (including an IEEE subcommittee on precision coaxial connectors [7]) tasked with producing standards for these precision connectors. Finally, by the late 1960s, the first fully automated vector network analyzers (VNAs) providing high-precision measurement capabilities were introduced (e.g., [8], [9]). The stage was now set for work to begin on introducing reliable measurement assurance techniques for measurements made using VNAs (Figure 2).

However, there were several other key developments that took place during the 1970s, 1980s, and 1990s that greatly improved the state of the art of measurements made using VNAs. These included the introduction of:

- smaller precision coaxial connectors (beginning with the 3.5-mm connector [10] and ending with the 1-mm connector [11]), enabling measurements to be made over wider bandwidths
- VNA calibration and verification kits containing high-precision devices suitable for calibrating and/ or verifying the performance of the VNAs
- reliable VNA calibration techniques [including thru-reflect-line (TRL) [12], line-reflect-line (LRL) [13], etc.)

Andrej Rumiantsev is with SUSS MicroTec Test Systems GmbH, Germany. Nick Ridler is with the National Physical Laboratory, United Kingdom.

Digital Object Identifier 10.1109/MMM.2008.919925

1527-3342/08/$25.00©2008 IEEE

- six-port VNAs [14] used by national measurement standards laboratories [such as the National Institute of Standards and Technology (NIST) in the United States and the National Physical Laboratory (NPL) in the United Kingdom, etc.] to provide an independent measurement method to verify the performance of the commercially available VNAs.

Finally, also by the late 1980s and early 1990s, national measurement standards laboratories (i.e., NIST, NPL, etc.) began turning their attention to demonstrating the reliability of VNA measurements made on planar circuits (such as on-wafer measurements) to support the rapidly developing microelectronics industry. Both NIST and NPL produced standard wafers [15], [16] that contained the planar circuit equivalent to the coaxial air line—i.e., precision sections of coplanar waveguide and/ or microstrip transmission line. These lines provided the reference standards for calibrating VNAs for on-wafer measurements.

All of the above activities greatly improved the state of the art for practitioners and users of VNA measurements. Also, in addition to all these activities, much was done by measurement experts working in industrial, academic, and government laboratories to establish traceability and other quality assurance mechanisms for these VNA measurements. These topics are discussed in "What is Traceability?" and "Measurement Assurance."

Systematic Measurement Errors

What Is Calibration and Error Correction?

Calibration is defined as the "set of operations that establish, under specified conditions, the relationship between values of quantities indicated by a measuring instrument or measuring system, or values represented by a material measure or a reference material, and the corresponding values realized by standards" [17]. As such, calibration traditionally involves having an instrument or component sent away periodically to a standards and/ or calibration laboratory, who then undertake the calibration process. This often results in a certificate of calibration being issued that demonstrates the current condition of the instrument or component.

However, in the context of a VNA, the term calibration can have at least two different meanings. First, the traditional concept of calibration can still be applied, with the VNA being sent away for calibration, typically every year or so. (Alternatively, some companies offer periodic on-site calibration, performed by a visiting calibration specialist.) However, of more relevance to this article is another form of calibration that is performed locally, usually each time the instrument is set up and configured for a given series of measurements. This second form of calibration is intended to remove systematic errors from the instrument hardware (and to take into account the presence of any accessories that may have been added to enable specific measurements to be performed) at the required frequencies for the measurements. For example, measurements may be required to be made in an on-wafer environment. In which case, first cables need to be connected to the VNA front-panel connectors, followed by coaxial adaptors, and finally on-wafer probes (Figure 3). This second form of calibration will correct for the effects of these added components as well as correct the systematic errors in the VNA. This is why this type of calibration is often referred to as *error correction*, and it is this type of calibration that will be discussed in this article.

The demand for increased measurement accuracy from the VNA can be achieved by improving the hardware, the models used for characterizing measurement errors, the calibration methods used for calculating these errors, and the definitions of calibration standards. For S-parameters, the systematic errors are often represented using so-called error models of the measurement system (i.e., VNA). The number of error coefficients included in the error model, as well as the type of error model, depends on

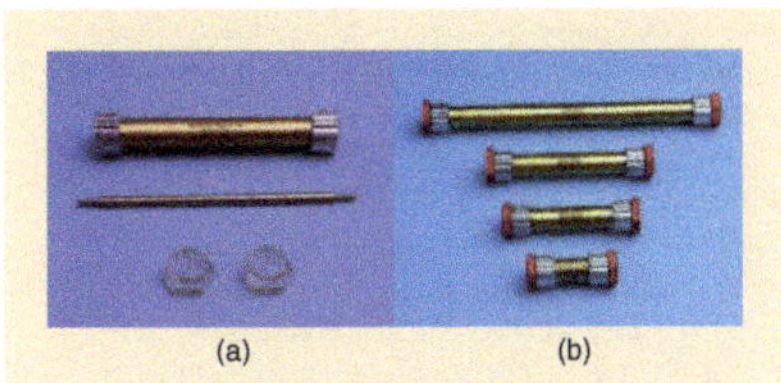

Figure 1. *An example of precision reference coaxial air lines of different length.*

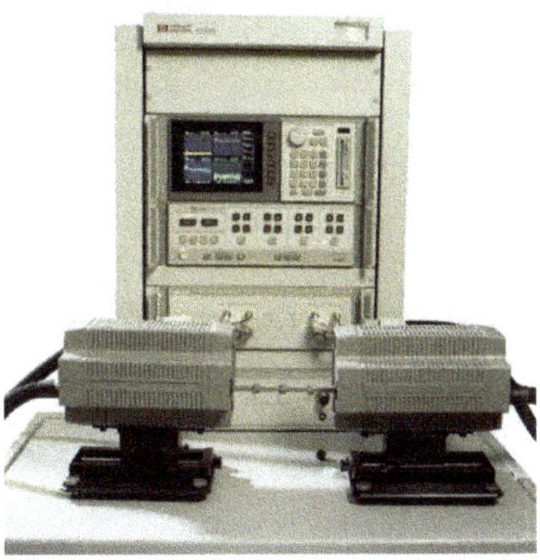

Figure 2. *A coaxial mm-wave measurement bench based on the Agilent 8510 VNA. This analyzer was the industry reference for microwave measurements for many years.*

What Is Traceability?

Traceability, in the context of a measurement, is defined as the "property of the result of a measurement or the value of a standard whereby it can be related to stated references, usually national or international standards, through an unbroken chain of comparisons all having stated uncertainties" [17]. Applying this concept to a VNA measurement, the stated references could be precision air lines (or their equivalent), the VNA is the transfer device used as part of the unbroken chain of comparisons, and the precision connectors enabling these comparisons to be made within acceptable limits to the uncertainty of measurement.

The benefit of having a measurement that is traceable stems from the fact that it can be used to demonstrate the equivalence of measurements made independently of one another. This is of paramount importance in a customer/supplier relationship where a common understanding is needed of the parameters that define (or specify) the performance of the device being bought or sold. Therefore, if two measurements of a quantity are made independently, and these measurements are both traceable, then their values will agree to within the stated uncertainties of the measurements. This is, therefore, an extremely valuable process that can provide the necessary underpinning assurance that is needed when operating within a truly global marketplace where the customer and supplier may be located in different parts of the world.

The vital role that traceability can play was recognized long ago and led to the introduction of national measurement accreditation schemes so that customers and suppliers could fully demonstrate the quality of their measurements to an independent third party (i.e., the accreditation body). These days, such accreditation processes are controlled by international standards (e.g., [72]), thus ensuring that the accreditation process is itself applied uniformly across all types of measurements and at all locations around the world. Most countries maintain a national accreditation body for this purpose, and these bodies are themselves linked through international accreditation organizations such as the International Laboratory Accreditation Cooperation (ILAC, www.ilac.org).

When traceability is harmonized within an agreed system of units (e.g., the international system of units, SI), then not only is it possible to demonstrate equivalence between measurements of the same quantity, but it also becomes possible to demonstrate the equivalence of measurements of different quantities. This is achieved through the relationship of these quantities to the so-called base quantities within the system of units. (In SI, the seven base quantities are length, mass, time, electric current, thermodynamic temperature, amount of substance, and luminous intensity.)

By following the traceability path of a measurement back to its fundamental base quantities, it is possible to demonstrate the harmonization of the measurement within the system of units. For example, a reflection measurement made along a transmission line can usually be traced back to dimensional measurements, since it is the dimensions of the transmission line that determine the impedance and therefore the amount of signal that is reflected by the line. The base quantity for dimensional measurements is length. Similarly, for power and noise measurements, these can usually be related back to heating effects. Therefore, the base quantity is thermodynamic temperature. In just about all microwave measurements, the frequency of the measurement needs to be known. Since frequency is the reciprocal of periodic time, the base quantity is time.

A key role of a national measurement standards laboratory (such as NIST, NPL, etc.) is to maintain primary reference standards of measurement. For example, at microwave frequencies, these are usually standards of power, impedance, attenuation, noise, etc. In addition, the national measurement standards laboratory is tasked with realizing the seven SI base quantities. By linking these two roles, the national measurement standards laboratory is able to deliver a wide range of traceable measurements to industry that are also harmonized within the SI.

The subsequent 'linking' of the capabilities of one national measurement standards laboratory to others is achieved through participation in international measurement comparison programs conducted under the auspices of organizations such as the International Bureau of Weights and Measures (BIPM, www.bipm.org) and their consultative committees. The results from these comparison exercises are analyzed and placed on a database maintained by the BIPM that demonstrates the capability of each national laboratory.

Finally, it is worth mentioning that these days, with the global accessibility of the Internet, measurement services that make extensive use of the Internet are beginning to be developed. These services are starting to play a role in providing traceable measurements in a highly efficient manner. For example, a system has recently been put in place by NPL that uses the Internet to provide traceability for high-precision measurements using VNAs at any location around the world [73].

- the hardware topology of the VNA,
- the number of VNA ports and measurement receivers
- the required measurement accuracy.

The following section presents commonly used models of S-parameter systematic measurement errors.

Flow-Graph S-Parameter Representation

The first error models used for automated S-parameter error correction were introduced at the end of the 1960s. They addressed the bidirectional two-port system and defined the influence of system imperfection on reflection (S_{11}, S_{22}) and transmission (S_{21}, S_{21}) measurements. These models were developed to represent systematic measurement errors using imaginary two-port error networks. They were described by S-parameters and were included in the measurement signal paths [8]. The model for a reflection (one-port) measurement consisted of only one error network. Originally, this network was described as a matrix of four S-parameters.

Measurement Assurance

Although traceability provides arguably the most acceptable method for assuring a given measurement, it is not always possible to provide such traceability for all types and ranges of measurement. This is particularly true of modern VNAs that offer many different measurement formats (e.g., logarithmic or linear; single-ended or differential; frequency- or time-domain; etc.) often over very wide dynamic ranges (sometimes up to 100 dB or more). Under these circumstances, the measurement community benefits from the use of additional assurance techniques to validate results from VNAs.

The first major contribution to this area was the introduction of verification standards and kits for VNAs [74], [75]. These verification kits can be measured routinely by the end-user and compared with reference values supplied by the manufacturer. The kits can also be returned periodically to the manufacturer, who checks the reference values. This provides a high degree of measurement assurance for the end-users. Verification kits have since been produced in many of the different types of connectors used by the industry, as well as in waveguide.

Another activity that has been very valuable to measurement practitioners in our industry is the user groups that have been set up over the years. These groups have enabled the key measurement issues, at any given time, to be identified, discussed, and resolved. Probably the first such user group set up by RF and microwave specialists was the Automatic RF Techniques Group (ARFTG), www.arftg.org , which was established back in 1972 [76]. ARFTG is a technical organization interested in all aspects of RF and microwave test and measurement. The group is still very active today, and continues to evolve in response to the many developing needs of the RF and microwave community. For example, a recent development within ARFTG has been the establishment of a Nonlinear Vector Network Analyser (NVNA) Users' Forum. This informal group meets three times each year - during the Spring ARFTG conference (which is itself part of Microwave Week), the Fall ARFTG Symposium, and European Microwave Week.

Other user groups of interest to the VNA community include ARMMS (www.armms.org)—the RF and Microwave Society—and ANAMET (www.npl.co.uk/anamet)—the RF and Microwave Metrology Club. Like ARFTG, these groups meet twice each year to discuss issues of relevance to each group.

An activity that some of these user groups undertakes is to provide the opportunity to participate in measurement comparison programs (MCPs). These are programs that allow many participants to make measurements of the same devices that travel between the participating laboratories [77], [78] (see Figure A). The results of the measurements of these traveling standards are compared to indicate the overall equivalence (or not) of the results. Such exercises are extremely useful for identifying serious errors that may be present in measurements made by any of the participants. Comparisons can also be undertaken in areas of measurement where traceability may not yet exist (e.g., time-domain measurements [79]).

All of the above processes—local auditing using verification kits, interactions with user groups, participation in MCPs—provide measurement assurance which complements that provided by classical traceability processes. Ultimately, for the very highest level of measurement assurance, one should consider traceability of measurement along with one or more of these other processes.

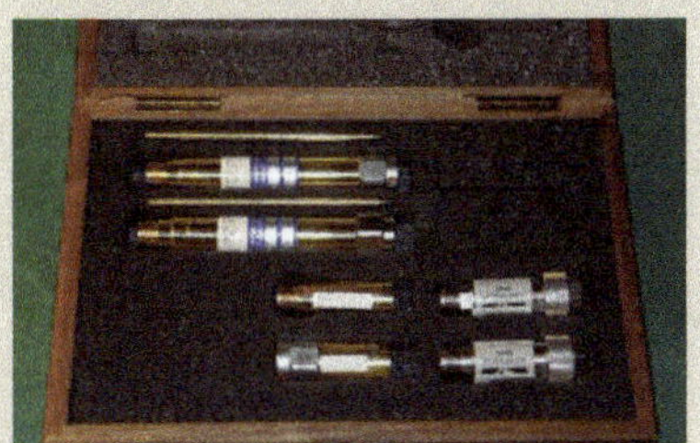

Figure A. *Type-N travelling standards used for the ARFTG MCP.*

However, it turned out that only the coefficients S_{11} and S_{22} and the product $S_{21}S_{12}$ were needed for further error correction. As a result, the three-term error model replaced the matrix consisting of four S-parameters where the coefficients e_{00}, e_{11}, and e_{01} are E_D (directivity), E_S (source match), and E_R (reflection tracking), respectively (Figure 4) [18]. Today, the three-term error model is still the most common representation of one-port calibration and error correction procedures.

Following from the above, the eight-term model represented the bidirectional system for automated measurements of two-port devices under test (DUTs) (Figure 5). The S-parameter-based model [Figure 5(a)] required all four coefficients (S_{11}, S_{12}, S_{21}, and S_{22}) to be known for each error adapter. The error correction of the transmission measurements included two factors $S_{21}^{(1)}S_{12}^{(2)}$ and $S_{21}^{(2)}S_{12}^{(1)}$ for the forward and reverse directions, respectively [8]. These factors were addressed as coefficients E_T in the error terms representation [Figure 5(b)] [19].

Alternative unidirectional test sets did not include an internal switch for redirecting the incident measurement signal between measurement ports. They allowed the DUT to be characterized in one direction only (for its S_{11} and S_{21} parameters). As introduced in [18], such a system can be described by only five error terms. An additional term represents the signal leakage between the measurement ports, thus extending the model to six parameters [18], [20] (see Figure 6).

The leakage terms (also called *crosstalk terms*) were later added to the eight-term model, one for each measurement direction, increasing the number of the error coefficients in general to ten [21].

The 8(10)- and 5(6)-term error models were in use for almost ten years without significant modification. [Note that here and elsewhere in the article, the number in parentheses represents the number of error terms, including any leakage terms (E_X). These terms are optional parameters that may not fully represent the crosstalk (as discussed further in this article) and thus we do not count it in our nomenclature.] Within any model, the error terms need to be defined for each measurement frequency point and saved in the VNA memory. Therefore, an extension of the error model, including the use of additional error terms, or development of a unified error model for different test sets were not commercially viable options. (At that time, the cost of computer memory was still a major design consideration.)

Rapid progress in semiconductor technologies at the end of the 1970s significantly expanded the availability of low-cost read/ write memory modules as well as mass storage devices embedded in measuring instruments. This greatly extended the capabilities of VNA error modeling. The measurement system description was unified and the 10(12)-term model was introduced for commercial VNAs independent of the test set configuration [19] (see Figure 7). This error model became the standard model for the description of systematic measurement errors of a

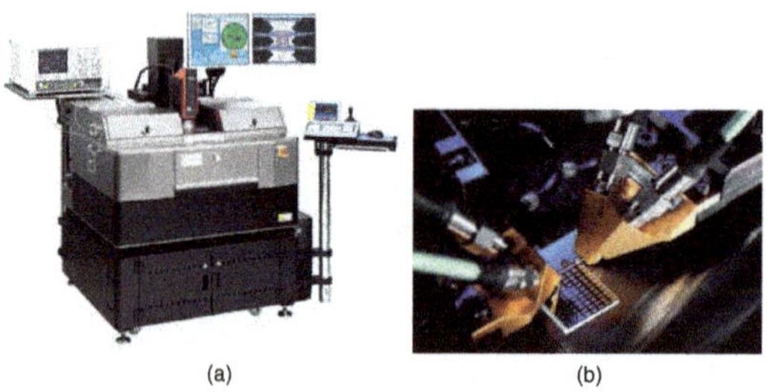

Figure 3. *(a) An example of a state-of-the-art 300-mm RF and microwave wafer-level measurement system. The system includes: the EMI-shielded and light-tight automated probe system with integrated thermal management and automated RF calibration, a VNA, RF cables, and RF wafer probes. (b) The set of coplanar calibration standards (a calibration substrate) is used for the calibration of the system.*

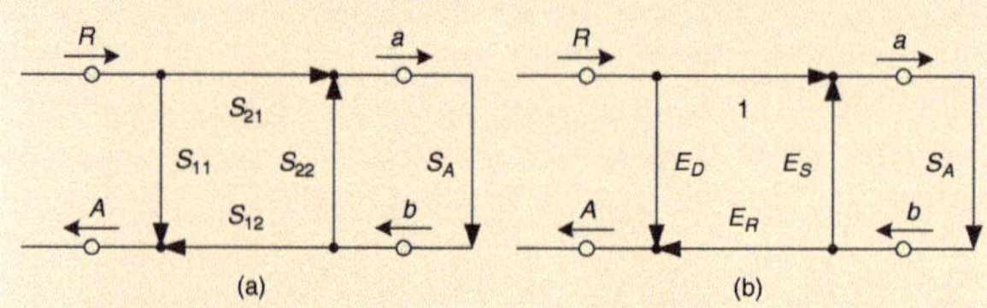

The relationship between the actual S_A and the measured S_M=A/R S-parameter of the DUT is given by

$$S_{11A} = \frac{S_{11M} - E_D}{E_S(S_{11M} - E_D) + E_R}$$

Figure 4. *The one-port three-term error model in (a)* **S**-*Parameter and (b) error terms representation.*

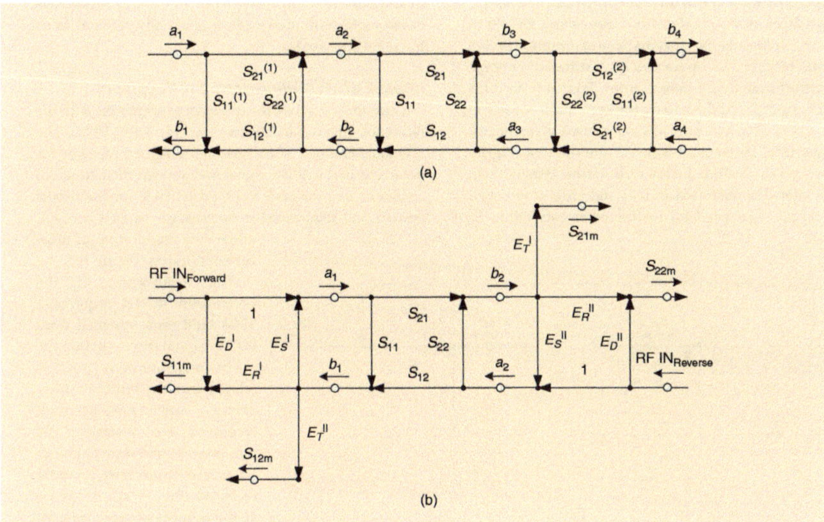

Figure 5. *The eight-term error model of a two-port VNA in (a)* **S**-*Parameter and (b) error terms representations. The unknown DUT* **[S]** *is connected between the error adapters. Prime and double-prime parameters correspond to the forward and reverse measurement directions, respectively.*

two-port VNA. It is implemented in all modern measurement instruments.

The equations for the relationship between the measured and actual S-parameter of a two-port DUT were given in [19] and [22]. However, these equations are somewhat bulky. An alternative simplified approach was introduced in [23]. For the measurement system, the relationship between the measured, m, waves and the incident, a, and reflected/ transmitted waves, b, at the DUT can be found using the scattering parameter definition:

$$\begin{pmatrix} m_2^I \\ a_1^I \end{pmatrix} = \begin{pmatrix} E_D^I & E_R^I \\ 1 & E_D^I \end{pmatrix} \begin{pmatrix} m_1^I \\ b_1^I \end{pmatrix}. \quad (1)$$

From (1) and Figure 7, the incident a_1^I, a_2^I, reflected b_1^I, and transmitted b_2^I waves at the DUT are

$$a_1^I = m_1^I + \frac{E_S^I}{E_R^I}\left(m_2^I - E_D^I m_1^I\right),$$
$$b_1^I = \frac{1}{E_R^I}\left(m_2^I - E_D^I m_1^I\right),$$
$$b_2^I = \frac{m_4^I}{E_T^I}, \quad a_2^I = \frac{E_L^I m_4^I}{E_T^I}. \quad (2)$$

The parameters $a_1^{II}, a_2^{II}, b_1^{II}$, and b_2^{II} can be found in a similar way, taking into account the switch in its other position. Once the wave parameters a and b are defined, the following matrix can be formed:

$$\begin{pmatrix} b_1^I & b_1^{II} \\ b_2^I & b_2^{II} \end{pmatrix} = \begin{pmatrix} S_{11} & S_{12} \\ S_{21} & S_{22} \end{pmatrix} \begin{pmatrix} a_1^I & a_1^{II} \\ a_2^I & a_2^{II} \end{pmatrix}, \quad (3)$$

or, in short,

$$[K] = [Sx][L]. \quad (4)$$

Finally, the S-parameters of the DUT can be found by

$$[Sx] = [K][L]^{-1}. \quad (5)$$

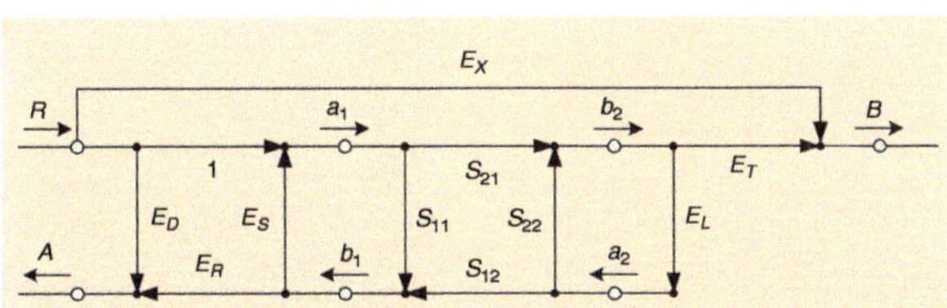

Figure 6. *The five-term unidirectional error model, represented by the error coefficients* $\mathbf{E_D}, \mathbf{E_S}, \mathbf{E_R}, \mathbf{E_L}$, *and* $\mathbf{E_T}$. *The leakage coefficient* $\mathbf{E_X}$ *is an optional parameter.*

Cascade Matrix T-Parameter Representation

The ten-term model, as described above and shown in Figure 8, represents the systematic measurement errors in terms of effective S-parameters. A different concept was introduced by engineers from Tektronix in 1975 [24]. They proposed describing the systematic measurement errors of a two-port system as two error boxes, characterized by transmission (T) parameters (Figure 9). Their model included eight error terms. However, as it was later demonstrated in [12] and [25], only seven error terms are needed for further error correction. To distinguish this approach from the old S-parameter-based eight-term model [8], it is usually referred to as the *seven-term model*.

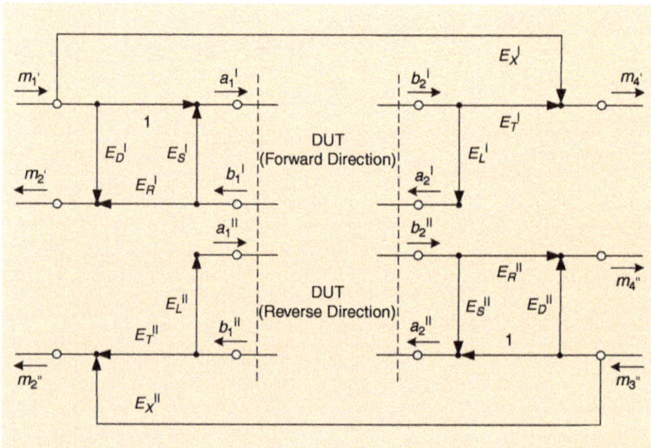

Figure 7. *The 10(12)-term error model for two-port bidirectional* **S***-parameter measurements. The error coefficients* **E** *represent the relationship between waves,* **m***, measured by the ideal VNA receivers and incident,* **a***, and transmitted/reflected waves,* **b***, at the DUT plane. Prime and double-prime parameters correspond to the forward and reverse measurement directions, respectively.*

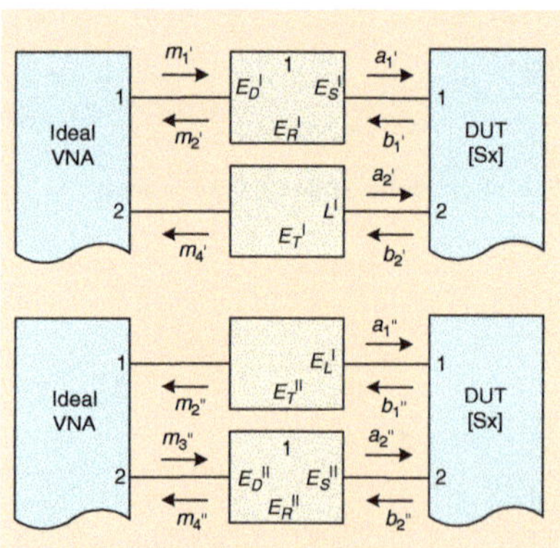

Figure 8. *Block diagram of a two-port VNA described by the ten-term model for the first and second state of the switch.*

Impact of VNA Measurement Receivers

It is common to relate the ten-term error model with the reference channel hardware concept of the VNA. The reference channel VNA has one reference receiver for detecting the incident signal and several measurement receivers, one for each VNA port. Thus, for the n-port system, the total number of receivers, k, is: $k = n + 1$, where n is the number of measurement ports (Figure 10).

The application of the seven-term model requires a VNA built on a so-called double-reflectometer principle: every measurement port is related with the individual reference and measurement receivers. For instance, the two-port double-reflectometer VNA uses four measurement receivers (Figure 11). In general, the number of measurement receivers k for a multiport double-reflectometer is $k = 2n$, where n is the number of system measurement ports.

Figure 11 shows a physical model of the systematic errors for a four-receiver VNA where $[Tx]$ is a measured DUT and $[A]$ and $[B]$ are the error boxes. The latter describe measurement systematic errors. The values $m_1 \dots m_4$ represent waves measured by ideal receivers.

It is straightforward to show that the relationship between $m_1 \dots m_4$, incident (a_1, a_2), and reflected or transmitted (b_1, b_2) signals is:

$$\begin{pmatrix} m_1' & m_1'' \\ m_2' & m_2'' \end{pmatrix} = \begin{pmatrix} A_{11} & A_{12} \\ A_{21} & A_{22} \end{pmatrix} \begin{pmatrix} T_{11} & T_{12} \\ T_{21} & T_{22} \end{pmatrix} \times \begin{pmatrix} B_{11} & B_{12} \\ B_{21} & B_{22} \end{pmatrix}^{-1} \begin{pmatrix} m_3' & m_3'' \\ m_4' & m_4'' \end{pmatrix}, \quad (6)$$

where: $m_1' \dots m_4'$ and $m_1'' \dots m_4''$ are the measured values in forward and reverse directions, respectively. $T_{11} \dots T_{22}$ are defined as the transmission parameters of a measured DUT. Alternatively, in shortened form,

$$M = ATB^{-1}, \quad (7)$$

where measurement matrix M is

$$M = \begin{pmatrix} m_1' & m_1'' \\ m_2' & m_2'' \end{pmatrix} \begin{pmatrix} m_3' & m_3'' \\ m_4' & m_4'' \end{pmatrix}^{-1}. \tag{8}$$

Finally, the T-parameters of the DUT are given by

$$T_X = A^{-1} M_x B. \tag{9}$$

Conversion of Error Models

Both seven-term and ten-term error models are used to describe the double-reflectometer VNA. If required, a seven-term model can be converted into a ten-term model. Several approaches have been published giving different conversion equations [22], [26]–[28]. These equations are slightly different, but are based on the same physical principle. The differences stem from the authors' notation used for the seven-term model elements, e.g., using the inverse of matrix $[B]$. Such conversion techniques are implemented in many double-reflectometer VNAs today.

There were also attempts to apply the seven-term model for the reference receiver VNA [29]. In fact, this assumes that the source match equals the load match of the test set, which holds only in the case of an ideal test set switch. For a real system, this assumption may lead to intolerable measurement inaccuracy, especially for highly reflective DUTs [30]. Only the ten-term model can guarantee the entire description of the reference receiver VNA.

Multiport Measurements and Signal Leakage Problems

As noted above, even the first error models of a VNA included special error term(s) to address the influence of one system measurement port on another (i.e. the so-called leakage term, E_X). The leakage term was simply defined as a transmission coefficient between VNA ports being perfectly matched. This definition holds only for those cases when the DUT has input and output impedances equal to the system impedance. When measuring other devices, the application of a leakage term defined in such a way degrades the measurement accuracy.

Further measurement experiments and practical experiences revealed that the leakage can have a very complicated nature. It is generally insufficient to use just one or two error terms to correctly represent this phenomenon. Clearly, another description of systematic measurement errors was required.

Such a concept was introduced by Speciale and Franzen in 1977 [31]. The systematic measurement errors of the n-port VNA were represented by a $2n$-port virtual error network, connected with its n-ports to the DUT and its other n-ports to the ideal, error-free VNA. The error network consists of $(2n)^2$ coefficients and describes all possible influences of the measurement ports on each other. In fact, one error term can be set to be a free parameter and the error model can be normalized with respect to this term. That is, only $4n^2 - 1$ coefficients are linearly independent from each other. Thus, these error terms completely describe such a system [32].

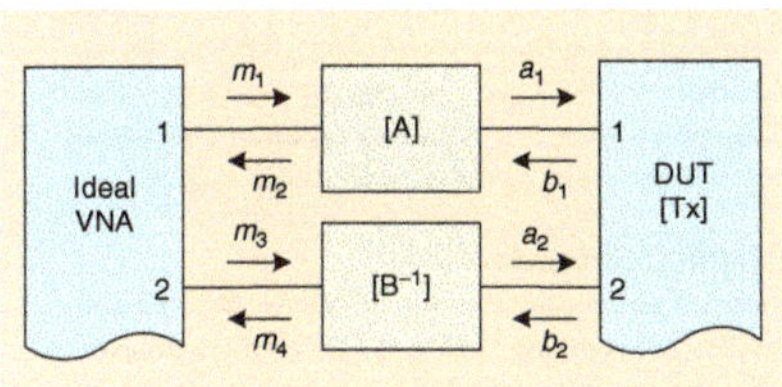

Figure 9. *Block diagram of a two-port VNA described by the cascade matrix representation (seven-term model).*

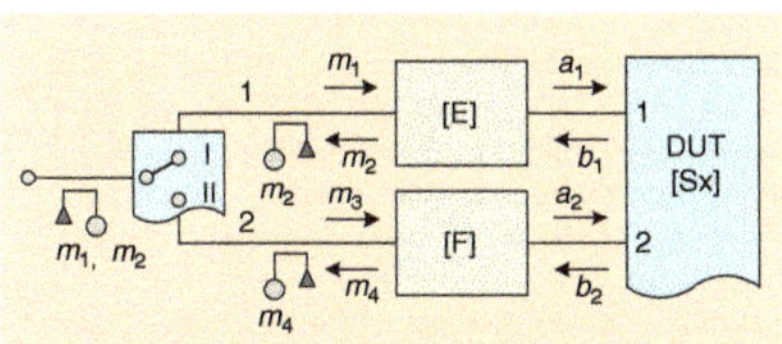

Figure 10. *Block diagram of VNA based on the reference channel architecture. It shows one reference receiver for incident signals* $\mathbf{m_1}$ *and* $\mathbf{m_3}$*, the signal source switch, the measurement receivers for signals* $\mathbf{m_2}$ *and* $\mathbf{m_4}$*, and the ten-term error model matrices* **[E]** *and* **[F]**.

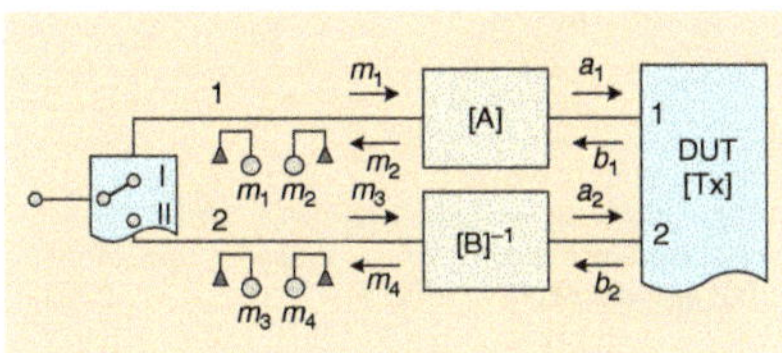

Figure 11. *Block diagram of VNA based on the double-reflectometer architecture. It shows the reference receivers,* $\mathbf{m_1}, \mathbf{m_3}$*; the signal source switch; the measurement receivers,* $\mathbf{m_2}$ *and* $\mathbf{m_4}$*; and the seven-term error model matrices* **[A]** *and* **[B]**.

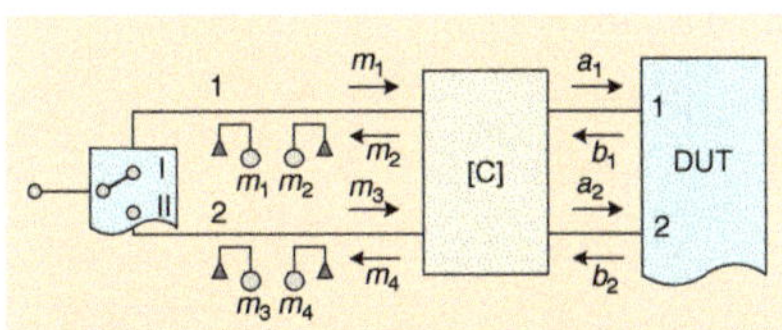

Figure 12. *Block diagram of the leaky VNA based on the double-reflectometer architecture. For the two-port system, the matrix* **[C]** *includes 15 error coefficients.*

The $4n^2 - 1$ model is only valid for VNAs built upon the double-reflectometer concept (with $2n$ measurement receivers, Figure 12). However, it was demonstrated much later that the full error model of a reference channel VNA (with $n + 1$ reference receivers) can also be defined (Figure 13). This includes significantly more error terms: e.g., 22 coefficients for a two-port VNA, compared with 15 coefficients for a two-port double-reflectometer VNA [33].

The error models including crosstalk describe the measurement system in a more general way. They can be easily transformed to their equivalent, crosstalk-free models by setting the crosstalk error coefficients to zero. Thus, the reduction of the 22-term model (for an $n + 1$ measurement receiver VNA) leads to a $(2n^2 + n)$-term crosstalk-free model (i.e., a ten-term model for the two-port case). Omitting the influence of the crosstalk from the $2n$ measurement receiver VNA ($4n^2 - 1$-term model) gives the $(4n - 1)$-term model (i.e., a seven-term model for the two-port case).

Partly Leaky Model

For some applications, the signal leakage between different measurement ports of a multiport system is not the same. For example, the multiport wafer-level measurement system configured with dual wafer probes (two ports per probe) shows a strong crosstalk between in-side (in-probe) ports, whereas the side-by-side (probe-to-probe) port influence is much lower. For such cases, it is feasible to include only those crosstalk coefficients in the system error model that most affect the measurement results.

The solution for the four-port measurement system was introduced in [34]. In this case, the error network is split into two parts. Each covers the in-side ports only (e.g., one network [C1] for Ports 1 and 2 and a separate network [C2] for Ports 3 and 4, as shown in Figure 14). This approach significantly simplifies the description of the measurement system by reducing the number of error terms from $4n^2 - 1$ to $2n^2 - 1$, where n is the number of VNA ports. Thus, only 31 error coefficients (for the partly leaky model) are needed, instead of 63 error coefficients (for the fully leaky model), when describing a four-port VNA.

Once the error model is known, the error terms can be calculated with the help of the calibration procedure. Various calibration methods have been developed over the 40-year history of vector network analysis. Some of these methods became standard de facto methods, while others were just intermediate steps towards improving the accuracy of S-parameter measurements.

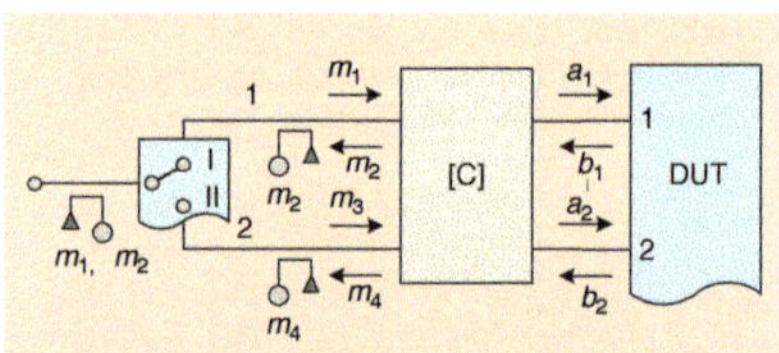

Figure 13. *Block diagram of the leaky VNA based on the reference channel architecture. For the two-port system, the matrix* [C] *includes 22 error coefficients.*

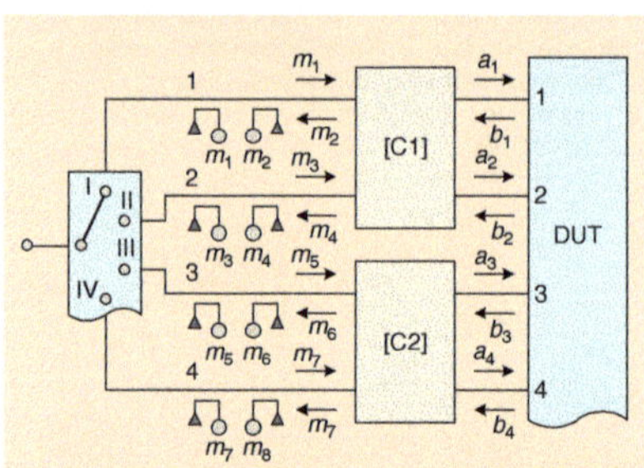

Figure 14. *A model of the double-reflectometer VNA allowing leakage between Ports 1 and 2 and between Ports 3 and 4.*

Calibration Procedures

First Iteration Solutions

Calibrating early VNAs was a lengthy and tedious process. Straightforward calculations of the required error terms as well as the error correction of the measured S-parameters of a DUT were not readily available at that time. Engineers were forced to rely on numerous variations of numerical and iterative procedures, e.g., [8].

First Explicit Solution

A significant advance was made in 1971 by Kruppa and Sodomsky [35]. For the first time, an explicit solution for calibrating a two-port VNA described by the eight-term model was introduced. This solution used three reflection standards (open, short, and match load termination) at each VNA port and two ports connected together (thru standard). Using the measurements of the open, short, and load at individual VNA ports, the three error terms S_{11}, S_{22}, and $S_{21}S_{12}$ (E_D, E_S and E_R) were defined for each port. The T_{21} and T_{12} terms were calculated from the forward and reverse transmission measurements of the thru standard, respectively (as shown in Figure 5).

The same work also introduced simple equations to perform a straightforward correction of the DUT's four S-parameters for the systematic measurement errors. Therefore, the need for lengthy iterative numerical calculations of error terms and error-corrected S-parameters was resolved.

This explicit approach was further modified for different test sets (error models) [20], [21] and, finally, the ten-term explicit calibration solution was introduced commercially by Hewlett-Packard in 1978 [19]. Since that time, this calibration procedure has become very popular under the name short-open-load-thru (SOLT) or thru-open-short-match (TOSM). Today, the SOLT calibration is a well-established technique that is implemented on all modern VNAs.

The accuracy of the SOLT procedure depends critically on the fabrication and modeling tolerances of the calibration standards (i.e., the lumped open, short, and load elements). Since the accuracy of these standards degrades with frequency, it remained a challenge to achieve reliable measurement results at high frequencies. Additional procedures, such as improving the calibration standard models (i.e., [36], [37]) or the use of standards initially characterized with respect to the reference calibration [38], can enhance the accuracy of the SOLT method.

Self-Calibration–TRL

The introduction of the TRL calibration (another variant of this is LRL) procedure by Engen and Hoer in 1974 was the next significant step in the development of VNA calibration theory [12]. For the first time, there was a method not requiring all standards to be either ideal or fully known. Using the redundancy of measurement results (an advantage of the double-reflectometer VNA and seven-term error model), TRL was able to define the originally unknown parameters of calibration standards like the reflection coefficient of the reflection standard and the propagation constant of the line standard. This new principle of calibrating a VNA with partly known standards was later called *self-calibration*.

Another advantage of the TRL technique is that it becomes possible to achieve real calibration and measurement traceability using well-defined air-isolated line standards. However, TRL is frequency limited. This restriction can be overcome by including additional line standards and applying a statistical analysis of the redundant measurement information (similar statistical techniques, such as weighted least squares [39] and generalized distance registration [40], have also been applied to one-port VNA calibration schemes resulting in a significant improvement in the overall accuracy of measurement), making TRL the accuracy benchmark per se [41]–[43].

TABLE 1. General requirements for the calibration standards.

Standard	Requirements
N1	Four known parameters (fully known)
N2	Minimum two known parameters
N3	Minimum one known parameter

Self-Calibration–Further Developments

After the introduction of the TRL self-calibration methods, many other different self-calibration procedures were developed. The measurement information redundancy obtained from the double-reflectometer VNA and its seven-term error model gives some calibration freedom: one or more standards may be partly unknown. This useful feature helps to define new calibration methods and optimize them for different applications.

For instance, the calculation of matrices $[A]$ and $[B]$ in Figure 9 can be performed by measuring three different two-port standards N_1, N_2, and N_3 instead of the DUT $[T]$ in (7)

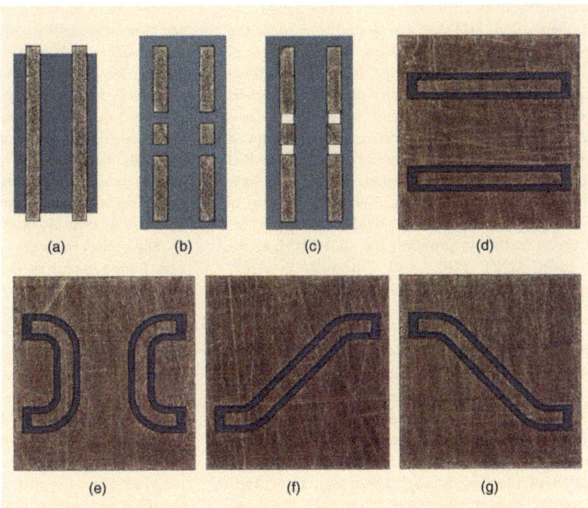

Figure 15. *Example of the commercially available (CSR) coplanar calibration standards: (a) paired shorts, (b) paired opens, (c) paired loads, (d) dual in-line thru lines, (e) dual loop-back thru lines, and, (f)–(g) cross-over thru lines. Such standards are used for most popular wafer-level calibration procedures.*

$$M_i = AN_iB^{-1}, (i = 1 \ldots 3). \quad (10)$$

To characterize the system completely [as in (6)], only seven unknowns have to be found from the 12 equations in (9). This redundancy produces general requirements to the calibration standards (Table 1) and makes it possible to derive many different calibration procedures [25], [44]–[46].

Reflection and transmission standards are addressed by the self-calibration procedure in two ways:

- one measurement of one known parameter (e.g., the reflection coefficient of a standard defines one error term)
- two measurements of the one unknown parameter taken under different conditions (e.g., the reflection coefficient of the same one-port standard measured at two VNA ports) give one error term.

Self-calibration requires seven error terms to be defined. In general cases, this can be met by any arbitrary combination of known and partly known standards (Figure 15). Today, TRL, line-reflect-match (LRM) [also often called thru-reflect-match (TRM) or thru-match-reflect (TMR)], short-open-load-reciprocal two-port (SOLR), quick-short-open-load-thru (QSOLT), and line-reflect-reflect-match (LRRM) are the most popular self-calibration procedures covering a very wide variety of applications.

Conventional and Improved LRM Procedures

The LRM method [47] was developed to resolve the frequency bandwidth limitation of conventional TRL. Instead of the line standard (or a set of different lines), it employed two one-port match (load) elements. Theoretically, LRM can be considered as a broadband calibration procedure. However, good calibration accuracy of commercially available LRM can be guaranteed only if using purely resistive, highly symmetrical 50Ω loads. This requirement is very difficult to achieve, especially at the wafer level. Some further improvements—like LRM, available from NIST [48], and line-reflect-match, advanced (LRM+) [49]—addressed this main drawback of conventional LRM.

SOLR

The SOLR method does not require the complete knowledge of the thru standard [50]. In fact, any passive two-port element providing a symmetrical (forward/ reverse) transmission coefficient (reciprocal) can be used for the calibration. SOLR is very helpful for setups where implementation of the thru is impractical: e.g., for coaxial applications when measurement ports have the same sex, or rectangular port configurations at the wafer-level. The accuracy of the SOLR method strongly depends on the one-port standards (open, short, load), which have to be either ideal or fully known.

QSOLT

Like SOLT, the QSOLT procedure expects all standards to be fully known. However, it removes the need to measure the one-port standards at the second VNA port [51], [52]. This feature dramatically reduces the time spent on reconnecting and remeasuring the standards. However, it should be noted that a VNA calibrated with the QSOLT method exhibits significant measurement errors at its second port, i.e., the port that did not have the one-port standards connected to it during calibration [53].

LRRM

The LRRM procedure was the first method that was developed explicitly to address the needs of wafer-level applications. It was designed to resolve the restrictions of the planar lumped load, such as potential asymmetry and the frequency dependence of its impedance [54]. However, like QSOLT, it measures the load standards at only one VNA port. For some applications, this may lead to less reliable measurement results at the second VNA port [55].

Table 2 gives a brief comparison of these popular self-calibration procedures for the following criteria:

- type of calibration standards
- use of standards
- definition of error term (ET) from reflection and transmission measurements
- products obtained from the redundancy information.

Calibration of the Leaky System

Obviously, calibrating a leaky system (e.g., described by the 15-term models) requires an extended number of standards and/ or calibration measurements. An iterative solution for the 15-term model was presented in [56]. It proposed four fully known two-port standards: one standard was the thru, while the remaining three standards were combinations of match-match, open-short, and short-open elements. As shown later in [57], the use of only four fully known two-port standards leads to an undetermined system of equations and, ultimately, a reduction in calibration accuracy. At least five such standards are required.

The explicit calibration and some self-calibration solutions for the 15-term model have been presented [57]–[60]. Also, the work in [33] gave a solution for the reference channel system (i.e., the 22-term model). Finally, the general self-calibration match-unknown-reflect-network (MURN) method for a leaky system was presented with eight unknown parameters of standards [58].

Multiport Cases and Hybridization

The fact that both ten-term and seven-term system descriptions can be applied to the multiport reflectometer VNA gives the user enough flexibility in choosing the most appropriate calibration method for his or her measurement system applications. Since the seven-term calibration procedures are insensitive to inaccuracies in some standards, this often makes them the preferred choice (e.g., [61], [62]).

When calibrating the seven-term system, selected error terms can be calculated using different methods. For instance, one can perform a hybrid calibration with a combination of SOLR and LRM [63] or another method [64]. This approach has benefits when some thru standards are difficult to characterize (e.g., at wafer level). However, hybrid methods may have limitations concerning the calibration dynamic range because they are based on the seven-term model [65].

An alternative way of integrating the advantages of different calibration procedures has been proposed by [66] and [67] with the generalized reflect-reflect-match-thru, advanced (GRRMT+) multiport solution. In contrast to hybrid calibrations, the GRRMT+ procedure uses the seven-term-based self-calibration LRM+ and SOLR procedures to calculate the accurate behavior of the partly known standards (i.e., the reflects and the thrus). Once all calibration standards are fully known, error terms are

TABLE 2. Comparison of TRL, LRM, SOLR, QSOLT, and LRRM calibration procedures.

	TRL		LRM/LRM+		SOLR		QSOLT		LRRM	
	Port 1	Port 2	Port 1	Port 2	Port 1	Port 2	Port 1	Port 2	Port 1	Port 2
Transmission Standards										
THRU Four Known S-parameters	☑ (4 ET)		☑ (4 ET)		×		☑ (4 ET)		☑ (4 ET)	
LINE Known: S_{11}, S_{22} Unknown: S_{21}, S_{12}	☑ (2 ET)		×		×		×		×	
RECIPROCAL Known: $S_{21} = S_{12}$, Known for $+/-90$ Degree Unknown: S_{11}, S_{22}	×		×		☑ (1ET)		×		×	
Sum of Error Terms Defined from Transmission Measurements	6		4		1		4		4	
Reflection Standards										
	Port 1	Port 2	Port 1	Port 2	Port 1	Port 2	Port 1	Port 2	Port 1	Port 2
OPEN One Known Per Port	×	×	×	×	☑ (1 ET)	☑ (1 ET)	☑ (1 ET)	×	×	×
SHORT One Known Per Port	×	×	×	×	☑ (1 ET)	☑ (1 ET)	☑ (1 ET)	×	×	×
LOAD One Known Per Port	×	×	☑ (1 ET)	☑ (1 ET)	☑ (1 ET)	☑ (1 ET)	☑ (1 ET)	×	☑ (1 ET)	×
REFLECT $S_{11}, = S_{22}$, Known for $+/-90$ Degree One Known for Two Ports	☑ (1 ET)		☑ (1 ET)		×		×		☑ (as open) (1 ET) ☑ (as short) (1 ET)	
Sum of Error Terms Defined From Reflection Measurements	1		3		6		3		3	
Self-Calibration Product	Reflection Coefficient of the Reflect Propagation Constant of the Line		Reflection Coefficient of the Reflect		S-Parameters of the Reciprocal		No		Reflection Coefficient of both Reflects	

calculated by the modified GSOLT procedure with nonideal, but known, standards. Thus, the limitations of the multiport ten-term, multiport seven-term, and hybrid methods are overcome all in one procedure.

Future Perspectives

In the last four decades, we have observed remarkable advances in microwave measurement instrumentation as well as in calibration and error correction methodologies. This significantly influenced the evolution of high-frequency semiconductor devices. Precise measurement results are crucial for understanding the real performance of a DUT, verifying its model, and improving its design. Thus, progress in the *S*-parameter measurement methods accelerated the development of, for example, high-performance telecommunication and defense systems.

Today's progress in wireless technologies and high-frequency broadband applications and the requirements for low power consumption, reduced electro-magnetic interferences, increased sensitivity, and increased data transfer rates drive the development of high-frequency passive and active differential devices. Therefore, the improvement of measurement systems is integral for providing broadband differential driving signals.

The first multiport VNAs enabling true differential measurement are already commercially available [68], [69]. Some methods for correcting systematic measurement errors have recently been published [70], [71]. These methods represent modifications of existing approaches for single-ended systems. The next significant step in calibration and error correction theory could well be the introduction of true-differential error models and calibration standards. New straightforward true-differential calibration methods will drastically simplify the calibration process. It will bring the accuracy of measurement and characterization of differential devices to new levels.

References

[1] B.O. Weinschel, "Air-filled coaxial lines as absolute impedance standards," *Microwave J.*, pp. 47–50, vo, 7, Apr. 1964.

[2] I.A. Harris and R.E. Spinney, "The realization of high-frequency impedance standards using air-spaced coaxial lines," *IEEE Trans. Instrum. Meas.*, vol. 13, no. 4, pp. 265–272, 1964.

[3] L. Essen and K.D. Froome, "The refractive indices and dielectric constants of air and its principal constituents at 24,000 Mc/ s," *Proc. Phys. Soc.*, vol. 64B, no. 10, 1951, pp. 862–875.

[4] K.H. Wong, "Using precision coaxial air dielectric transmission lines as calibration and verification standards," *Microwave J.*, vol. 31, pp. 83–92, Dec. 1988.

[5] A.E. Sanderson, "A radically new coaxial connector for high-precision measurements," *GR Experimenter*, vol. 37, pp. 1–6, Feb.–Mar. 1963.

[6] F.R. Huber and H. Neubauer, "The Dezifix connector—A sexless precision connector for microwave techniques," *Microwave J.*, vol. VI, pp. 79–85, June 1963.

[7] G-IM Subcommittee, "IEEE standard for precision coaxial connectors," *IEEE Trans. Instrum. Meas.*, vol. 17, no. 3, pp. 204–204, 1968.

[8] R.A. Hackborn, "An automatic network analyzer system," *Microwave J.*, vol. 11, pp. 45–52, May 1968.

[9] S.F. Adam, "A new precision automatic microwave measurement system," *IEEE Trans. Instrum. Meas.*, vol. 17, no. 4, pp. 308–313, 1968.

[10] S.F. Adam, G.R. Kirkpatrick, N.J. Sladek, and S.T. Bruno, "A high performance 3.5 mm connector to 34 GHz," *Microwave J.*, vol. 19, pp. 50–54, July 1976.

[11] K. Howell and K. Wong, "DC to 110 GHz measurements in coax using the 1 mm connector," *Microwave J.*, vol. 42, pp. 22–34, July 1999.

[12] G.F. Engen and C.A. Hoer, "Thru-reflect-line: An improved technique for calibrating the dual six-port automatic network analyzer," *IEEE Trans. Microwave Theory Tech.*, vol. 27, no. 12, pp. 987–993, 1979.

[13] C.A. Hoer and G.F. Engen, "On-line accuracy assessment for the dual six-port ANA: extension to nonmating connectors," *IEEE Trans. Instrum. Meas.*, vol. 36, pp. 524–529, June 1987.

[14] G.F. Engen, "The six-port reflectometer: An alternative network analyzer," *IEEE Trans. Microwave Theory Tech.*, vol. 25, no. 12, pp. 1075–1080, 1977.

[15] D. Williams, R. Marks, K. Phillips, and T. Miers, "Progress toward MMIC on-wafer standards," in *Proc. 36th ARFTG Microwave Measurements Conf.-Fall*, 1990, pp. 73–83.

[16] D.J. Bannister and M. Perkins, "Traceability for on-wafer S-parameter," *IEE Proc. Science, Measurement and Technology*, vol. 139, 1992, no. 5, pp. 232–234.

[17] *International Vocabulary of Basic and General Terms Used in Metrology*, 2nd Ed. International Organization for Standardization, Geneva, Switzerland, 1993.

[18] B.P. Hand, "Developing accuracy specifications for automatic network analyzer systems," *Hewlett-Packard J.*, vol. 21, pp. 16–19, Feb. 1972.

[19] J. Fitzpatrick, "Error models for system measurement," *Microwave J.*, vol. 21, pp. 63–66, May 1978.

[20] S. Rehnmark, "On the calibration process of automatic network analyzer systems (short papers)," *IEEE Trans. Microwave Theory Tech.*, vol. 22, no. 4, pp. 457–458, 1974.

[21] H.V. Shurmer, "Calibration procedure for computer-corrected s parameter characterisation of devices mounted in microstrip," *Electronics Lett.*, vol. 9, no. 14, pp. 323–324, 1973.

[22] D. Rytting, "Advances in microwave error correction techniques," in *Proc. Hewlett-Packard RF and Microwave Measurement Symp. and Exhibition*, June 1987, pp. 6201-6302.

[23] H. Heuermann, "GSOLT: The calibration procedure for all multi-port vector network analyzers," in *MTT-S Int. Microwave Symp. Dig.*, 2003, pp. 1815–1818.

[24] N.R. Franzen and R.A. Speciale, "A new procedure for system calibration and error removal in automated S-parameter measurements," in *Proc. 5th European Microwave Conf.*, 1975, pp. 69–73.

[25] H.J. Eul and B. Schiek, "A generalized theory and new calibration procedures for network analyzer self-calibration," *IEEE Trans. Microwave Theory Tech.*, vol. 39, no. 4, pp. 724–731, 1991.

[26] R.B. Marks, "Formulations of the basic vector network analyzer error model including switch-terms," in *Proc. 50th ARFTG Microwave Measurements Conf.-Fall*, 1997, pp. 115–126.

[27] S. Vandenberghe, D. Schreurs, G. Carchon, B. Nauwelaers, and W. De Raedt, "Identifying error-box parameters from the twelve-term vector network analyzer error model," in *Proc. 60th ARFTG Microwave Measurements Conf.-Fall*, 2002, pp. 157–165.

[28] H.J. Eul, "Methoden zur Kalibrierung von heterodynen und homodynen Netzwerkanalysatoren," Ruhr-Universitaet, Bochum, 1990.

[29] "In-fixture microstrip device measurements using TRL* calibration," Hewlett-Packard Company, Santa Clara, CA, 1991.

[30] D. Zelinka and M. Shaw, "A comparative study of TOSL, TRL, and TRL* network analyzer calibration techniques, using microstrip test fixtures," in *Proc. 46th ARFTG Microwave Measurements Conf.-Fall*, 1995, pp. 9–18.

[31] R.A. Speciale and N.R. Franzen, "Super-TSD, a generalization of the TSD network analyzer calibration procedure, covering n-port measurements with leakage," in *MTT-S Int. Microwave Symp. Dig.*, 1977, pp. 114–117.

[32] V. Teppati and A. Ferrero, "On-wafer calibration algorithm for partially leaky multiport vector network analyzers," *IEEE Trans. Microwave Theory Tech.*, vol. 53, no. 11, pp. 3665–3671, 2005.

[33] H. Heuermann and B. Schiek, "Results of network analyzer measurements with leakage errors-corrected with direct calibration techniques," *IEEE Trans. Instrum. Meas.*, vol. 46, no. 5, pp. 1120–1127, 1997.

[34] V. Teppati, A. Ferrero, D. Parena, and U. Pisani, "A simple calibration algorithm for partially leaky model multiport vector network analyzers," in *Proc. 65th ARFTG Microwave Measurements Conf.-Spring*, 2005, pp. 5-9.
[35] W. Kruppa and K.F. Sodomsky, "An explicit solution for the scattering parameters of a linear two-port measured with an imperfect test set," *IEEE Trans. Microwave Theory Tech.*, vol. MTT-19, no. 1, pp. 122–123, 1971.
[36] S. Padmanabhan, P. Kirby, J. Daniel, and L. Dunleavy, "Accurate broadband on-wafer SOLT calibrations with complex load and thru models," in *Proc. 61st ARFTG Microwave Measurements Conf.-Spring*, 2003, pp. 5–10.
[37] D. Blackham and K. Wong, "Latest advances in VNA accuracy enhancements," *Microwave J.*, vol. 48, pp. 78-94, July 2005.
[38] N. Ridler and N. Nazoa, "Using simple calibration load models to improve accuracy of vector network analyser measurements," in *Proc. 67th ARFTG Microwave Measurements Conf.-Spring*, 2006, pp. 104–110.
[39] D. Blackham, "Application of weighted least squares to OSL vector error correction," in *Proc. 61st ARFTG Microwave Measurements Conf.-Spring*, 2003, pp. 11-21.
[40] M.J. Salter, N.M. Ridler, and P.M. Harris, "Over-determined calibration schemes for RF network analysers employing generalised distance regression," in *Proc. 62nd ARFTG Microwave Measurements Conf.-Fall*, 2003, pp. 127–142.
[41] R.B. Marks, "A multiline method of network analyzer calibration," *IEEE Trans. Microwave Theory Tech.*, vol. 39, no. 7, pp. 1205–1215, 1991.
[42] D.F. Williams, J.C.M. Wang, and U. Arz, "An optimal vector-network-analyzer calibration algorithm," *IEEE Trans. Microwave Theory Tech.*, vol. 51, no. 12, pp. 2391–2401, 2003.
[43] D.F. Williams, R.B. Marks, and A. Davidson, "Comparison of on-wafer calibrations," in *Proc. 38th ARFTG Microwave Measurements Conf.-Fall*, 1991, pp. 68–81.
[44] K.J. Silvonen, "A general approach to network analyzer calibration," *IEEE Trans. Microwave Theory Tech.*, vol. 40, no. 4, pp. 754–759, 1992.
[45] H. Heuermann and B. Schiek, "A generalization of the Txx network analyzer self-calibration procedure," in *Proc. 22nd European Microwave Conf.*, 1992, pp. 907–912.
[46] H. Heuermann and B. Schiek, "Robust algorithms for Txx network analyzer self-calibration procedures," *IEEE Trans. Instrum. Meas.*, vol. 43, no. 1, pp. 18–23, 1994.
[47] H.J. Eul and B. Schiek, "Thru-match-reflect: One result of a rigorous theory for de-embedding and network analyzer calibration," in *Proc. 18th European Microwave Conf.*, 1988, pp. 909–914.
[48] D.F. Williams and R.B. Marks, "LRM probe-tip calibrations using nonideal standards," *IEEE Trans. Microwave Theory Tech.*, vol. 43, no. 2, pp. 466–469, 1995.
[49] R. Doerner and A. Rumiantsev, "Verification of the wafer-level LRM+ calibration technique for GaAs applications up to 110 GHz," in *Proc. 65th ARFTG Microwave Measurements Conf.-Spring*, 2005, pp. 15-19.
[50] A. Ferrero and U. Pisani, "Two-port network analyzer calibration using an unknown 'thru'," *IEEE Microwave Guided Wave Lett.*, vol. 2, no. 12, pp. 505–507, 1992.
[51] A. Ferrero and U. Pisani, "QSOLT: A new fast calibration algorithm for two port S parameter measurements," in *Proc. 38th ARFTG Microwave Measurements Conf.-Fall*, 1991, pp. 15–24.
[52] H.J. Eul and B. Schiek, "Reducing the number of calibration standards for network analyzer calibration," *IEEE Trans. Instrum. Meas.*, vol. 40, no. 4, pp. 732–735, 1991.
[53] J.A. Jargon, R.B. Marks, and D.K. Rytting, "Robust SOLT and alternative calibrations for four-sampler vector network analyzers," *IEEE Trans. Microwave Theory Tech.*, vol. 47, no. 10, pp. 2008–2013, 1999.
[54] A. Davidson, K. Jones, and E. Strid, "LRM and LRRM calibrations with automatic determination of load inductance," in *Proc. 36th ARFTG Microwave Measurements Conf.-Fall*, 1990, pp. 57–63.
[55] R. Doerner, "Evaluation of wafer-level LRRM and LRM+ calibration techniques," in *Proc. 69th ARFTG Microwave Measurements Conf.-Spring*, 2007, pp. 86-89.
[56] J.V. Butler, D.K. Rytting, M.F. Iskander, R.D. Pollard, and M.A. Vanden Bossche, "16-term error model and calibration procedure for on-wafer network analysis measurements," *IEEE Trans. Microwave Theory Tech.*, vol. 39, no. 12, pp. 2211–2217, 1991.
[57] K.J. Silvonen, "Calibration of 16-term error model [microwave measurement]," *Electronics Lett.*, vol. 29, no. 17, pp. 1544–1545, 1993.
[58] H. Heuermann and B. Schiek, "15-term self-calibration methods for the error-correction of on-wafer measurements," *IEEE Trans. Instrum. Meas.*, vol. 46, no. 5, pp. 1105–1110, 1997.
[59] H. Heuermann and B. Schiek, "Results of network analyzer measurements with leakage errors corrected with the TMS-15-term procedure," in *IEEE MTT-S Int. Microwave Symp. Dig.*, 1994, vol. 3, pp. 1361–1364.
[60] K. Silvonen, "LMR 16—A self-calibration procedure for a leaky network analyzer," *IEEE Trans. Microwave Theory Tech.*, vol. 45, no. 7, pp. 1041–1049, 1997.
[61] A. Ferrero, U. Pisani, and K.J. Kerwin, "A new implementation of a multiport automatic network analyzer," *IEEE Trans. Microwave Theory Tech.*, vol. 40, no. 11, pp. 2078–2085, 1992.
[62] J. Martens, "Multiport SOLR calibrations: Performance and an analysis of some standards dependencies," in *Proc. 62nd ARFTG Microwave Measurements Conf.-Fall*, 2003, pp. 205–213.
[63] J. Martens, D. Judge, and J. Bigelow, "VNA calibration modifications and hybridizations for simplified high frequency multiport/ differential measurements," in *ARMMS Conf. Dig.*, Apr. 2005.
[64] L. Hayden, "A hybrid probe tip calibration of multiport vector network analyzers," in *Proc. 68th ARFTG Microwave Measurements Conf.-Fall*, 2006, pp. 176-183.
[65] H. Heuermann, "Multi-port calibration techniques for differential parameter measurements with network analyzers," in *Proc. European Microwave Conf., Rohde and Schwarz Workshop*, Oct. 2003, pp. 1-6.
[66] H. Heuermann, A. Rumiantsev, and S. Schott, "Advanced on-wafer multiport calibration methods for mono- and mixed-mode device characterization," in *Proc. 63rd ARFTG Microwave Measurements Conf.-Spring*, 2004, pp. 91–96.
[67] A. Rumiantsev, H. Heuermann, and S. Schott, "A robust broadband calibration method for wafer-level characterization of multiport devices," in *Proc. 69th ARFTG Microwave Measurements Conf.-Spring*, 2007, pp. 56-60.
[68] *ZVA Vector Network Analyzer*, Rohde and Schwartz, Munich, Germany, 2007.
[69] "Dual-source 4-port network analyzers true-mode stimulus application (TMSA)," Agilent, Santa Clara, CA, 2007.
[70] J. Simon, "True differential stimulus gives additional insight into nonlinear amplifier behavior," in *Proc. 69th Microwave Measurements Conf.-Spring*, 2007, pp. 33-41.
[71] J. Dunsmore, K. Anderson, and D. Blackham, "Complete pure-mode balanced measurement system," in *Proc. 2007 IEEE/MTT-S Int. Microwave Symp.*, 2007, pp. 1485–1488.
[72] *General Requirements for the Competence of Testing and Calibration Laboratories*, vol. ISO/ IEC 17025:2005, 2005.
[73] R.A. Dudley and N.M. Ridler, "Traceability via the Internet for microwave measurements using vector network analyzers," *IEEE Trans. Instrum. Meas.*, vol. 52, no. 1, pp. 130–134, 2003.
[74] R.W. Beatty, "Calculated and measured S11, S21, and group delay for simple types of coaxial and rectangular waveguide two-port standards," *NBS Tech. Note 657*, Boulder, CO, Dec. 1974.
[75] M.A. Maury Jr. and G.R. Simpson, "Two-port verification standards in 3.5 mm and 7 mm for vector network analyzers," *Microwave J.*, vol. 27, pp. 101–110, June 1984.
[76] R. Tucker, "The history of the Automatic RF Techniques Group," *IEEE Microwave Mag.*, vol. 8, no. 4, pp. 69–74, 2007.
[77] R.A. Ginley, "Confidence in VNA measurements," *IEEE Microwave Mag.*, vol. 8, no. 4, pp. 54–58, 2007.
[78] G.D. Jones and N.M. Ridler, "Comparison assesses the quality of network measurements," *Microwaves & RF*, vol. 34, pp. 101–104, Jan. 1995.
[79] M.J. Maddock and N.M. Ridler, "ANAMET-991: TDNA measurement comparison exercise," ANAMET, Teddington, U.K., Rep. 031, 2001.

12

Attachment to Chapter 3

12.1 Paper [49]: "Verification of the Wafer-Level LRM+ Calibration Technique for GaAs Applications up to 110 GHz"

R. Doerner and A. Rumiantsev, "Verification of the wafer-level LRM+ calibration technique for GaAs applications up to 110 GHz," in *ARFTG Microwave Measurements Conference-Spring, 65th*, 2005, pp. 15–19.

In this article the accuracy of the transfer TMR calibration was compared to that of the benchmark NIST multiline TRL procedure for the first time. The comparison was performed on NIST verified GaAs coplanar waveguide calibration reference material RM8130. The NIST calibration comparison method was used to quantify the difference between measured S-parameters corrected by NIST multiline TRL and the transfer TMR. The worst-case error bounds for TMR corrected S-parameter measurements were determined up to 110 GHz. It was demonstrated that the difference between the benchmark multiline TRL and transfer TMR was comparable with the measurement system drift.

In this paper, I developed the verification methodology, planned the work and wrote the main part of the paper. My coauthor Ralf Doerner from Ferdinand-Braun-Institut (FBH), Leibniz-Institut fuer Hoechstfrequenztechnik (Berlin, Germany) made all required measurements. We shared calculation and the data analysis.

Verification of the Wafer-Level LRM+ Calibration Technique for GaAs Applications up to 110 GHz

Ralf Doerner* and Andrej Rumiantsev**

*Ferdinand-Braun-Institut fuer Hoechstfrequenztechnik (FBH),
Gustav-Kirchhoff-Str. 4, Berlin, D-12489, Germany, ralf.doerner@ieee.org.
**SUSS MicroTec Test Systems GmbH, Suss-Str. 1, Sacka, D-01561, Germany, a.rumiantsev@ieee.org.

***Abstract* — In this article the accuracy of the LRM+ calibration is compared to that of the benchmark NIST multiline TRL procedure for the first time. The comparison is performed on NIST verified GaAs coplanar waveguide calibration reference material 8130. The NIST calibration comparison method is used to quantify the difference between measured S-parameters corrected by NIST multiline TRL and an advanced LRM+ calibration. The worst-case error bounds for LRM+ corrected S-parameter measurements are determined up to 110 GHz. It is demonstrated that the difference between benchmark multiline TRL and LRM+ is comparable with the measurement system drift. Verification results prove that LRM+ can be successfully used for accurate GaAs on-wafer calibration with customized standards. This overcomes some drawbacks of multiline TRL while providing the same calibration accuracy.**

***Index Terms* — calibration, error correction, calibration comparison, scattering parameters measurement.**

I. Introduction

First attempts to calibrate wafer-level RF measurement set-ups were performed in the beginning of the 1980's. Yet, the verification of wafer-level calibration accuracy has remained a critical issue. Different calibration procedures have been developed in the past years (i.e. [1]–[7]). All of them rely on ideal, fully or partly known reference elements (calibration standards), realized in planar design (microstrip or coplanar).

In contrast to coaxial and waveguide applications, a great variety of fabrication techniques makes it almost impossible to trace back planar calibration standards to a natural reference. This substantially complicates the task of specifying and verifying planar calibrations.

However, research undertaken by the National Institute of Standards and Technology (NIST) provided a procedure comparing wafer-level calibrations to those performed by NIST [8]. It can be used by industrial laboratories for verification purposes. With the help of this approach, the LRM+ [7] procedure has been compared to a well-defined benchmarking multiline TRL calibration developed by NIST [2].

As demonstrated in [7], accurate measurement results can be achieved up to 110 GHz on conductive wafers using LRM+ and a customized set of standards. However, although results presented in [7] verify the calibration accuracy qualitatively, the quantitative verification of the LRM+ procedure is still a challenge.

Several accuracy verification procedures for on-wafer calibration are currently in use, e.g. [9], [10]. But almost all of them rely on error-corrected measurements of well-known reference elements. The difficulty to realize an ideal verification element on the test wafer reduces the accuracy of such approaches.

In contrast to this, a method to define error boundaries of error-corrected measurements, in relation to a well-defined calibration, was proposed in [8] and realized in a software package developed by NIST.

This method is used here in order to assess accuracy of the LRM+ procedure quantitatively. The following section briefly describes the verification approach used while Section III presents experimental results.

II. Verification Procedure

A. Calibration Comparison Procedure

The procedure used for accuracy verification provides the worst-case deviations of the measured S-parameters of passive devices for the examined (first-tier) calibration with respect to the benchmark (second-tier) calibration. Deviations are treated as $|S_{ij}-S_{ij}'|$, for $ij \in \{11, 12, 21, 22\}$, where S_{ij}' is the S-parameter measured by the calibration to be tested, and S_{ij} is the S-parameter measured by the benchmark calibration.

NIST multiline TRL was selected as the benchmark calibration. In conjunction with methods proposed in [12] and [13], this procedure allows accurate setting of the measurement system reference impedance to 50 Ohm as well as a precise definition of the measurement reference plane.

Both the LRM+ and the benchmark multiline TRL calibration were performed on the semi-insulating GaAs reference material 8130 (RM 8130), provided by NIST.

B. Reference Material 8130

The RM 8130 consists of a coplanar wave guide (CPW) multiline TRL calibration set: a 550 μm long thru line, five lines with lengths of 2.685 mm, 3.75 mm, 7.115 mm, 20.245 mm, and 40.55 mm, and two offset shorts located in a distance of 225 μm from the beginning of the line. There are also additional 12 verification reference elements. For the LRM+ calibration we use the 550 μm thru line, the 225 μm offset short, and two offset loads. The paired load standard with the resistance of about 73 Ohm for both ports was used for the first LRM+ calibration, while the asymmetrical verification resistor with the port 1 resistance of about 46 Ohm and port 2 resistance of about 133 Ohm was used for the second LRM+ calibration.

According to the individual test results provided by NIST for every RM 8130, the actual line capacitance is 1.7877 pF/cm. This value is used by the benchmark multiline TRL calibration for the accurate definition of the characteristic impedance of the RM 8130 lines and the transformation of the measurement system reference impedance.

C. Wafer-Level Measurement Setup and Software

The experimental setup for the 110 GHz wafer-level measurements includes an Agilent 8510XF VNA, a manual wafer-probe station, and the 110 GHz wafer probe tips having a pitch of 125 μm. The examined first-tier LRM+ calibration was performed using external calibration software[1]. The second-tier benchmark multiline TRL calibration as well as the accuracy analysis was performed with the help of the MultiCal® software package[2].

III. Experimental Results

To avoid additional contact uncertainty, all RM 8130 calibration standards required for benchmarking multiline TRL and the LRM+ as well as the reference elements were measured in one measurement series in the frequency range from 150 MHz up to 110 GHz. At the end of the experiment, the multiline TRL calibration standards were re-measured providing the second measurement series.

A. Verification of the Measurement Setup Integrity

First, stability of the measurement instrument and its capability to reproduce NIST measurements was validated using a multiline TRL calibration, GaAs reference material RM 8130, and the software package MultiCal©. Acquired data were corrected externally, using the multiline TRL procedure and compared to the reference data provided with the RM 8130 by NIST. The second measurement series of TRL calibration standards was used to define the drift of the measurement setup within the experiment. Obtained results are presented in Fig. 1. They are limited to 40 GHz due to the frequency limitation of the original RM 8130 measurement reference data.

According to [11], this bound increases with frequency and should not exceed 0.1 up to 40 GHz. As demonstrated in Fig. 1, the measurement setup generally meets this requirement. Thus, it forms a reliable basis for calibration comparison purposes. However, some discontinuities detected around 35 GHz may deteriorate verification accuracy in the range 32…39 GHz. Also, it has to be noted that due to the hardware limitation of the VNA it was not possible to measure accurately below 500 MHz. So, measured and calculated data below 500 MHz were not taken into consideration.

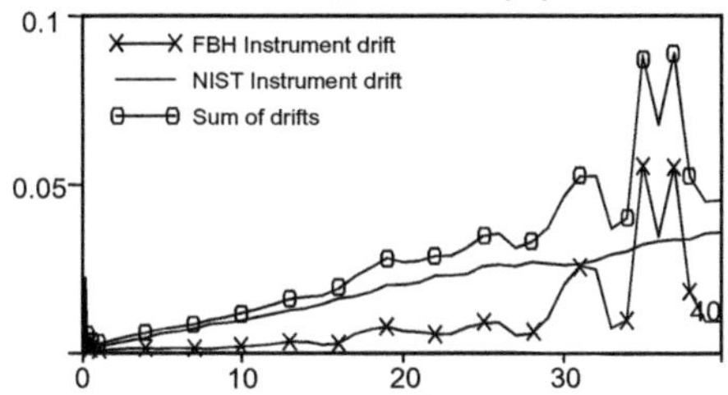

Fig. 1. Verification results on integrity of the wafer-level measurement setup used.

B. Accuracy and Repeatability Verification of the LRM+

In a second step, a multiline TRL calibration up to 110 GHz was performed for data obtained from the first and the second measurement series. The measurement system reference impedance was set to 50 Ohm and the measurement reference plane was defined at the center of the RM 8130 550 μm thru standard. The measurement system drift was determined up to 110 GHz from the second measurement series of all required multiline TRL standards by a second-tier multiline TRL.

Electrical parameters of a paired load standard (Load 1) were calculated for both cases. As shown in Fig. 3, the loads are almost symmetrical and have a resistance different from 50 Ohm (approximately 73 Ohm) and an additional reactive part. Loads are slightly dispersive. As discussed in [7], standards of this kind can be successfully used for the LRM+ calibration.

In the next step, two LRM+ calibrations were performed up to 110 GHz using the first and the second measurement series. The system reference impedance was set back to 50 Ohm by means of an LRM+ algorithm for each port individually. According to the calibration comparison technique proposed

[1] The LRM+ calibration procedure is implemented in the commercially obtainable software SussCal® from SUSS MicroTec.

[2] The MultiCal® package is provided by National Institute of Standards and Technology.

in [11], a second-tier multiline TRL calibration was used as benchmark calibration to determine the upper error bound for both cases.

The error bounds of these two LRM+ calibrations as well as a measurement system drift are shown in Fig. 2. It is obvious that the difference between the first and the second LRM+ calibration is marginal. Experimental results prove that both LRM+ calibrations provide the same measurement accuracy as the benchmarking NIST multiline TRL up to 110 GHz. The variation of the measurement accuracy is comparable with the measurement system drift over the whole frequency range.

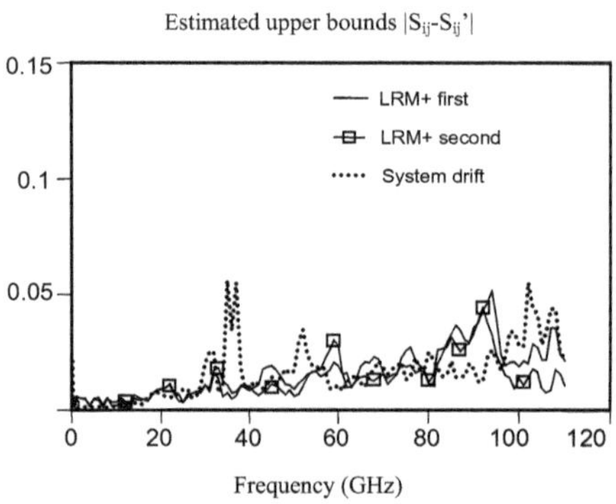

Fig. 2. The maximum possible differences from the benchmark NIST multiline TRL calibration for two 110 GHz LRM+ calibrations and actual system drift.

C. *LRM+ with Asymmetrical Load Standard*

A combination of two different load elements was used for the next experiment, namely the port-1 element of the RM 8130 verification resistor (see Fig. 3, Fig, 4, Load 2: Z11) and the port-2 element of the RM 8130 load standard used in the previous experiment (see Fig. 3, Fig. 4, Load 1: Z22). This combination artificially repeats the calibration conditions typically occurring in practice for customized wafer-embedded LRM calibration kits. The load model was defined based on measurement data corrected with the multiline TRL.

Using this load combination together with the 550 μm thru and the offset short, four different variants of the LRM calibration procedures can be performed: the simple LRM [3], the LRM, normalized to the port-2 load impedance [4], the LRM+ with the DC load (R1 = 45.92 Ohm, R2 = 72.71 Ohm) correction, and the LRM+ with the complete correction of the load standard imperfectness. The accuracy of each of them was compared to the benchmark multiline TRL. Fig. 5 presents the experimental results.

As expected, the simple LRM differs from the benchmarking multiline TRL by a value of about 0.6 over the whole frequency range. The estimated measurement error of a simple LRM procedure in the low-frequency range is caused by the deviation of the load resistance from 50 Ohm and its port asymmetry. Fig. 3 shows that the real part of the port-1 impedance is nearly constant over frequency. At the same time, the imaginary part increases slightly (Fig. 4). The real part of the port-2 load impedance decreases with frequency, while its imaginary part increases. This results in a nearly constant port asymmetry of the load standard used.

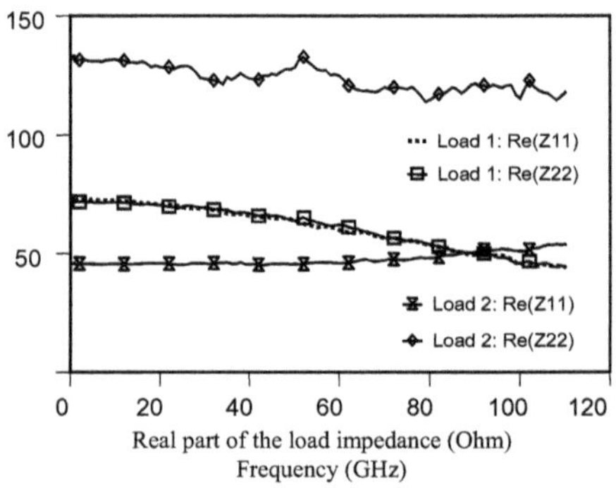

Fig. 3. The real part of the measured port 1 (Z11) and port 2 (Z22) impedance of the load standard (Load 1) and a resistor (Load 2), located on the RM 8130.

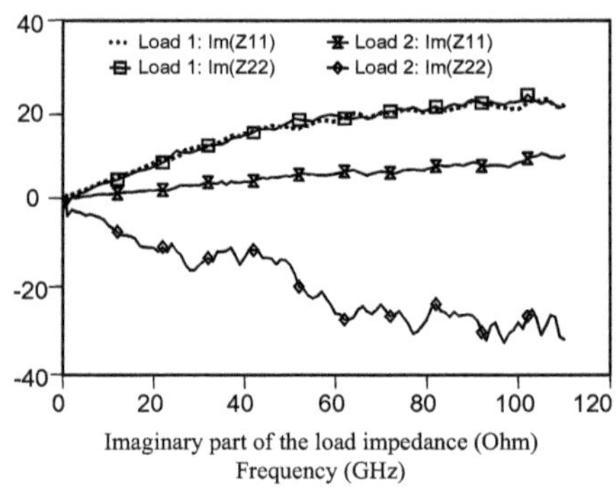

Fig. 4. The imaginary part of the measured port 1 (Z11) and port 2 (Z22) impedance of the load standard (Load 1) and a resistor (Load 2), located on the RM 8130.

Normalizing LRM to the impedance of the port-2 load provides better accuracy beyond 40 GHz than using the simple LRM. This arises from the decreasing difference in magnitude of the load impedances at both ports due to their reactive parts.

Results for the DC load corrected LRM+ generally correspond to those presented in [4]. However, in contrast to the experimental results, the procedure from [4] used the same load standard for both calibration ports.

The complete LRM+ calibration provides the best accuracy with full individual correction of the load imperfectness for each measurement port. The upper error bounds are comparable with the system drift. It has to be noted that the LRM+ error bounds of this experiment are comparable with those found in Section III.B (Fig. 2), while the used load standard was strongly asymmetrical.

Estimated upper bounds $|S_{ij}-S_{ij}'|$

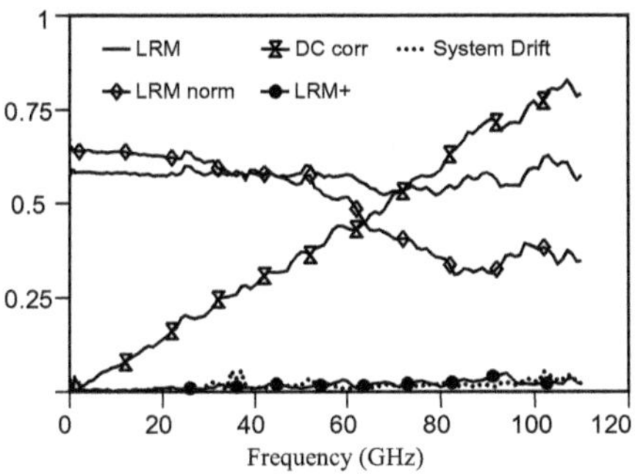

Fig. 5. The maximum possible differences between measurements of passive devices from NIST multiline TRL calibration for the simple LRM, the normalized LRM, the DC load corrected LRM+, and the complete LRM+ calibration.

D. Port Symmetry Verification

Accurate calibration aims for the exact definition of the reference impedance at both ports to the same value (i.e. 50 Ohm). As discussed in [5], LRM-like procedures set the reference impedance to the impedance of the load standard. Typically, on-wafer load standards are realized in pair. This should provide measurements of identical loads at both ports. Supposed that these loads are not equal (i.e. due to the fabrication tolerances), the reference impedance will involve errors, with regard to both the desired value and the difference between ports. The observed effect is called "port asymmetry": measurements of the same one-port device show different results at each port. The positive or negative offset from the expected value depends on the amount of the load asymmetry. To avoid this, either the load standard has to be realized symmetrically or its asymmetry should be accounted for by the calibration procedure.

The series attenuator embedded in RM 8130, symmetrical for return loss measurements, was used to verify the port symmetry of simple LRM, normalized LRM, DC load corrected LRM+, and the complete LRM+. The verification results are presented in Fig. 6. In contrast to other LRM procedures, the return loss measurements of the attenuator corrected by LRM+ calibration are highly symmetrical. The results of port symmetry verification obviously correspond with those obtained from the calibration verification (Fig. 5). Again, it is demonstrated clearly that conventional LRM-like procedures do not yield sufficient calibration accuracy for non-symmetrical non-ideal load standards.

The method [8], which is used in this work for the accuracy analysis of different LRM calibrations, also covers the measurement error caused by the "port asymmetry" problem. However, there are cases where this method is not applicable (i.e. for conductive wafers, or if the RM 8130 is not available). As demonstrated, the measurement of a suited symmetrical two-port element provides an alternative tool to verify calibration accuracy utilizing the port symmetry.

Return loss magnitude (dB)

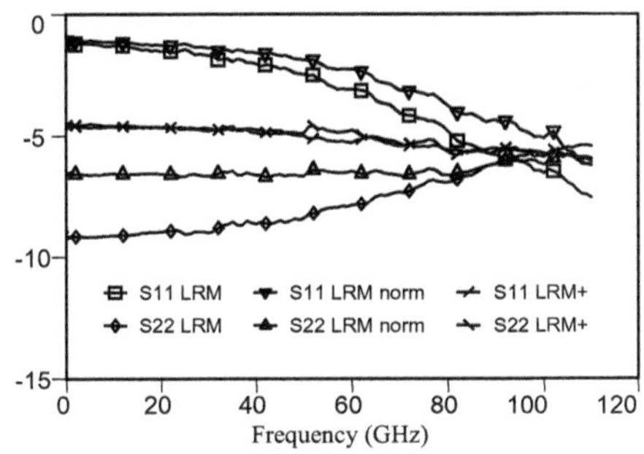

Fig. 6. Results of port symmetry verification for the LRM, the normalized LRM, and the LRM+ calibration using the RM 8130 series attenuator.

IV. Conclusions

Summarizing one can state that our experimental results demonstrate that the difference of benchmarking multiline TRL and LRM+ accuracy is comparable to the measurement system drift over the whole frequency range studied (500 MHz...110 GHz). This proves LRM+ to be a valuable tool as it overcomes the main drawback of multiline TRL: LRM+ does not require a large set of calibration standards but nevertheless provides calibration accuracy comparable to the NIST multiline TRL. LRM+, therefore, saves wafer space, minimizing the test chip size to only three standards, realized in the same design. Thus, a fully automated calibration is possible even when using a fixed wafer probes configuration. The required determination of the electrical model of the used load standards can be done easily, e.g., by means of the approaches presented in [7], [14]. As shown above, LRM+ can be successfully used for GaAs applications up to 110 GHz.

ACKNOWLEDGEMENT

The authors would like to thank Dr. W. Heinrich for his valuable comments.

REFERENCES

[1] G.F. Engen, C.A. Hoer, "Thru-Reflect-Line: An Improved Technique for Calibrating the Dual Six Port Automatic Network Analyzer", *IEEE Trans. Microwave Theory Tech.*, vol. 27, pp. 987–993, December 1979.

[2] R. Marks, "A Multiline Method of Network Analyzer Calibration", *IEEE Trans. Microwave Theory Tech.*, vol. 39, no. 7, pp. 1205–1215, July 1991.

[3] H.-J. Eul, B. Schiek, "Thru-Match-Reflect: One Result of a Rigorous Theory for De-embedding and Network Analyzer Calibration", *Proceedings of the 18th European Microwave Conference*, Stockholm, pp. 909–914, 1988.

[4] D. Williams, R. Marks, "LRM Probe-Tip Calibration using Nonideal Standards", *IEEE Trans. Microwave Theory Tech.*, vol. 43, no. 2, pp. 466–469, February 1995.

[5] R. Marks, D. Williams, "A General Waveguide Circuit Theory", *J. Res. Natl. Inst Stand. Technol.*, vol. 97, pp. 533–561, Sept.-Oct. 1992.

[6] H.-J. Eul, B. Schiek, "A generalized theory and new calibration procedures", *IEEE Trans. Microwave Theory Tech.*, vol. 39, no. 4, pp. 724–731, April 1991.

[7] R. Scholz, F. Korndoerfer, B. Senapati, A. Rumiantsev, "Advanced Technique for Broadband On-Wafer RF Device Characterization", *63rd ARFTG Conf. Digest*, pp. 83–90, June 2004.

[8] D. Williams, R. Marks, A. Davidson, "Comparison of On-Wafer Calibrations", *38th ARFTG Conf. Digest*, pp. 68–81, December 1991.

[9] B. Oldfield, "Wafer Probe Calibration Accuracy Measurements", *48th ARFTG Conf. Digest*, pp. 54-58, December 1996.

[10] C. Woodin, M. Goff, "Verification of MMIC On-Wafer Microstrip Calibration", *IEEE MTT-S Digest*, pp. 1029–1032, 1990.

[11] "Manual for Verifying Probe Station Integrity and On-Wafer Measurement Accuracy", *NIST/Industrial MMIC Consortium*, NIST, 24 August, 1998.

[12] D. Williams, R. Marks, "Transmission Line Capacitance Measurement", *IEEE Microwave and Guided Wave Lett.*, vol. 1, pp. 243–245, September 1991.

[13] R. Marks, D. Williams, "Characteristic Impedance Determination Using Propagation Constant Measurements", *IEEE Microwave and Guided Wave Lett.*, vol. 1, pp. 141–143, June 1991.

[14] D. Walker, D. Williams, J. Morgan, "Planar Resistors for Probe Station Calibration", *40th ARFTG Conf. Digest*, pp. 1-9, December 1992.

12.2 Paper [51]: "Advanced On-Wafer Multiport Calibration Methods for Mono- and Mixed-Mode Device Characterization"

H. Heuermann, A. Rumiantsev, and S. Schott, "Advanced on-wafer multiport calibration methods for mono- and mixed-mode device characterization," in *ARFTG Microwave Measurements Conference-Spring, 63rd*, 2004, pp. 91–96.

This paper outlined the theoretical background and the results of advanced RF calibration procedures especially suited to support wafer level RF device characterization with mono and/or mixed-mode interfaces. Based on the large number of possible methods to combine two-port calibration algorithms with multiport techniques, two methods were derived and investigated focusing on on-wafer devices. The results of our experiments demonstrated the effectiveness of the novel RRMT on-wafer calibration method for multiport devices.

In this work, I was responsible for collecting measurement data and data analysis. My co-author Prof. Holger Heuermann (Fachhochschule Aachen, Germany) developed the mathematical solution. Steffen Schott (SUSS MicroTec, Germany) organized the access to a 4-port VNA.

Advanced On-Wafer Multiport Calibration Methods for Mono- and Mixed-Mode Device Characterization

Holger Heuermann*, Andrej Rumiantsev**, and Steffen Schott**
* Univ. of Applied Sciences Aachen, Institute of High Frequency Tech., Germany, 52066 Aachen, Eupener Str. 70
** SUSS MicroTec Test Systems GmbH, Germany, 01561 Sacka, Sussstr. 1

***Abstract*— This paper outlines the theoretical background and the results of advanced RF calibration procedures specially suited to support wafer level RF device characterization with mono- and/or mixed-mode interfaces. The analysis of on-wafer measurement systems bring up the benefit of using load standards, reflectometers as well as the GSOLT-model to perform high quality device characterizations. Based on the large number of possible methods to combine two-port calibration algorithms with multiport techniques, two methods were derived and investigated focusing on on-wafer devices. The results of our experiments demonstrate the effectiveness of novel RRMT on-wafer calibration method for multiport devices.** [1]

I. Introduction

The accurate modeling of microwave devices is necessary to develop communication products on low cost semiconductors and MMICs. A correct modeling of semiconductors needs precise on-wafer measurement data. To ensure the validity of the measured data, the RF test equipment must be calibrated.

In the past decades, many RF calibration methods for two-port scattering parameter measurements have been published and many of them are established [1]–[6]. Meanwhile, for both network analyzer architectures, the reflectometer-concep and the reference-channel concept, multiport calibration methods are published [7], [8].

The purpose of the presented work was to design a multiport calibration procedure, which is optimized for on-wafer conditions. The algorithm must support mono- and mixed-mode interfaces, [9].

This paper describes the requirements of an on-wafer measurement system for multiport devices, including calibration standards, vector network analyzer (VNA) architectures as well as error-correction algorithms. A new and very short theory for calibration and error-correction of multiport VNAs based on the reflectometer-concep is presented. The results of sections IV and V introduce the multiport-LRRM[2]- and the novel RRMT[3]-procedures. Results representative of the RRMT-method show that this procedure is a very robust and accurate calibration method.

II. On-Wafer Measurement System

Users of VNAs for measurements of devices with coaxial standards are not familiar with on-wafer calibration and error-correction techniques. However, RF engineers, performing on-wafer measurements, are familiar with calibration methods and the necessary standards. The reasons being the complexity of on-wafer standards and the on-wafer devices under test (DUTs) which are critical components for accurate modeling of the devices. Therefore, calibration methods are established for one- and two-port on-wafer measurements. The most often used methods are SOLT[4], [19], and the self-calibration procedures TRL[5], [1], LRM[6], [3], and LRRM (Line/Reflect/Reflect/Match [4], [12].

SOLT relies on fully known standards, which is difficul to achieve at wafer level beyond 40 GHz, [13], [14]. In contrast to the SOLT, the advantage of the self-calibration procedures (e.g. TRL, LRM, etc.) is that the reflectio standard uncertainty (radiation, losses, and an imperfect phase behavior) can be ignored without any affect to the calibration accuracy.

The inability to compensate for imperfect transmission lines limits the on-wafer TRL frequency range [13]. Other TRL drawbacks include multiple transmission lines, repositioning the probes and the requirement of the very long transmission lines for low frequencies [20].

Conventional LRM calibration sets the reference impedance to the value of the ideal match standard. NIST LRM uses an arbitrary match, but the match must be similar for both ports [6]. LRRM [4], [12] like the QSOLT procedure, [17], relies on the measurements of the load standard at one port only. This approach is free of the possible port's asymmetry of the planar load standard, but it sets the reference impedance of the second measurement port inaccurately [18].

To perform a self-calibration without any systematic error, a VNA based on the reflectomete architecture is needed. This kind of VNA has, in the three-port case, six measurement channels as shown in Fig.1. The basics of a multiport calibration for this kind of VNA are based on the 7-term model and are published in [7].

[2]LRRM is the abbreviation for Line/Reflect/Reflec Match.

[3]RRMT is the abbreviation for Reflect/Reflect/Ma h/Thru.

[4]SOLT is the abbreviation for Short/Open/Line/Thru

[5]TRL is the abbreviation for Thru/Reflect/Lin

[6]LRM is the abbreviation for Line/Reflect/Matc

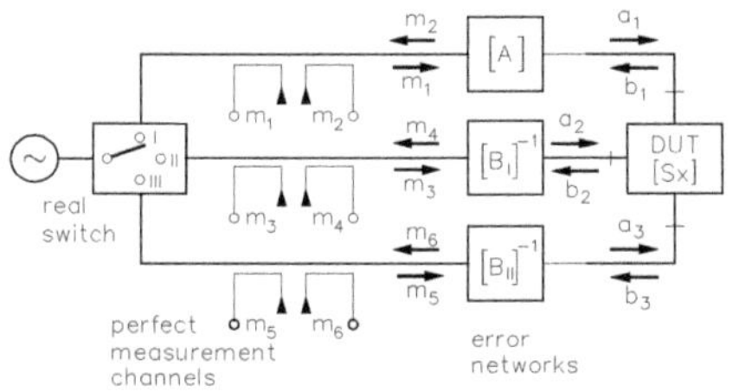

Fig. 1. Block diagram of a 3-port vector network analyzer with three ports and six measurement channels (triple-reflectometer

However, for the VNA architecture with less measurement channels, (Fig. 2) a different multiport method was published, [8], which is based on the 10-term model[7]. This GSOLT[8] method can also be used for VNAs, based on the reference-channel or reflectomete concept.

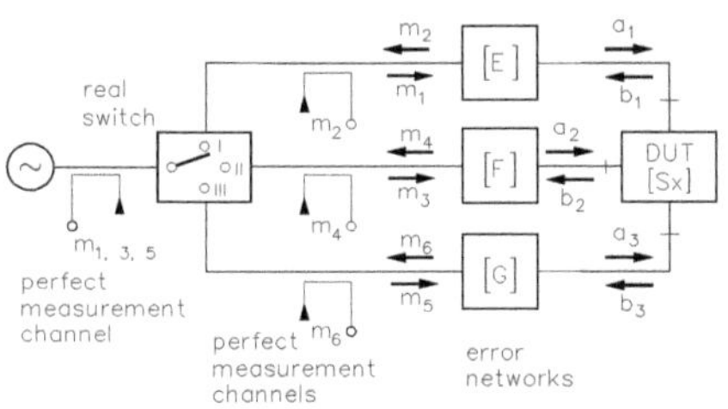

Fig. 2. Block diagram of a 3-port vector network analyzer with four measurement channels (reference-channel architecture)

In comparison between the two calibration procedures, the 7-term model and the 10-term model shows avowedly that the 7-term procedure has so-called bracings, [11]. The main dilemma of these bracings are, that the error-corrected measurements do not include the full transmission dynamic performance. Procedures using the 10-term model are much more robust.

Because of this deficien y, for multiport on-wafer calibrations we combined the GSOLT-technique and the LRM-procedure, which results in the RRMT-method. We compared RRMT with another multiport 7-term model calibration procedure, which also required two unknown reflectio standards but only one match.

Next a simplifie theory for the multiport 7-term model is presented.

[7]The 10-term model is the basic model of the 12-term SOLT-model.
[8]The name GSOLT is the short form of *General SOLT*.

III. Theory of GTXX

The way to calibrate or to error-correct the multiport VNA with the GTXX[9]-procedure is described by using a three-port triple reflectomete VNA. These results can easily be extrapolated to n-port VNAs. The block diagram with the error networks of a three-port triple-reflectomete VNA is shown in Fig. 1. The block diagram shown in Fig. 1 can be separated into two double-reflectomete VNAs: Using the firs and second state of the switch, the error matrices $[\mathbf{A}]$ and $[\mathbf{B_I}]^{-1}$ can be calculated with a standard two-port calibration TXX-procedure, [5], as well as any method which fit the 7-term model, e.g. TRL. With the firs and third state of the switch, the 7 parameters of the error boxes $\mathbf{A}$ and $\mathbf{B}_{II}$ can also be calculated by using a TXX-procedure.

The notation is using the transmission parameters

$$[\mathbf{Mx}] = [\mathbf{A}]\,[\mathbf{Nx}]\,[\mathbf{B_i}]^{-1} \tag{1}$$

with the measurement matrix $[\mathbf{Mx}]$ for the DUT $[\mathbf{Nx}]$, [5].

The number of error coefficient is 3 for the error box $\mathbf{A}$ and 4∗(n-1)[10] for the transmission calibration steps, i.e. 4∗n-1 error coefficient in total.

A. GTXX Error Correction Process

It is useful to invert the error matrices for the GTXX multiport error correction process.

$$[\mathbf{G}] = [\mathbf{A}]^{-1}\,, \quad [\mathbf{H_i}] = [\mathbf{B_i}]^{-1}\,, \quad (i = I, II) \tag{2}$$

Referring to Fig. 1, the relations

$$\begin{pmatrix} b_1 \\ a_1 \end{pmatrix} = [\mathbf{G}] \begin{pmatrix} m_1 \\ m_2 \end{pmatrix}\,, \tag{3}$$

$$\begin{pmatrix} b_2 \\ a_2 \end{pmatrix} = [\mathbf{H_I}] \begin{pmatrix} m_3 \\ m_4 \end{pmatrix}\,, \quad \begin{pmatrix} b_3 \\ a_3 \end{pmatrix} = [\mathbf{H_{II}}] \begin{pmatrix} m_5 \\ m_6 \end{pmatrix} \tag{4}$$

hold in every state of the switch for the inverted transmission parameters. That means that for a three-port, all three sets of error corrected incident and reflecte wave values

$$\begin{pmatrix} b_1 \\ b_2 \\ b_3 \end{pmatrix} = \begin{pmatrix} S_{11} & S_{12} & S_{13} \\ S_{21} & S_{22} & S_{23} \\ S_{31} & S_{32} & S_{33} \end{pmatrix} \begin{pmatrix} a_1 \\ a_2 \\ a_3 \end{pmatrix} \tag{5}$$

can be calculated with high accuracy.

Finally, we can combine the three sets of S-parameter equations to the matrix equation

$$\underbrace{\begin{pmatrix} b_1' & b_1'' & b_1''' \\ b_2' & b_2'' & b_2''' \\ b_3' & b_3'' & b_3''' \end{pmatrix}}_{=[\mathbf{K}]} = \underbrace{\begin{pmatrix} S_{11} & S_{12} & S_{13} \\ S_{21} & S_{22} & S_{23} \\ S_{31} & S_{32} & S_{33} \end{pmatrix}}_{=[\mathbf{Sx}]} \underbrace{\begin{pmatrix} a_1' & a_1'' & a_1''' \\ a_2' & a_2'' & a_2''' \\ a_3' & a_3'' & a_3''' \end{pmatrix}}_{=[\mathbf{L}]} \tag{6}$$

and use the equation

$$[\mathbf{Sx}] = [\mathbf{K}]\,[\mathbf{L}]^{-1} \quad . \tag{7}$$

to calculate the error-corrected S-parameters.

[9]GTXX stands for *General TXX*.
[10]n: Number of the measurement ports.

B. A Comparison of GSOLT and GTXX

The advantages of GTXX:

- Less calibration measurements.
- Many possible self-calibration standards.
- Easy to create calibration standards can be used.
- Switch can have poor isolation.

The advantages of GSOLT:

- Less measurement channels = cheaper VNAs.
- More robust for high dynamic measurements.

To overcome the drawback of GSOLT for on-wafer measurements which require perfect open and short standards, we have derived GSOLT in the most general way of three reflec tion standards with known but not ideal reflectio parameters. Using this common approach we are able to use the results of TXX self-calibration algorithms, which were already done for RRMT.

IV. THEORY OF MULTIPORT-LRRM-PROCEDURE

The theory of the two-port LRRM method is well-known, [4], [12], by using the standards shown in Fig. 3.

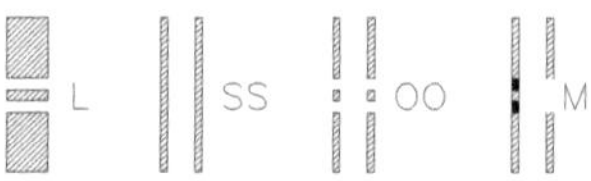

Fig. 3. Calibration standards of the LRRM-procedure

Since the match standard is only contacted at one port, the GSOLT-model can not be used for the multiport procedure. Thus we described in [10] how to realize the multiport-LRRM-procedure based on GTXX.

V. THEORY OF THE NOVEL RRMT-PROCEDURE

It is well known that the self-calibration process works with high accuracy for the reflectio standards. In addition, only a thru and a load are attractive candidates for on-wafer measurements. The assumption that the loads are identical and the used 7-term model are the reasons for the so-called bracings of the conventional LRM procedure. The traditional solution to overcome this problem the measurement of the load standard at one port only (e.g. LRRM procedure). However, as described in [10] and [21], this approach has many other drawbacks.

Our new solution to overcome the bracing problem of conventional LRM is the RRMT-procedure. The RRMT procedure consists of the following steps: We use the self-calibration process to calculate the accurate values of the open and the short standard (The bracing of the LRM-procedure have only negligible effect on the calculation of the highly reflect ve self-calibration values). We use, for the calculation of the R-standards, the most robust formula for 7-term procedures, [5]. We already have four well-known standards: R1=S, R2=O, M=L, T, all with non-ideal but known values. So, we derive the GSOLT-procedure with non-ideal values. The mathematical derivation is the same as the presented GSOLT, [2], [8], but the equations are much more complex, [10].

In this way, RRMT uses the benefit of the LRM-technique and the robustness of the GSOLT-procedure (10-term model) respectively. Errors in the match standards have no influenc on the measurement's performance apart from the fact that the real values of the match standards establish separate system impedances for the forward and the reverse direction scattering parameters. The last can be easily overcome definin the arbitrary match standard for each port independently according to [13], [15], [16], and [21].

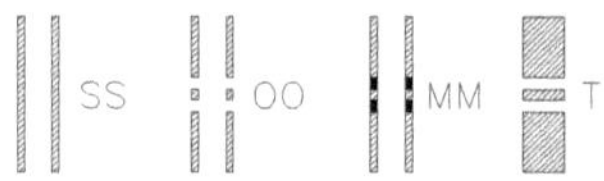

Fig. 4. Calibration standards of the RRMT-procedure

VI. MEASUREMENT RESULTS

We used the multiport network analyzer from Ballmann[11] over the frequency range of 1 GHz to 6 GHz (maximum) to measure the raw data in ASCII-SxP-format. The raw measurement data has been read out and processed on an external computer. A four-port RRMT-procedure was implemented in Matlab.

Fig. 5 shows the top-view of a used differential thru-standard and the dual |Z|-probes, manufactured by SUSS MicroTec.

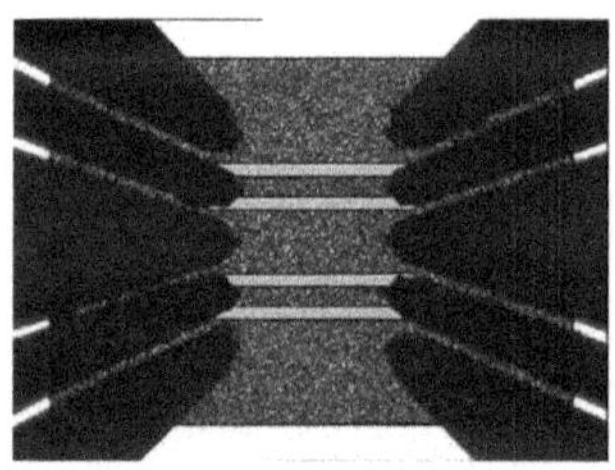

Fig. 5. On-wafer calibration standard for differential thru

Fig. 6 outlines a similar substrate to the one we used.

Fig. 7 shows the full set-up we used including the SUSS PA200HF Probe Station.

The firs example for self-monitoring values of the RRMT is given in Fig. 8 with an illustration of the self-calibration results of the short-standard.

However, this and some other figure show a measurement error at 3 GHz. This error was caused by the test-set of the used network analyzer.

[11]The Ballmann network analyzer family S20X is based on a X-reflectomete. We used an 8-port network analyzer.

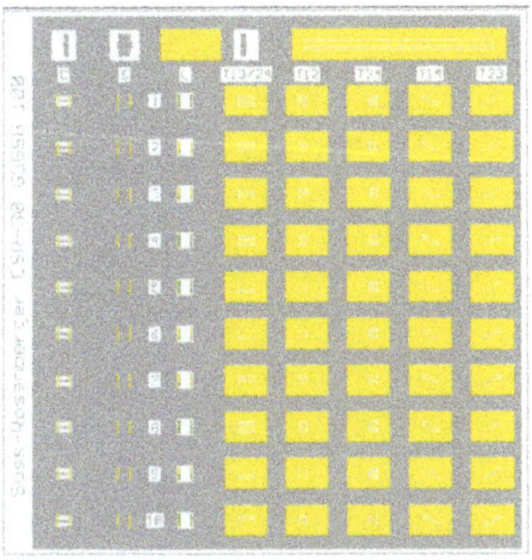

Fig. 6. On-wafer calibration and verificatio substrate

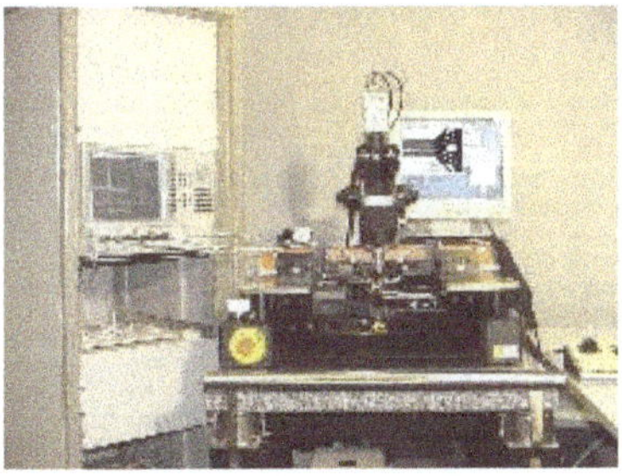

Fig. 7. Set-up for the on-wafer measurements

Fig. 9 shows the self-monitoring values of the open-standard calculated from the self-calibration process of the RRMT-procedure.

Both self-calibration results indicate that the set-up and self-calibration process works excellently.

The following results are error-corrected four-port RRMT-measurements.

However, a long matched transmission line with an open end is a very hard test (so-called ripple test) for a network analyzer set-up. Positive results of the magnitude at port 4 (Fig. 10) illustrate once more the robustness of the RRMT-procedure.

Also, the reflectio value of a long matched line is a good proportion for verfication Fig. 11 shows that the results are better than -30 dB.

The last measurement results in Fig. 12 illustrate the very accurate measurement performance of the RRMT-multiport procedure for transmission measurements. The results of a long line connected between port 1 and 4 were calculated by using the four-port RRMT-method.

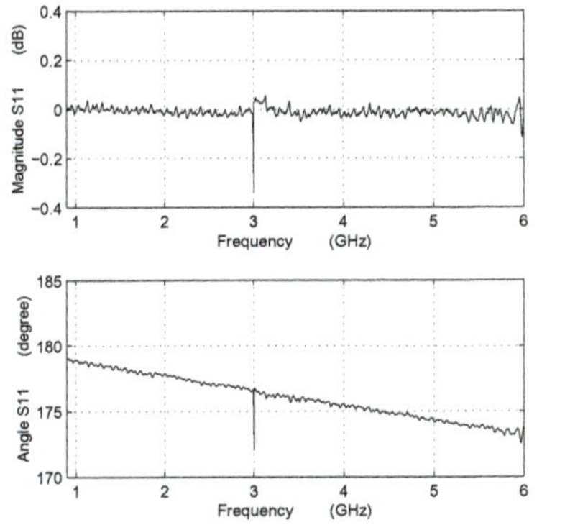

Fig. 8. RRMT-results in magnitude and phase of the self-calibration part for short-standard

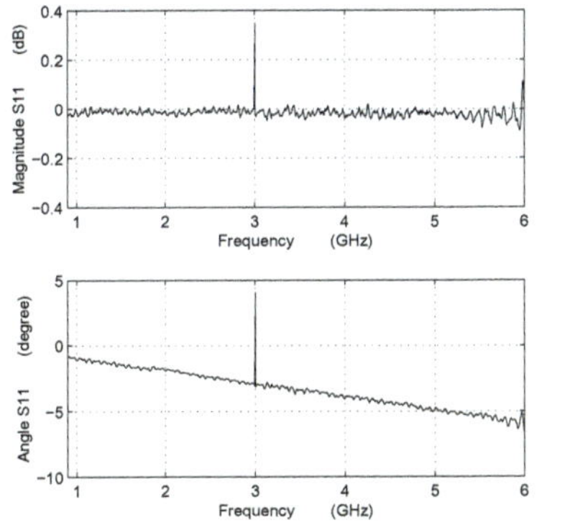

Fig. 9. RRMT-results in magnitude and phase of the self-calibration part for open-standard

VII. Conclusion

In this paper, advanced RF-techniques to perform accurate RF-on-wafer multiport measurements were presented. A large number of multiport calibration procedures can be designed. Two of them were derived and compared regarding the on-wafer conditions. A comparison was done with experimental measurements. The results clearly demonstrated proof of the advantages of the new multiport RRMT calibration procedure.

Acknowledgment

The authors are grateful to Mr. Ballmann, who provided the multiport network analyzer S208.

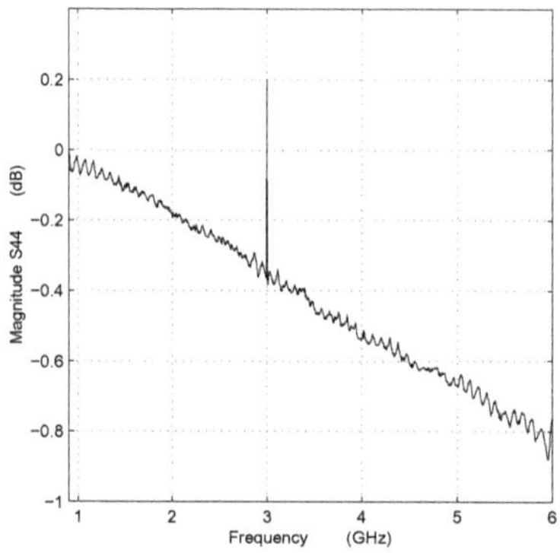

Fig. 10. Magnitude of RRMT error corrected measurement results of a long matched transmission line with an open end, connected at port 4

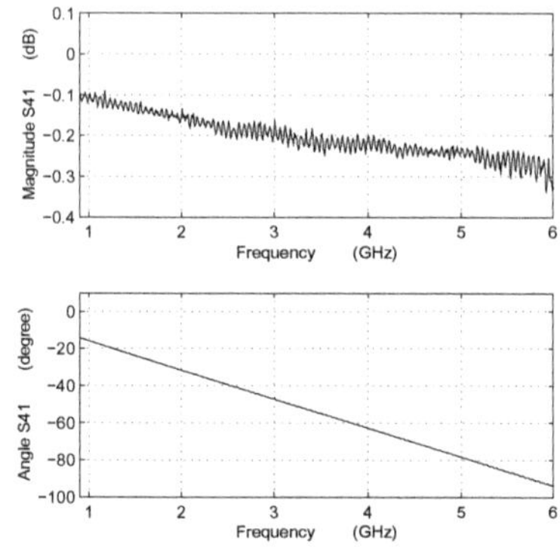

Fig. 12. Magnitude and phase of RRMT error corrected measurement results of a long transmission line, connected between port 1 and port 4

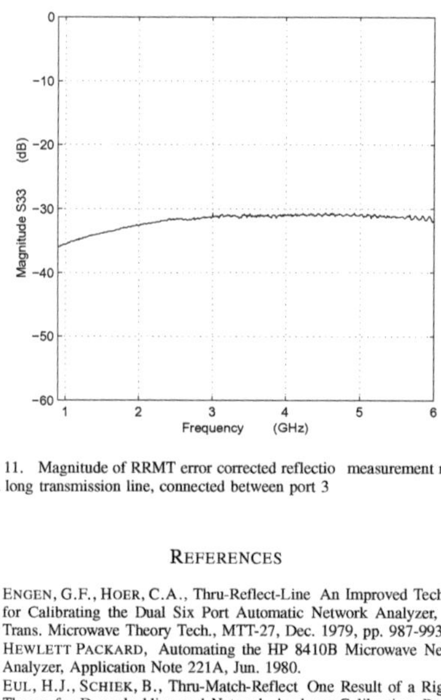

Fig. 11. Magnitude of RRMT error corrected reflectio measurement results of a long transmission line, connected between port 3

References

[1] Engen, G.F., Hoer, C.A., Thru-Reflect-Line An Improved Technique for Calibrating the Dual Six Port Automatic Network Analyzer, IEEE Trans. Microwave Theory Tech., MTT-27, Dec. 1979, pp. 987-993

[2] Hewlett Packard, Automating the HP 8410B Microwave Network Analyzer, Application Note 221A, Jun. 1980.

[3] Eul, H.J., Schiek, B., Thru-Match-Reflect One Result of a Rigorous Theory for De-embedding and Network Analyzer Calibration, Proceedings of the 18th European Microwave Conference, Stockholm, 1988, pp. 909-914

[4] Davidson, A., Jones, K., Strid, E., LRM and LRRM Calibrations with Automatic Determination of the Load Inductance, ARFTG Digest Dec. 1990, pp. 57-64

[5] Heuermann, H., Schiek, B., Robust Algorithms for Txx Network Analyzer Procedures, IEEE Trans. Instrum. Meas., Feb. 1994, pp. 18-23

[6] Williams, D., Marks, R., LRM Probe-Tip Calibration using Nonideal Standards, IEEE Trans. Microwave Theory Tech., vol. 43, No. 2, 1995, pp. 466-469

[7] Ferrero, A., Pisani, U., Kerwin, K.J., A New Implementation of a Multiport Automatic Network Analyzer, IEEE Trans. Microwave Theory Tech., vol. 40, Nov. 1992, pp. 2078-2085

[8] Heuermann, H., GSOLT: The Calibration Procedure for all Multi-Port Vector Network Analyzers, MTT-S International Microwave Symposium, Philadelphia, 2003, pp. 1815-1818

[9] Stengel, B., Mixed-Mode S-Parameters Measurements and Applications, MTT-S International Microwave Symposium, Workshop WMO, Seattle, 2002

[10] Heuermann, H., Kalibrierverfahren zur Durchfuehrung von Mehrtormessungen auf Halbleiterscheiben, Invention P 103 14 463.3 from 25.03.03

[11] Heuermann, H., Schiek, B., 15-Term Self-Calibration Methods for the Error-Correction of On-Wafer Measurements, IEEE Trans. Instrum. Meas., IM-5, Oct. 1997, pp. 1105-1110

[12] Purroy, F., Pradell, L., New Theoretical Analysis of the LRRM Calibration Technique for Vector Network Analyzers, IEEE Trans. Instrum. Meas, IM-5, Oct. 2001, pp. 1307-1314

[13] Williams, F., Characterization of Thin-Film Calibration Elements, 38th ARFTG Conf. Digest, 1991, pp. 25-36

[14] Imperato, M., Weller, T., Dunleavy, L., On-Wafer Calibration Using Space-Conservative (SOLT) Standards, MTT-S International Microwave Symposium, June 1999, pp. 1643-1646

[15] Walker, D., Kaiser, R., Williams, D., Coakley K., Lumped-Element Models for On-Wafer Calibration, 56th ARFTG Conf. Digest, 2000, pp. 89-92

[16] Kirby, P., Dunleavy, L., Weller T., Load Models for CPW and Microstrip SOLT Standards on GaAs, 56th ARFTG Conf. Digest, 2000

[17] Ferrero, A., Pisani, U., QSOLT: A New Calibration Algorithm for Two Port S-Parameter Measurements, 38th ARFTG Conf. Dig., San Diego, Dec. 1991, pp. 5-6

[18] Jargon, J., Marks, R., Rytting, D., Robust SOLT and Alternative Calibrations for Four-Sampler Vector Network Analyzers, IEEE Trans. Microwave Theory Tech., vol. 47, No. 10, 1999, pp. 2008-2013

[19] Kruppa, W., Sodomsky, K., An Explicit Solution for the Scattering Parameters of a Linear Two-Port Measured with an Imperfect Test Set, IEEE Trans. Microwave Theory Tech., Jan. 1971, pp. 122-123

[20] Marks, R., A Multiline Method of Network Analyzer Calibration, IEEE Trans. Microwave Theory Tech., Vol. 39, No. 7, 1991, pp. 1205-1215

[21] SUSS MicroTec, Application Overview: High Frequency Probing, December 2002

12.3 Paper [53]: "A Robust Broadband Calibration Method for Wafer-Level Characterization of Multiport Devices"

A. Rumiantsev, H. Heuermann, and S. Schott, "A robust broadband calibration method for wafer-level characterization of multiport devices," in *ARFTG Microwave Measurements Conference-Spring, 69th*, 2007, pp. 56–60.

This paper described the theory and practical results of the new multiport calibration procedure especially suited for wafer-level device characterization over a wide frequency range. An analysis of the currently available multiport calibration approaches was carried out. The advantages and drawbacks of each approach were demonstrated. It was shown that a robust wafer-level multiport calibration procedure should combine the strengths of both 7-term and 10-term based algorithms. Corresponding to these requirements, the definition of the advanced multiport RRMT+ algorithm was given. The results of a practical experiment proved the theory and demonstrated the advantages of the new multiport RRMT+ calibration procedure.

For this paper, I planned the entire work, derived the mathematical solution, acquired measurement results, did the calculations and wrote the paper. My two co-authors, Prof. Holger Heuermann (Fachhochschule Aachen, Germany) and Steffen Schott (SUSS MicroTec, Germany) provided scientific and management supervision respectively.

A Robust Broadband Calibration Method for Wafer-Level Characterization of Multiport Devices

Andrej Rumiantsev [1], Holger Heuermann [2], and Steffen Schott [1]

[1] SUSS MicroTec Test Systems GmbH, Suss-Str. 1, D-01561 Sacka, Germany
[2] University of Applied Sciences Aachen, Eupener Str. 70, D-52066 Aachen, Germany

Abstract **— This paper describes the theory and practical results of the new multiport calibration procedure especially suited for wafer-level device characterization over a wide frequency range. An analysis of the currently available multiport calibration approaches was carried out. The advantages and drawbacks of each approach are demonstrated. It is shown that a robust wafer-level multiport calibration procedure should combine the strengths of both 7-term and 10-term based algorithms. It should also provide reference-match measurements on each VNA measurement port and be insensitive to the behavior of highly-reflective standards and the design of transmission standards.**

Corresponding to these requirements, the definition of the advanced multiport RRMT+ algorithm is given. The results of a practical experiment proved the theory and demonstrated the advantages of the new multiport RRMT+ calibration procedure.

Index Terms **— multiport calibration, error correction, scattering parameters measurement.**

I. Introduction

The progress in wireless technologies and high-frequency broadband applications, the requirements for low power consumption, reduced electro-magnetic interferences, increased sensitivity, and data transfer rates drive the development of high-frequency, multiport mono- and mixed-mode devices, and thus improvement of multiport measurement systems.

As an example, the first 4-port 40 GHz vector network analyzer (VNA) based on a true $2n$-receivers concept has been recently reported [1]. The hybrid multiport systems already provide 4-port measurement capability up to 67 GHz [2, 3]. The expansion of the frequency range of multiport VNAs requires new approaches for their calibration, particularly for wafer-level applications.

Many different calibration procedures were developed and implemented in commercial VNAs in the past. These can be separated into two groups based on the 10-term or 7-term model. 10-term approaches can be used for low-end VNAs (with $n+1$ measurement receivers, where n is number of VNA ports) and demonstrate the highest potential calibration dynamic range [4]. However, the requirement to have either ideal of fully known calibration standards significantly limits the application of these procedures. Therefore, it is not recommended to use 10-term based calibrations for wafer-level measurements beyond 20 GHz.

Many 7-term procedures are not sensitive to the non-ideal calibration standards short and open. Variants of these procedures (such as 2-port and multiport TRL[1], LRM[2], SOLR[3], LRM+[4], or TXX for general) are widely used for the calibration of broadband wafer-level measurement systems [5-8, 10, 12-14,]. However, as it was reported in [10], potential limitation of the calibration dynamic range of the 7-term-based methods may reduce the measurement accuracy, especially for differential devices under high-rejection conditions.

Additionally, multiport TXX-procedures may lead to significant calibration errors in the following cases (typical for many wafer-level setups):

- partly-known reflection (R) standards are not symmetrical (for instance, when measuring at a packaged-level);
- the rectangle, loop-back, or cross-over thru (Fig. 1) introduces a resonance at certain frequencies within the measurement range;
- the open, short, and load standards are not ideal or insufficiently modeled (for SOLR-like procedures)
- the load standard is not purely resistive and its impedance is frequency dependent;
- the load impedance is asymmetrical;

Fig. 1. Configuration of thru standards for a 4-port wafer-level calibration: straight thrus (port 1 – port 2, port 3 – port 4), loop-back thrus (port 1 – port 3, port 2 – port 4), cross-over thrus (port 1 – port 4, port 2 – port 3).

---

[1] Thru-Reflect-Line
[2] Line-Reflect-Match
[3] Short-Open-Load-Reciprocal
[4] Advanced Line-Reflect-Match

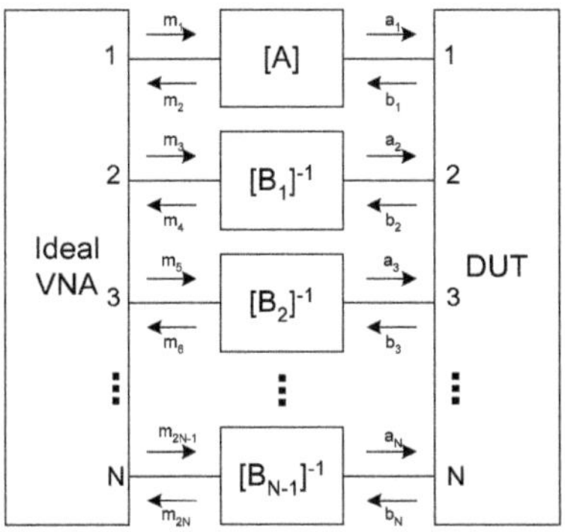

Fig. 2. Block diagram of n-port VNA that represents the ideal VNA, the matrices of error terms $[A]$ and $[B_i]$, and the DUT.

Therefore, it is necessary to develop a new multiport calibration procedure that will be free of the mentioned limitations.

The next chapter of this paper briefly describes the multiport measurement system and discusses different calibration approaches. Then, the novel generalized RRMT+ is introduced and experimental results verifying the application of this method are presented.

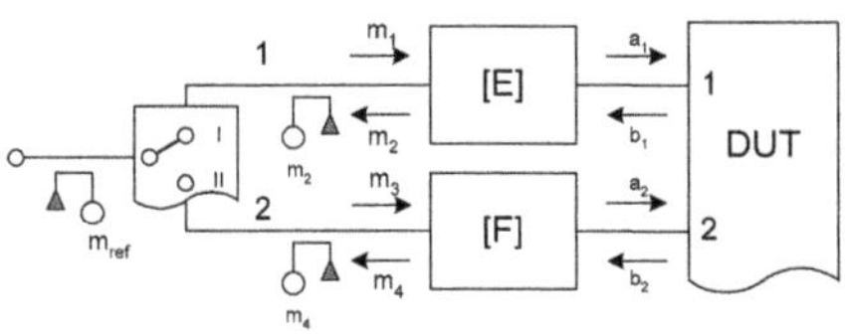

Fig. 3. Block diagram of the n-port VNA based on the reference channel architecture. For simplification, the diagram is reduced to its 2-port part. It shows the reference receiver m_{ref}, the signal source switch, the measurement receivers m_2 and m_4, and the 10-term model error matrices $[E]$ and $[F]$.

II. The Theory of the Generalized RRMT+

A. Multiport System

A multiport VNA can be described in general as shown in Fig. 2. The schematic diagram consists of an ideal VNA, the matrices $[A]$, $[B_i]$ of systematic measurement errors, and the DUT. An example of an n-port VNA with a reference channel architecture (with n+1 measurement receivers) is given in Fig. 3. Its error model is based on the 10-term representation: for every state of the system switch, 5 error terms can be defined for each pair of VNA ports. For instance, the two-port case gives terms: E_D^I, E_R^I, E_S^I, F_L^I, and F_T^I for the first state of the switch; and F_D^{II}, F_R^{II}, F_S^{II}, E_L^{II}, and E_T^{II}, for the second state of the switch (Fig. 4). Introducing matrices $[G]$, $[H]$, and so on, this description can easily be extended to the n-port case with $2*n^2+n$ error coefficients.

The multiport reflectometer VNA can be described by both the 10-term and the 7-term based models. The 7-term based n-port VNA error models consists of the $4*(n-1)+3$ coefficients.

Crosstalk between calibrating ports is not addressed in these models. Crosstalk affected systems require more sufficient modeling and include additional error terms.

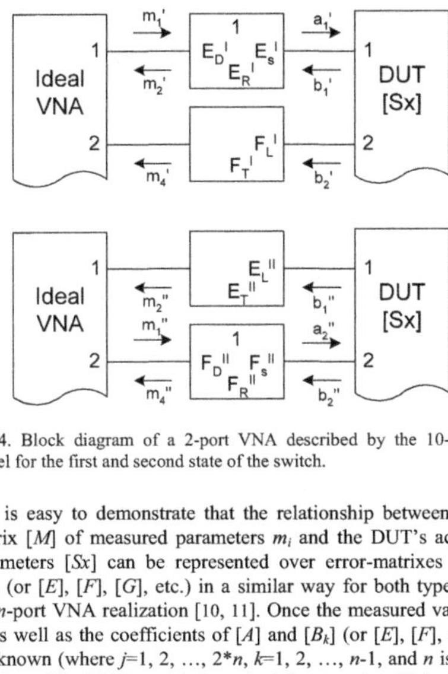

Fig. 4. Block diagram of a 2-port VNA described by the 10-term model for the first and second state of the switch.

It is easy to demonstrate that the relationship between the matrix $[M]$ of measured parameters m_i and the DUT's actual parameters $[Sx]$ can be represented over error-matrixes $[A]$, $[B_i]$, (or $[E]$, $[F]$, $[G]$, etc.) in a similar way for both types of the n-port VNA realization [10, 11]. Once the measured values m_j as well as the coefficients of $[A]$ and $[B_k]$ (or $[E]$, $[F]$, etc.) are known (where j=1, 2, …, $2*n$, k=1, 2, …, n-1, and n is the number of the VNA ports), the actual S-parameters $[Sx]$ of the unknown DUT can easily be found [e.g. 10].

B. Generalized RRMT+ Calibration Procedure

The fact that both 10-term and 7-term system descriptions can be applied to the multiport reflectometer VNA, gives the user enough flexibility in choosing an appropriate calibration procedure that fits to the measurement system applications in the best way possible. Since the 7-term calibration procedures are not sensitive to inaccurate reflection calibration standards, they are used more often.

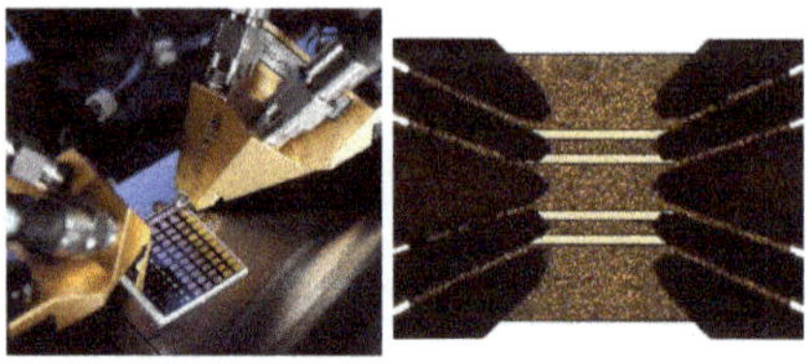

Fig. 5. Configuration of a 4-port wafer-level measurement and calibration: two Dual |**Z**| Probes and a differential (dual) straight calibration thru standard.

Calibrating the 7-term system, some selected error terms can be calculated by different variants of the TXX calibration within a single procedure: for instance, combining LRM and SOLR (a "hybrid calibration") [14, 15]. This approach has benefits when some thru standards are difficult to characterize (e.g. at wafer level). However, hybrid calibrations are neither free of possible "port bracing" nor the limitation of the calibration dynamic range because they are based on the 7-term model [10].

The novel RRMT+[5] multiport calibration approach is free from these limitations as well as from drawbacks discussed in the introduction. RRMT+ integrates the advantages of the 10-term and 7-term calibration algorithms into one procedure and extends the algorithm reported in [11, 16] for non-ideal thru standards.

In contrast to hybrid calibrations, the RRMT+ procedure uses the 7-term-based self-calibration process to calculate the accurate behavior of partly know standards (reflects and thrus) and the 10-term process to calibrate the system. Reflect standards (partly-known open and short) are characterized by the very robust LRM+ method [8]; reciprocal thrus (loop-back, cross-over, Fig. 1) are defined with the help of the SOLR algorithm [5]. Additionally, introducing at least one straight line in the measurements allows the characteristic impedance Z_0, phase, and attenuation constants of the straight thru to be found [17]. If necessary, the load can be specified.

Once all calibration standards are fully known, all error terms are calculated by the modified GSOLT procedure with non-ideal, but know standards [11]. Therefore the limitations of the multiport 10-term (SOLT), multiport 7-term (TXX) conventional and hybrid methods are overcome in one procedure.

[5] Advanced Reflect-Reflect-Match-Thru

C. RRMT+ Requirements for Calibration Standards

Depending on the number of DUT ports and their design, different configurations of the wafer-level measurement system are possible. For example, Fig. 5 shows two dual wafer probes with the same pitch. Calibrating the system like this, the set reflection (open, short, load) and transmission (thru, lines) standards can be measured. Reflection standards are typically grouped into one element consisting of four same-type standards. Thus, all required S-parameters (return loss) can be obtained after just one movement and contact. All elements in this group are equal to each other.

Full 4-port calibration requires 6 thru standards, as shown in Fig. 1. However, the electrical characteristics of only 3 standards are required: straight (port 1 – port 2), loop back (port 1 – port 3), and cross over (port 1 – port 4). The others are symmetrical.

Therefore, RRMT+ requires at minimum that:

1. open and short for all four ports are partly known (only within +/- $\pi/2$ their phase);
2. load port 1 and port 2 (or port 3 and port 4) are known, but they can be represented by any impedance element and different from each other;
3. straight thru port 1 – port 2 (or port 3 – port 4) is known.

Alternatively, requirements 2 and 3 can be replaced by adding an additional straight line with known physical length.

These requirements can vary depending on the system configuration and application type.

III. Experimental Results

The theory of the novel RRMT+ calibration approach was proven by experimental results. The experimental setup included a Rohde and Schwarz 40 GHz 4-port ZVA network analyzer; the semiautomatic PA200 probe system, the Dual |**Z**| Probe 500 µm pitch GSGSG configuration, and the GSGSG CSR-34 calibration substrate from SUSS MicroTec. Some additional test devices were used for the calibration comparison and calibration accuracy verification.

It is important to note that the probes and standards included in this setup had a very large pitch (500 µm). This presents a significant challenge for accurate broadband calibration (beyond 20 GHz).

To avoid the influence from contact repeatability error, the measured data were acquired in one measurement series in raw format and saved to an external PC. The calibration and the error correction were performed offline with the help of SussCal Professional calibration software[6].

The experimental results demonstrated a significant improvement in the measurement accuracy when the new RRMT+ calibration algorithm was used (Fig. 6-8).

[6] Available from SUSS MicroTec.

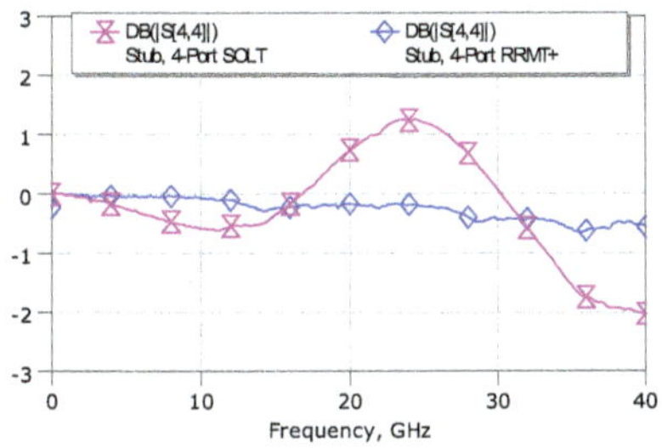

Fig. 6. Return loss measurements of the dual 6000 µm stub line on port 4 with respect to the 4-port SOLT and 4-port RRMT+ calibrations.

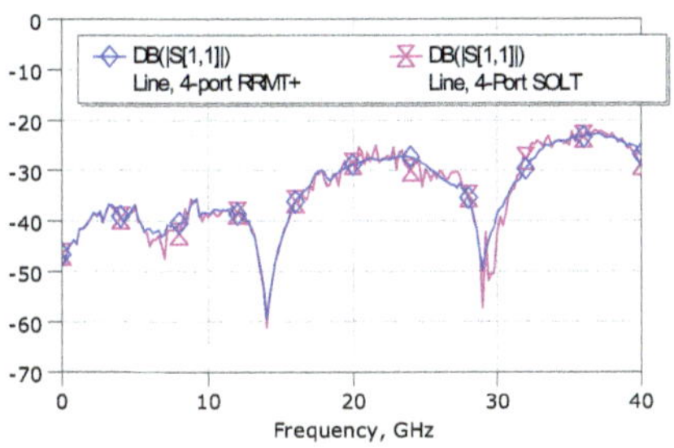

Fig. 7. Return loss measurements of the dual 6000 µm line on port 1 with respect to the 4-port SOLT and 4-port RRMT+ calibrations.

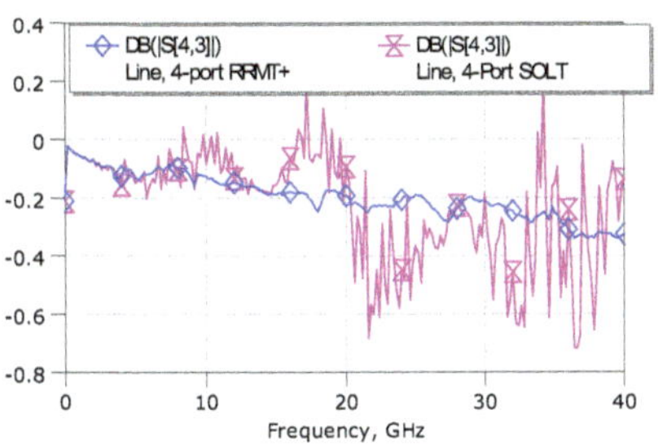

Fig. 8. Insertion loss measurements of the dual 6000 µm line between port 3 and port 4 with respect to the 4-port SOLT and 4-port RRMT+ calibrations.

IV. Conclusion

The analysis of conventional multiport calibration procedures was fulfilled. Their advantages and drawbacks for wafer-level applications were discussed and the novel RRMT+ calibration procedure was introduced.

It was demonstrated that the RRMT+ approach is not sensitive to conventional wafer-level multiport calibration problems: in general, it does not require accurate modeling of all reflection and transmission standards. This fact guarantees that the RRMT+ method is significantly less sensitive to typical calibration problems in a multiport wafer-level setup. Thus it is ideal for broadband characterization of multiport and differential devices without limitations of their sizes, frequency and design.

Acknowledgement

The authors wish to acknowledge excellent assistance and support of Frieder Simon, Christian Evers, and Klaus Beister of Rohde and Schwarz for providing the 40 GHz 4-port ZVA used for this experiment.

References

[1] Rohde and Schwarz, "Vector network analyzer R&S ZVA", *Product Brochure*, June 2006.

[2] Agilent Technology, "Agilent solutions for multiport and balanced devices", *Product Brochure*, March 2006.

[3] Anritsu, "Lightning™ 37000D vector network analyzers", *Product Brochure*, June 2005.

[4] H. Heuermann, "GSOLT: The calibration procedure for all multi-port vector network analyzers", *MTT-S Digest 2003*, pp. 1815-1818.

[5] A. Ferrero, U. Pisani, "Two-port network analyzer calibration using an unknown 'thru'", *IEEE Microwave and Guided Wave Letters*, Vol. 2, Issue 12, pp. 505-507, December 1992.

[6] H.-J., Eul, B. Schiek, "A generalized theory and new calibration procedures", *IEEE Trans. Microwave Theory Tech.*, Vol. 39, no.4, pp. 724-731, April 1991.

[7] H. Heuermann, B. Schiek, "Robust Algorithms for Txx network analyzer self-calibration procedures", *IEEE Trans. Instrum. Meas.*, IM-1, Feb. 1994, pp. 18-23.

[8] R. Doerner, A. Rumiantsev, "Verification of the wafer-level LRM+ calibration technique for GaAs applications up to 110 GHz", *65th ARFTG Conference Digest*, June 2005.

[9] A. Ferrero, F. Sanpietro, U. Pisani, "Multiport vector network analyzer calibration: a general formulation", *Trans. Microwave Theory Tech.*, vol. 42, no. 12, pp. 2455–2461, December 1994.

[10] H. Heuermann, "Multi-port calibration techniques for differential parameter measurements with network analyzers", *European Microwave Conference*, October 2003, Rohde and Schwarz Workshop.

[11] H. Heuermann, A. Rumiantsev, S. Schott, "Advanced on-wafer multiport calibration methods for mono- and mixed-mode device characterization", *63rd ARFTG Conference Digest*, June 2004.

[12] J. Martens, "Multiport SOLR calibrations: performance and an analysis of some standards dependencies", *62nd ARFTG Conference Digest*, pp. 205 – 213, December 2003.

[13] T. Buber, A. Rodriguez, A. Jenkins, and others, "Multimode TRL and LRL calibrated measurements of differential devices", *64th ARFTG Conference Digest*, pp. 157 – 166, June 2004.

[14] J. Martens, D. Judge, J. Bigelow, "VNA calibration modifications and hybridizations for simplified high frequency multiport/differential measurements, *ARMMS Conference* Digest, April 2005.

[15] L. Hayden, "A hybrid probe tip calibration of multiport vector network analyzers", *68th ARFTG Conference Digest*, Fall 2006.

[16] "Calibration method for carrying out multiport measurements on semiconductor wavers", US Patent 7,130,756.

[17] A. Rumiantsev, R. Doerner, S. Thies, "Calibration standards verification procedure using the calibration comparison technique", *36th European Microwave Conference Digest*, September 2006.

13

Attachment to Chapter 5

13.1 Paper [76]: "Applying the Calibration Comparison Technique for Verification of Transmission Line Standards on Silicon up to 110 GHz"

A. Rumiantsev, P. L. Corson, S. L. Sweeney, and U. Arz, "Applying the calibration comparison technique for verification of transmission line standards on silicon up to 110 GHz," in *ARFTG Microwave Measurements Conference-Spring, 73rd*, Boston, MA, 2009, pp. 1–6.

This paper presented the results of extracting the electrical characteristics of planar lines using the calibration comparison method for standards realized in IBM's advanced 0.13 μm RF CMOS 8SF process. For the first time, this method was applied to characterizing the customized standards on silicon up to 110 GHz. Additionally, this paper considered the influences of the reference benchmark calibration standards, on the characterization accuracy of silicon wafer-embedded lines at mm-wave frequencies.

Here, I developed the entire verification methodology, did all calculations (except extraction of the DUT parameters), and wrote the major part of the conference paper. The design of the custom calibration standards, the electromagnetic simulations and the acquisition of measurement data and the extraction of the DUT parameters were done by Philip Corson and Susan Sweeney from IBM Microelectronics (Essex Junction, VT, USA). The analysis of the results was performed together with all co-authors, including Uwe Arz from Physikalisch-Technische Bundesanstalt (PTB) (Braunschweig, Germany).

Applying the Calibration Comparison Technique for Verification of Transmission Line Standards on Silicon up to 110 GHz

Andrej Rumiantsev [1], Phillip L. Corson [2], Susan L. Sweeney [2], and Uwe Arz [3]

[1] SUSS MicroTec Test Systems GmbH, Suss-Str. 1, Sacka, D-01561, Germany, a.rumiantsev@ieee.org;
[2] IBM Microelectronics, Essex Junction, VT 05452, USA, corson@us.ibm.com; slsweene@us.ibm.com
[3] Physikalisch-Technische Bundesanstalt (PTB), Bundesallee 100, Braunschweig, D-38116, Germany, uwe.arz@ptb.de

Abstract **— This paper will present the results of extracting the electrical characteristics of planar lines using the calibration comparison method for standards realized in IBM's advanced 0.13 µm CMOS process. For the first time, this method is applied to characterizing the customized standards on silicon up to 110 GHz. Additionally, this paper considers the influences of the reference benchmark calibration standards, included with GaAs reference material RM8130, on the characterization accuracy of silicon wafer-embedded lines at mm-wave frequencies.**

I. Introduction

Accurate on-wafer measurement of high frequency scattering (S-) parameters have already become an essential step of characterizing devices used in high frequency applications, such as advanced communications circuits. As technology scaling advances and device performance increases, it is becoming essential to improve the calibration and ultimately the measurement accuracy of an on-wafer S-parameter characterization system. Accurate on-wafer network analyzer calibration and establishment of the measurement reference plane close to the wafer-embedded device under test (DUT) at mm-wave frequencies is now necessary, reducing the reliance on inaccurate de-embedding methods of the probe padset parasitics.

The multiline TRL[1] calibration [1] procedure together with methods for characterizing planar lines [2, 3] developed by the National Institute of Standards and Technology (NIST) has already established a solid reference in the on-wafer calibrations for semi-insulating wafers. However, even more challenging and urgently required, is the application of these techniques and the analysis of their accuracy for semi-conductive wafers [4].

A great variety of fabrication techniques complicates the task of specifying and verifying planar calibrations. Research undertaken by NIST provided a procedure for comparing wafer-level calibrations, identifying setup drift, as well as verifying calibration standards [5]. Furthermore, the calibration comparison procedure can be successfully used to characterize the electrical properties of unknown standards on alumina [6] as well as on silicon [7-9]. However, to the author's knowledge, this method has not been verified yet for characterizing calibration standards on lossy substrates above 50 GHz.

[1] Thru-Reflect-Line

This paper will present the results of extracting the electrical characteristics of planar lines using the calibration comparison method for standards realized in an IBM advanced 0.13 µm CMOS process. For the first time, this method is applied to characterize customized standards on silicon up to 110 GHz. Additionally, this paper considers the influences of wafer probes and the reference benchmark calibration standards on the characterization accuracy of lines on silicon.

II. On-Wafer Calibration Standards and Characterization Methods

A. Calibration Standard on Silicon

The experimental calibration set included all standards required for the multiline TRL calibration (with three lines). Standards were fabricated within a typical 2-port GSG on-wafer probe padset. The ground pads, signal pad, and signal launch lines were placed on an upper metal level layer, while a meshed ground plane is placed under the signal geometries on the first metal level layer. The transmission lines were fabricated as extensions of the signal launch lines. The line design addresses the specifics of the bulk CMOS process. Line lengths were adjusted to cover a frequency band from 1 GHz to 110 GHz. The test set included various lumped standards, such as Load, and different Open and Short elements.

Three different techniques were applied to characterize the electrical properties of the test set lines: the lumped load method, the calibration comparison method, and EM simulations.

B. Lumped Load Method

The lumped load method was originally introduced in [3]. It extracts the characteristic impedance Z_0 of a quasi-TEM line fabricated on a semi-insulating substrate from its measured capacitance per unit length C in three steps. First, the multiline TRL calibration is performed with the examined lines setting the calibration reference impedance Z_r to the characteristic impedance of the line Z_l. Second, the reflection coefficient Γ_{load} and the DC resistance $R_{load.dc}$ of a reference load element are measured. Finally, the line capacitance is calculated from:

$$C\left[1-j\frac{G}{\omega C}\right] \approx \frac{\gamma}{j\omega R_{load.dc}} \frac{1+\Gamma_{load}}{1-\Gamma_{load}} \tag{1}$$

where: γ is the line propagation constant, measured from the multiline TRL calibration.

C. Calibration Comparison Method

The calibration comparison method described in [8] relies on a two-tier calibration process. The first-tier calibration is performed in a set of well-defined reference lines. In this calibration, the system reference impedance Z_r is set to 50 Ω, and the position of the measurement reference plane is moved to the probe tips. The second-tier calibration is performed in the transmission lines one wishes to characterize. Here, the reference plane is also set to the probe tips. The calibration comparison procedure yields error boxes that can be modeled by a simple equivalent circuit. Finally, a robust estimate of the measured line's characteristic impedance can be derived, which is insensitive to the contact pad parasitics.

D. EM Simulation

The thru line was simulated with a simplified model using the 3D full-wave electromagnetic field simulation package Ansoft HFSS®[4]. The 3D structure included all the passivation and metal layers with interconnects simplified to solid bars. Simulation results yield the electrical parameters of the transmission line including gamma and the characteristic impedance.

III. Measurement Setup

The experimental setup included an Agilent 8510XF VNA, a semi-automatic wafer-probe station, and 110 GHz wafer probe tips with a pitch of 100 µm. To establish a reference calibration, two different sets of standards were used: a semi-insulating GaAs reference material 8130[2] (RM8130), and a commercially available alumina calibration substrate recommended by its vendor for application up to 110 GHz.

Measurement data from the VNA were recorded with SussCal®[3]. Off-line calibration, error correction, calibration comparison, and the line parameter extraction were performed using MultiCal©, LINEPAR, CAP, and ExtractZo from NIST.

A. Reference Material RM8130

The RM8130 consists of a coplanar wave guide (CPW) multiline TRL calibration set: a 550 µm long thru line, five lines with lengths of 2685 µm, 3750 µm, 7115 µm, 20245 µm, and 40550 µm, and two offset shorts located at a distance of 225 µm from the beginning of the line. Additionally, there are 12 verification reference elements.

According to the individual test results provided by NIST for each RM8130, the actual line capacitance of RM CPW line used for this investigation is 1.7877 pF/cm. This value is used by the benchmark multiline TRL calibration for the accurate definition of the characteristic impedance of the RM8130 lines and the transformation of the measurement system reference impedance to 50 . RM8130 is only specified for use up to 40 GHz [10].

[2] Available from NIST

[3] SussCal is available from SUSS MicroTec.

[4] HFSS is commercially-available from ANSYS Inc.

B. Alumina Calibration Substrate

The alumina calibration substrate consists of lumped standards (open, short, and load) as well as CPW line sets: a 150 µm long thru line, four lines with lengths of 6550 µm, 1450 µm, 950 µm, and 500 µm. Its estimated effective dielectric constant is about 5.26. The substrate is 625 µm thick.

The CPW lines of the alumina calibration substrate are not characterized by the vendor. Thus, the line characteristic impedance and the propagation constant must be computed for the alumina benchmark probe tip calibration [6].

We supposed that the measured characteristics of alumina CPW lines can vary significantly due to two factors: the fabrication tolerances of the lines and the influence of different probe tip design. Thus, we measured two alumina substrates of the same type using two differently designed wafer probes.

IV. Experimental Results

A. Verification of CPW Lines from Alumina Substrate

The CPW lines from both alumina substrates were characterized by the lumped load and the calibration comparison method. For the lumped load method, the DC resistance of the load standard from each substrate was accurately measured. The line capacitance per unit length C was extracted from CAP program and the $G/(C)$ component was neglected. The effective dielectric constant was measured by the multiline TRL procedure and the line characteristic impedance Z_0 was extracted using the method of [2].

For the calibration comparison method, the first-tier benchmark multiline TRL was performed on the reference material RM8130. The reference plane was set to the probe tip end and the reference impedance was accurately transformed to 50 using the capacitance value of the RM lines as specified by NIST. The characteristic impedance Z_0 of the CPW lines on alumina was extracted from the second-tier multiline TRL error boxes with the help of the ExtractZo program. Finally, inductance L, resistance R, capacitance C, and conductance G per unit length of the alumina CPW lines were estimated from the LINEPAR program (Fig. 1).

Conductance G and resistance R per unit length extracted from both methods are in good agreement for both measured setups below 40 GHz. The variation of about 7% for the capacitance C and about 5% for the inductance L per unit length can be attributed to tolerances in the fabrication of the alumina substrates.

Fig. 1 also shows that the calibration comparison results strongly deviate above 40 GHz for both setups. This indicates the cross section of the RM8130 CPW lines was not optimized for frequencies above 40 GHz.

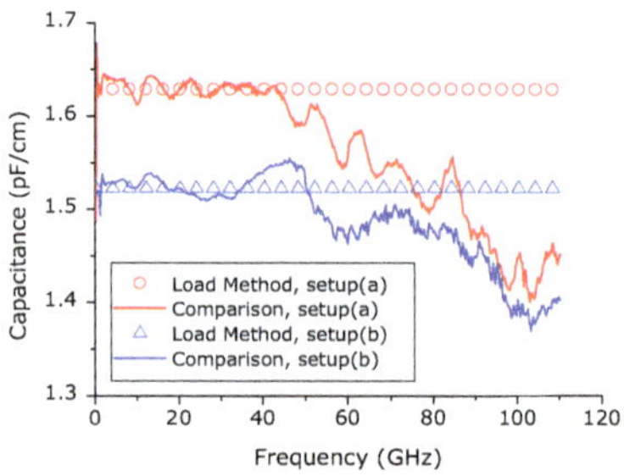

a)

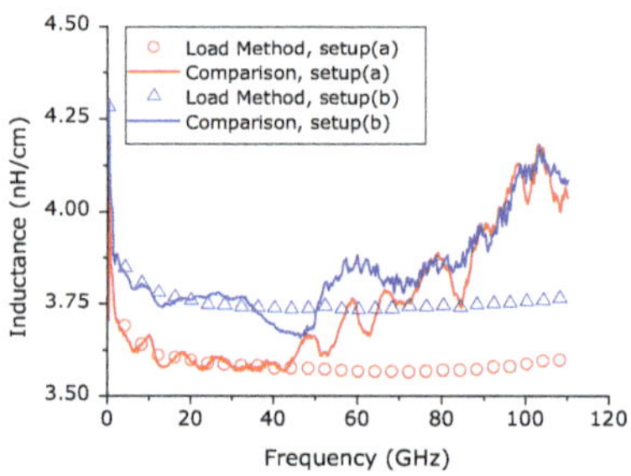

b)

Fig. 1. Parameters of the alumina CPW line extracted from the calibration comparison to the reference RM8130 calibration and the load method for two setups.

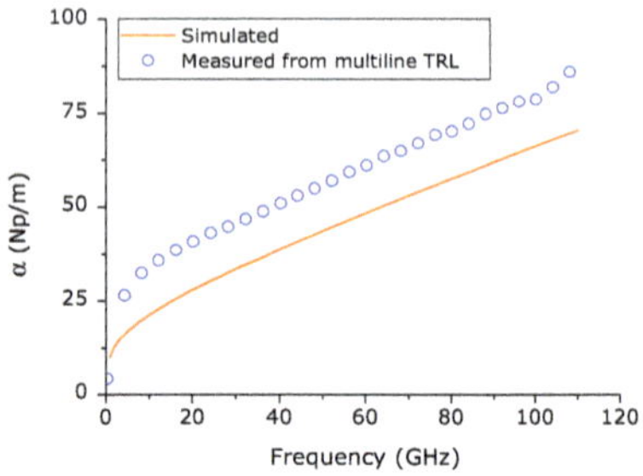

a)

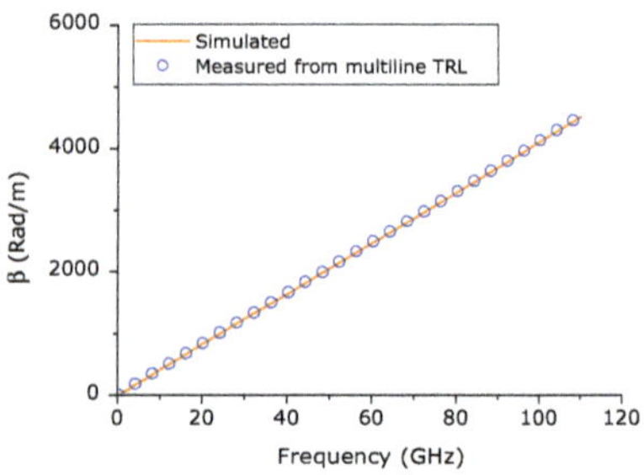

b)

Fig. 2. Simulated and measured attenuation (a) and phase (b) constants of examined silicon line.

Comparing results for the conductances per unit length of both methods up to 40 GHz, we concluded that the evaluated alumina CPW lines have negligible transverse losses.

B. Lines on Silicon

The propagation constant of the examined lines on silicon was measured using multiline TRL and compared to the simulation result (Fig. 2). The line characteristic impedance Z_0 was extracted from three approaches: calibration comparison method referenced to the RM8130 (up to 40 GHz), calibration comparison method referenced to alumina substrate (up to 110 GHz), and the lumped load method (up to 110 GHz), as shown in Fig. 3. From the results obtained (Fig. 4), line capacitance *C*, inductance *L* per unit length as well as coefficients *G/(C)* and *L/(R)* were calculated. Fig. 4, c) shows the transverse losses of the silicon lines used in this study can be neglected.

C. Extraction of DUT Parameters

An on-wafer NFET transistor, with a channel length of 0.12 μm for which the wiring to the transistor was carefully designed to be consistent with the on-wafer calibration standards, was also measured. The raw, uncalibrated measured data were corrected to a calibration reference plane occurring in the middle of the via stack at the end of the signal launch line of the probe padset, as shown in [4]. A small amount of parasitics remaining after calibration, occurring between the calibration reference plane and the device, were de-embedded using standard Open-Short de-embedding methods.

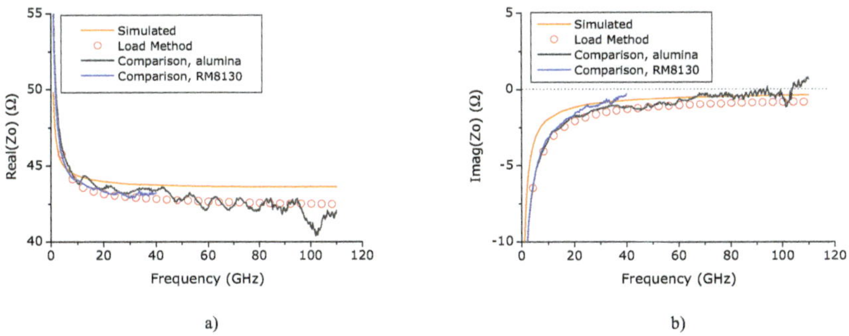

Fig. 3. Simulated and extracted real (a) and imaginary (b) parts of the examined silicon line characteristic impedance Z_0. Extracted data were obtained from the load method and the calibration comparison method to reference material RM8130 and alumina substrate. Comparison to RM8130 is limited in frequency to 40 GHz due specification of the reference material used.

Figure 5 shows the extracted drain-to-gate capacitance versus frequency for the transistor. In addition to the excellent flatness versus frequency, the data also shows very little variation and sensitivity to the difference in the various methods of extracting the electrical characteristics of the calibration standards.

V. Discussion

Both the calibration comparison method and the lumped load method have proven to be valid methods for extracting the electrical characteristics of planar transmission line standards. From Fig. 1 we conclude that the use of the NIST RM8130 cannot be recommended above 40 GHz. However, at low frequencies the calibration comparison method using the RM is in good agreement with the lumped load method. Furthermore, as shown in Fig. 3 and Fig. 4, both methods are in acceptable agreement up to 110 GHz. The difference in the substrate material between the GaAs of the RM, the alumina, and the silicon appears to be successfully compensated for in the calibration comparison method.

We observed slight rippling of the extracted parameters of alumina CPW lines (Fig. 1), as well as for the silicon lines measured with the respect to the first-tier reference calibration on alumina (Fig. 3 and Fig. 4). Possible reasons for such effects include the influence of the high-resistivity absorbing layer that extends under the ground conductors of the alumina CPW lines, non-ideal calibration boundary conditions, and the impact of the design of the wafer probes on the calibration standards characteristics [11]. Further detailed investigations of these phenomena are required.

As shown in Fig. 1, our results indicate that there are some fabrication variations (such as thickness, surface roughness and the roughness of the edges between ground and signal conductors, signal gap, and the substrate dielectric), or probe-type dependencies, in commercially offered alumina substrates. Therefore, in order to use the alumina substrate for reference calibrations, it is mandatory to characterize the electrical properties of the substrate together with the probes, using a well established benchmark calibration. This issue highlights the need for a reference material that is valid to 110 GHz. To the author's knowledge, such a reference material is not available yet.

Thus, the lumped load method of measuring line capacitance per unit length is currently recommended for alumina. Also, it was proven that the alumina CPW lines have negligible loss and the extraction of the coefficient $G/(\ C)$ is not needed.

Most importantly however, the lumped load method demonstrates acceptable accuracy for determining the electrical characteristics of the transmission line standards on silicon investigated here up to 110 GHz. The success shown in this study is attributable to the design of the line standards on the silicon. The first metal level ground plane under the signal provides a dominant microstrip propagation mode. This ground plane is common practice in the industry. Due to fabrication rules, the ground plane must be a mesh, and further studies are needed to evaluate the geometries where the microstrip mode dominates and the lumped load method is valid.

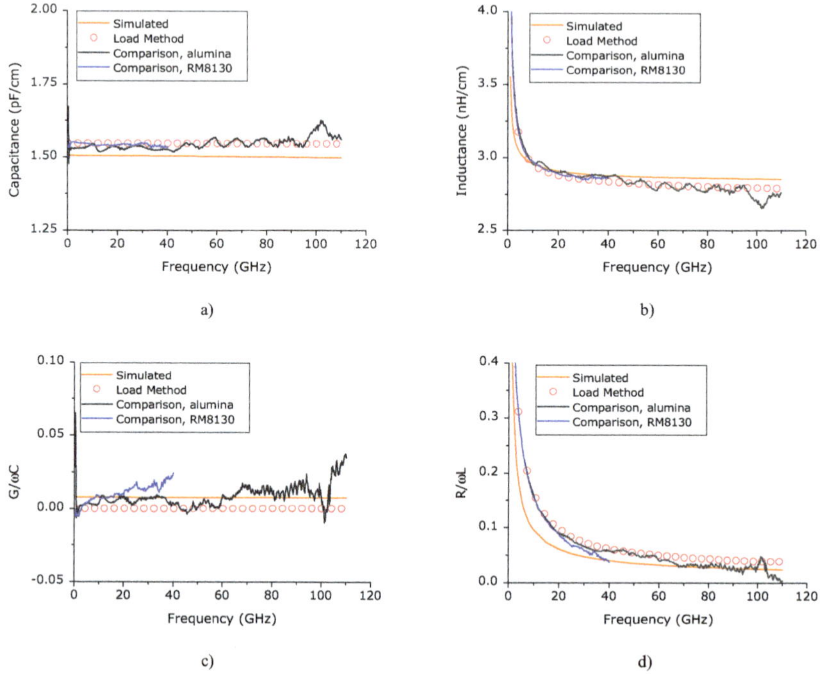

Fig. 4. Comparison of the simulated and extracted parameters of the examined silicon line. Parameters are extracted from the lumped load method as well as from the calibration comparison to the reference material RM8130 and alumina substrate. Comparison to RM8130 is limited in frequency to 40 GHz due specification of the reference material used.

Careful characterization of the thru and the load is essential for developing an accurate LRM-based on-wafer calibration method for use in silicon processes [4]. This work has shown successful characterization of the thru line, providing a basis for a successful LRM on-wafer calibration.

All three methods for determining the electrical characteristics of the calibration standards studied here, and using the resulting TRL calibration to correct the raw transistor measurements, are yielding tight agreement for extracted transistor parameters, as shown in Fig. 5. This result reveals the device insensitivity to the method of line standard characteristics extraction. In addition, it indicates that the successful development of an on-wafer calibration method on silicon for device measurement up to 110 GHz is possible.

VI. Conclusion

Summarizing, for the first time the calibration comparison method was applied to extract the electrical characteristics of planar CMOS lines up to 110 GHz. This was done using commercially available alumina calibration substrate, as well as the GaAs reference material RM8130. We demonstrated that commercially available calibration substrates can be used for characterizing customized standards on silicon at mm-wave frequencies with acceptable accuracy.

As a result, the implementation of wafer-embedded calibration techniques on lossy substrates at mm-wave frequencies, such as multiline TRL, can be significantly simplified. Thus, the required improvement in the measurement accuracy of advanced RF CMOS components at frequencies up to 110 GHz can be achieved.

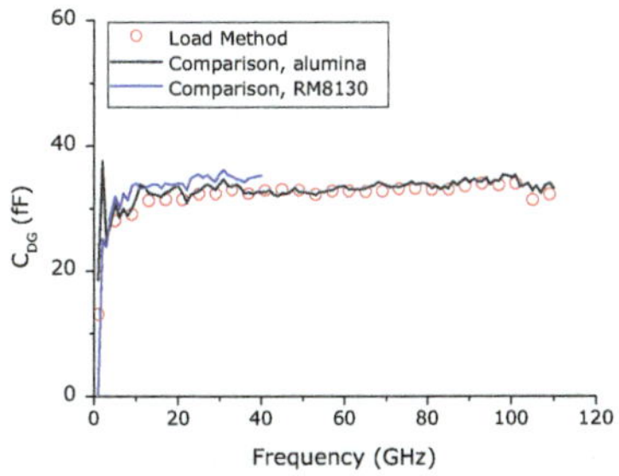

Fig. 5 Extracted drain-to-gate capacitance of a of 0.12 μm channel length NFET with respect to the load and calibration comparison methods.

ACKNOWLEDGMENT

The authors wish to acknowledge Ralf Doerner from Ferdinand-Braun-Institut fuer Hoechstfrequenztechnik (FBH), for support of 110 GHz measurements of an alternative alumina substrate.

REFERENCES

[1] R. B. Marks, "A multiline method of network analyzer calibration," *Microwave Theory and Techniques, IEEE Transactions on,* vol. 39, pp. 1205-1215, 1991.

[2] R. B. Marks and D. F. Williams, "Characteristic impedance determination using propagation constant measurement," *IEEE Microwave and Guided Wave Letters,* vol. 1, pp. 141-143, June 1991.

[3] D. F. Williams and R. B. Marks, "Transmission line capacitance measurement," *Microwave and Guided Wave Letters, IEEE,* vol. 1, pp. 243-245, 1991.

[4] A. Rumiantsev, S. L. Sweeney, and P. L. Corson, "Comparison of on-wafer multiline TRL and LRM+ calibrations for RF CMOS applications," *ARFTG Microwave Measurements Conference-Fall, 72nd,* pp. 132-135, 2008.

[5] D. F. Williams, R. B. Marks, and A. Davidson, "Comparison of on-wafer calibrations," in *ARFTG Microwave Measurements Conference-Fall, 38th.* vol. 20, 1991, pp. 68-81.

[6] A. Rumiantsev, R. Doerner, and S. Thies, "Calibration standards verification procedure using the calibration comparison technique," in *36th European Microwave Conference*, 2006, pp. 489-491.

[7] D. F. Williams, U. Arz, and H. Grabinski, "Characteristic-impedance measurement error on lossy substrates," *Microwave and Wireless Components Letters, IEEE,* vol. 11, pp. 299-301, 2001.

[8] U. Arz, D. F. Williams, and H. Grabinski, "Characteristic impedance measurement of planar transmission lines," in *URSI-GA*, 2002.

[9] M. F. Ktata, U. Arz, H. Grabinski, and H. Fischer, "Influence of the ground line position on the signal integrity of product-related interconnects in the frequency and time domain," *Advanced Packaging, IEEE Transactions on,* vol. 28, pp. 152-159, 2005.

[10] "Reference Material 8130. Coplanar waveguide calibration set", *Report of Investigation*, NIST, Gaithersburg, MD 20899, USA, 1998.

[11] A. Rumiantsev, R. Doerner, and E. M. Godshalk, "Influence of calibration substrate boundary conditions on CPW characteristics and calibration accuracy at mm-wave frequencies," *ARFTG Microwave Measurements Conference-Fall, 72nd,* pp. 168-173, 2008.

13.2 Paper [22]: "Calibration Standards Verification Procedure Using the Calibration Comparison Technique"

A. Rumiantsev, R. Doerner, and S. Thies, "Calibration standards verification procedure using the calibration comparison technique," in *36th European Microwave Conference*, 2006, pp. 489–491.

This article examined the verification procedure of coplanar calibration standards based on the calibration comparison technique. For the first time, different commercially available alumina calibration substrates were compared using the NIST multiline TRL calibration procedure, calibration comparison approach, and the NIST GaAs CPW calibration reference material RM8130. The worst case error bounds for the measurement of passive devices compared to the reference multiline TRL on the RM8130 were calculated for each tested substrate.

This work was co-authored by Ralf Doerner from Ferdinand-Braun-Institut (FBH), Leibniz-Institut fuer Hoechstfrequenztechnik (Berlin, Germany) and Steffen Thies from Rosenberger (Fridolfing, Germany). For this work, I developed the verification methodology, planned the work, defined the measurement campaign, and wrote the conference paper. I did calculations and the data analysis together with Ralf Doerner. The measurement results were acquired by Ralf Doerner at FBH facility.

Calibration Standards Verification Procedure Using the Calibration Comparison Technique

Andrej Rumiantsev*, Ralf Doerner**, Steffen Thies***

*SUSS MicroTec Test Systems GmbH, Suss-Str. 1, Sacka, D-01561, Germany

**Ferdinand-Braun-Institut fuer Hoechstfrequenztechnik (FBH), Gustav-Kirchhoff-Str. 4, Berlin, D-12489, Germany

*** Rosenberger Hochfrequenztechnik GmbH & Co. KG, Hauptstrasse 1, Fridolfing D-83413, Germany

***Abstract* — This article examines the verification procedure of coplanar calibration standards based on the calibration comparison technique. For the first time, different commercially available alumina calibration substrates were compared using the NIST multiline TRL calibration procedure, calibration comparison approach, and the NIST GaAs CPW calibration reference material RM8130. The characteristic impedance of the verified calibration line standards is extracted using the multiline TRL procedure and the definition of the load standard resistance. The error circuit, describing difference of the substrate medium and the standard design is extracted. Finally, the worst case error bounds for the measurement of passive devices compared to the reference multiline TRL on the RM8130 are calculated for each tested substrate.**

***Index Terms* — calibration, error correction, calibration comparison, scattering parameters measurement.**

I. Introduction

Significant progress in modern 2.5- and 3-dimensional microwave simulation tools provides quick and accurate design and optimization of microwave components. However, variations in the fabrication process raise the necessity of the accurate post-production measurement and characterization. It is especially a problem for the CPW calibration standards: the design and fabrication process makes it impossible to trace the mechanical and electrical properties of the standards to the natural constants directly. The alternative verification approaches are still a challenge. The uncertainty introduced during the manufacturing of standards influences the accuracy of the calibration, and it can be used for indirectly verifying the standards.

The research undertaken by the National Institute of Standards and Technology (NIST) provided a technique for comparing different wafer-level calibrations to those performed by NIST, which have well-defined boundaries (the system reference impedance and the position of the measurement plane) [1]. This technique can be successfully used by industrial laboratories for calibration verification purposes, e.g. [2]. This paper presents the application and results of using the calibration comparison technique for verifying the accuracy of commercially-available CPW calibration standards.

II. Verification Procedure

The calibration comparison technique provides the worst-case deviations of the measured S-parameters of passive devices for the examined (first-tier) calibration with respect to the benchmark (second-tier) calibration. Deviations are treated as $|S_{ij}-S_{ij}'|$, for $ij \in \{11, 12, 21, 22\}$, where S_{ij}' is the S-parameter measured by the calibration to be tested, and S_{ij} is the S-parameter measured by the benchmark calibration. The deviations are calculated from the obtained error matrices or "error boxes" [A] and [B] (Fig. 1), and represent three sources of possible error: differing calibration methods, medium and launch geometries, and fabrication error. In the ideal case, both calibrations, benchmark and examined, are equal, [A] and [B] are unity matrices, and the error bounds $|S_{ij}-S_{ij}'|$ are zero.

Selecting the same calibration method for the first-tier and the second-tier calibration eliminates the first error source. Using the procedure outlined in [3], the error resulting from the difference of the medium and launch geometries for the reference and test calibration can be excluded. Finally, the error matrices [A] and [B] will quantitatively represent the fabrication uncertainty of the examined set of standards.

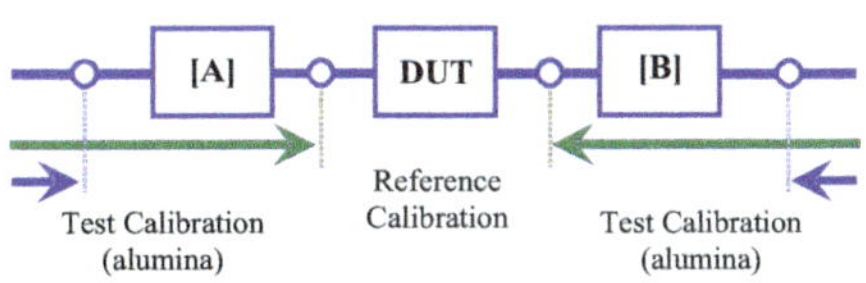

Fig. 1. The error matrixes [A] and [B] obtained by the calibration comparison method.

III. Experimental Results

For the independent evaluation, an experimental setup providing 110 GHz measurement capabilities was used. The setup for wafer-level measurements included an Agilent PNA 110 GHz network analyzer, a PM8 manual wafer-probe station from SUSS MicroTec, and 110 GHz wafer probe tips with a 125 μm pitch. The GaAs CPW lines (the reference material RM8130) fabricated and specified by NIST were used as a reference calibration set.

The integrity of the setup (Fig. 2) was verified and the measurement system drift within the experiment was defined acc. to [4] using the RM8130, the reference data, the multiline TRL calibration method, and the MultiCal® software package provided by NIST.

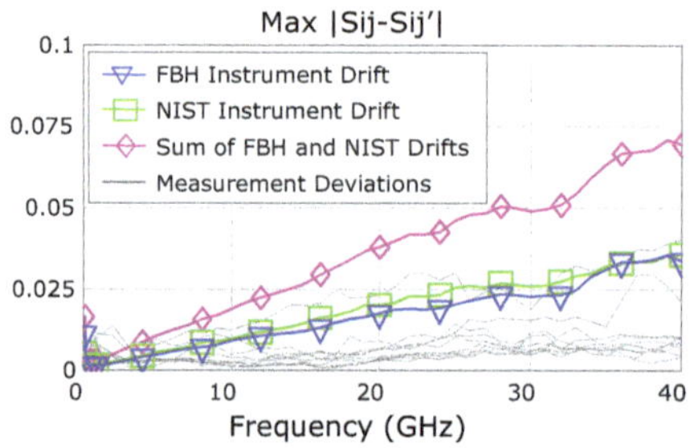

Fig. 2. Verification results on integrity of the 110 GHz wafer-level measurement setup used. Due to the frequency limitation of the reference data for the RM8130 provided by NIST, the results for the setup are limited to 40 GHz.

NIST multiline TRL [5] was selected as the calibration method. In conjunction with methods proposed in [6] and [7], this procedure allows accurate setting of the measurement system reference impedance to 50 Ohm as well as a precise definition of the measurement reference plane. To avoid additional uncertainty due to contact repeatability, all required data were acquired in one measurement series in raw-data format with the help of the calibration software SussCal®[1] and saved externally for further analysis.

If the characteristic impedance Z_0 and the physical length l are well known, a calibration based on the distributed standards (TRL-like) provides very accurate results [5]. Then, the lumped standards (Open, Short, and Load) can easily be specified [8]. Therefore, the characteristic impedance Z_0 of the alumina CPW lines was selected as verification criteria. The examined substrates were the commercially available from SUSS MicroTec CSR-3, CSR-8, and comparable substrate from another vendor (referenced further as the Sub A).

It was found that the line Z_0 definition method based on the direct measurement of its capacitance C per unit length [6] of the alumina CPW lines does not provide the required accuracy. The commercially available alumina substrates typically have lines shorter then 7 mm. This significantly decreases the measurement accuracy of the line's C. Thus the found variation range of C was 1.651…1.337 pF/cm. It corresponds to the characteristic impedance Z_0 variation from 46.34 Ohm to 57.01 Ohm and demonstrates the significant measurement error.

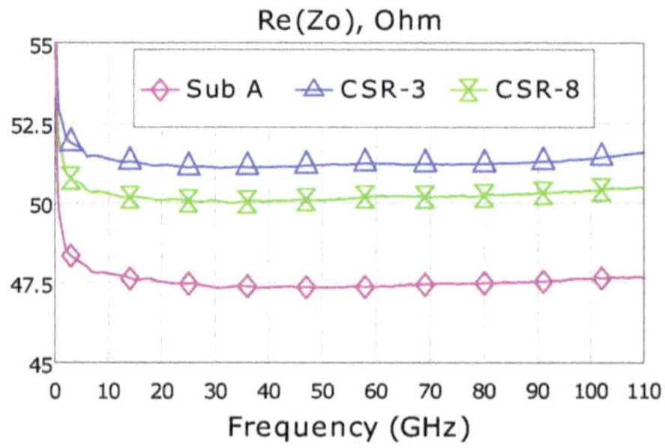

Fig. 3. Extracted value (real part) of the line characteristic impedance Z_0 for CSR-3, CSR-8, and Sub A calibration substrates.

Consequently, an alternative procedure was used to obtain the line capacitance C and, finally, the characteristic impedance Z_0 of the tested alumina lines. The resistance R of the load standard located on the same substrate was accurately measured using the 4-terminal method. Then, the first tier multiline TRL was performed setting the calibration reference impedance to the characteristic impedance Z_0 of the alumina lines used. The frequency range where the tested lines are no longer dispersive and, on the other hand, the load reactance is negligible was used to extract the value of the line capacitance C acc. to [6]. The procedure was repeated using other alumina substrates. The Table I shows the extracted values of the line capacitance C for each substrate respectively.

TABLE I
Extracted Line Capacitance

Substrate	C (pF/cm)
A	1.57
CSR-3	1.51
CSR-8	1.5

Based on the results, the reference plane of the examined calibration was accurately set to the probe tip end and the measurement reference impedance was transformed to 50 Ohm. The line characteristic impedance Z_0 was extracted

[1] Available from SUSS MicroTec.

for each calibration substrate with the help of MultiCal® software package (Fig. 3).

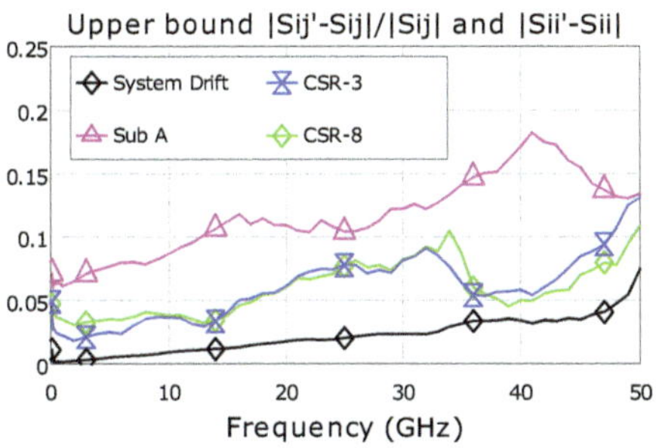

Fig. 4. Results of the accuracy verification for the multiline TRL calibration on CSR-3, CSR-8, and Sub A calibration substrates without correction for the line impedance.

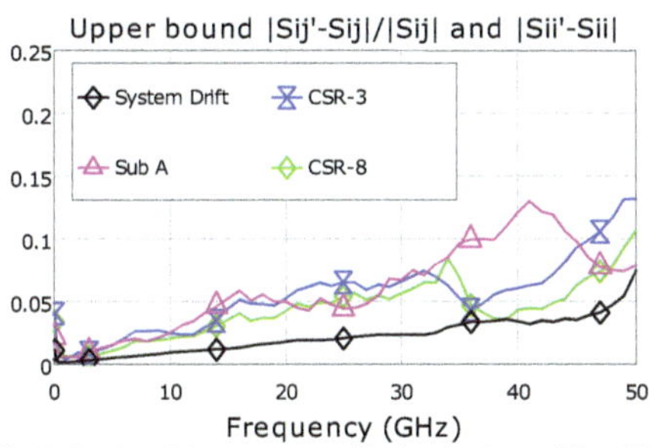

Fig. 5. Results of the accuracy verification for the multiline TRL calibration on CSR-3, CSR-8, and Sub A calibration substrates with correction for the line impedance.

The second-tier reference calibration was done. Results for tested alumina substrates are presented in Fig. 4 and Fig. 5 and validate the extraction method.

Finally, the same validation procedure was performed with the respect to the first-tier lumped LRM+ calibration [2] on each substrate proving its accuracy (Fig. 6).

IV. Conclusion

Different commercially available alumina calibration substrates were verified with the help of the calibration comparison procedure. The line characteristic impedance Z_0 was extracted. The found values are within tolerance of 3.6% for the CSR-3, 0.8% for the CSR-8, and 4.9% for the Sub A calibration substrates, respectively (for the frequency range of 10 GHz–110 GHz).

It has been demonstrated that the maximum error bound of 0.13 for the CSR-3, 0.11 for CSR-8, and 0.18 for the Sub A can be obtained up to 50 GHz. Obviously, additional correction techniques will be required to achieve the same results as for the CSR substrate family if the other commercially available alumina substrate is used.

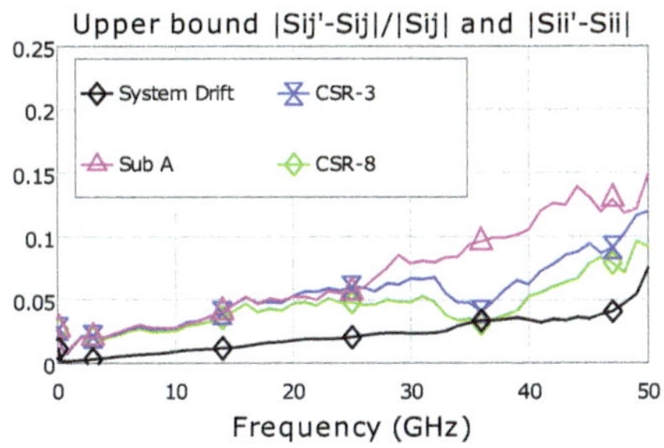

Fig. 6. Results of the accuracy verification for the LRM+ calibration procedure on CSR-3, CSR-8, and Sub A calibration substrates.

References

[1] D. Williams, R. Marks, A. Davidson, "Comparison of on-Wafer Calibrations", *38th ARFTG Conference Digest*, pp. 68–81, December 1991.

[2] R. Doerner, A. Rumiantsev, "Verification of the Wafer-Level LRM+ Calibration Technique for GaAs Applications up to 110 GHz", *65th ARFTG Conference Digest*, June 2005.

[3] D.K. Walker, D.F. Williams, "Compensation for Geometrical Variations in Coplanar Waveguide Probe-Tip Calibration", *IEEE Microwave and Guided Wave Letters*, vol. 7, no. 4, pp. 20–30, 1997.

[4] "Proposed Procedures for Verifying Probe Station Integrity and On-Wafer Measurement Accuracy", *NIST/Industrial MMIC Consortium*, NIST.

[5] R. Marks, "A Multiline Method of Network Analyzer Calibration", *IEEE Trans. Microwave Theory Tech.*, vol. 39, no. 7, pp. 1205–1215, July 1991.

[6] D. Williams, R. Marks, "Transmission Line Capacitance Measurement", *IEEE Microwave and Guided Wave Lett.*, vol. 1, pp. 243–245, September 1991.

[7] R. Marks, D. Williams, "Characteristic Impedance Determination Using Propagation Constant Measurements", *IEEE Microwave and Guided Wave Lett.*, vol. 1, pp. 141–143, June 1991.

[8] F. Williams, "Characterization of Thin-Film Calibration Elements", *38th ARFTG Conference Digest*, pp. 25–36, December 1991.

14

Attachment to Chapter 6

14.1 Paper [55]: "Comparison of On-Wafer Multiline TRL and LRM+ Calibrations for RF CMOS Applications"

A. Rumiantsev, S. L. Sweeney, and P. L. Corson, "Comparison of on-wafer multiline TRL and LRM+ calibrations for RF CMOS applications," in *ARFTG Microwave Measurements Conference-Fall, 72nd*, pp. 132–135, 2008.

This paper presented a quantitative comparison of the reference multiline TRL and the transfer TMR for a customized set of standards in a CMOS process using IBM's 0.13 μm technology. This comparison was undertaken for the first time and covered a frequency range from 1 GHz to 110 GHz. It was demonstrated that the accuracy of the on-wafer multiline TRL and the transfer TMR calibration were in very good agreement. Both methods outperformed the conventional off-wafer calibration with the DUT contact pad parasitics de-embedded.

For this paper, I planned the work, framed the design requirements to the transfer TMR standards, developed the accuracy verification method, did all offline calibration calculations and wrote related section of the paper. The standards were optimized, laid out, and measured by Philip Corson and Susan Sweeney from IBM Microelectronics (Essex Junction, VT, USA). Philip Corson provided EM simulation analysis. Susan Sweeney extracted the parameters of the test DUT. The analysis of the results was performed together with all co-authors.

Comparison of On-Wafer Multiline TRL and LRM+ Calibrations for RF CMOS Applications

Andrej Rumiantsev [#], Susan L. Sweeney [*], Phillip L. Corson [*]

[#] SUSS MicroTec Test Systems GmbH, Suss-Str. 1, Sacka, D-01561, Germany, a.rumiantsev@ieee.org
[*] IBM Microelectronics, Essex Junction, VT 05452, USA, slsweene@us.ibm.com; corson@us.ibm.com

***Abstract* — This paper presents a quantitative comparison of the reference multiline TRL and LRM+ for a customized set of standards in a CMOS process using IBM's 0.13 μm technology. This comparison was undertaken for the first time and covered a frequency range from 1 to 110 GHz. It was demonstrated that the accuracy of the on-wafer multiline TRL and LRM+ calibration were in very good agreement. Both methods outperform the conventional off-wafer calibration with the DUT contact pad parasitics de-embedded.**

***Index Terms* — calibration, error correction, calibration comparison, scattering parameters measurement, CMOS, de-embedding.**

I. Introduction

As technology scaling advances and device performance increases, it is becoming essential to obtain accurate high frequency scattering parameters of on-wafer devices used in applications such as communications circuits. Accurate on-wafer network analyzer calibration and the establishment of the measurement reference plane close to the device under test (DUT) is now necessary to reduce the reliance on inaccurate de-embedding methods of the probe padset parasitics.

Today, the commonly-used method for calibrating a wafer-level measurement setup for RF CMOS application consists of two steps:

- Step 1: off-wafer calibration to the probe tip performed on a commercially-available alumina calibration substrate;
- Step 2: translation of the measurement reference plane to the DUT using contact pad de-embedding techniques.

Such an approach was introduced for the first time at the end of the 1980s [1]. This approach recommended using Open and Short elements as the RF de-embedding test structures. This method provided a reasonable alternative to the on-wafer calibration challenges for silicon at that time, in contrast to the high-frequency GaAs applications [2]. Method [1] required neither a perfectly matched load nor the calibration lines embedded on the test chip.

However, further improvement of RF CMOS device performance and increasing operation frequencies to 110 GHz and beyond requires improved measurement accuracy. The de-embedding procedure has become more and more complicated [3, 4], and finally, the need to de-embed the load element has been reported [5, 6].

This work is a further development of wafer-level calibration techniques with a customized set of lumped standards for mm-wave frequency applications. The earlier investigations reported in [7] demonstrated implementation of a lumped LRM+[1] calibration methodology on a conductive (SiGe:C) wafer up to 110 GHz. However, results presented in [7] verify the calibration accuracy qualitatively, but the quantitative verification of the LRM+ procedure remained a challenge.

The achievements made in fabricating and characterizing planar distributed standards in conductive wafer processes observed over the last several years has established a solid background for accurate on-wafer calibration on silicon, e. g. [8-10]. This paper presents a quantitative comparison of the reference multiline TRL[2] [11] and lumped LRM+ [7] for a customized set of standards in a CMOS process using IBM's 0.13 μm technology, with an NFET transistor of gate length 0.12 μm as the DUT. This comparison was undertaken for the first time and covered a frequency range from 1 to 110 GHz.

Additionally, this paper considers proper design of planar structures and presents simulated and measured results of wafer-embedded standards.

Finally, key model parameters of an active DUT are extracted with respect to different wafer-embedded calibration schemes.

II. On-Wafer calibration Comparison

Different calibration procedures have been developed in the past years [12]. All of them rely on ideal, fully or partly known reference elements (calibration standards), realized in planar design (microstrip or coplanar).

In contrast to coaxial and waveguide applications, a great variety of fabrication techniques makes it almost impossible to trace planar calibration standards to a primary reference. This substantially complicates the task of specifying and verifying planar calibrations. However, research undertaken by the National Institute of Standards and Technology (NIST) provided a procedure for comparing wafer-level calibrations,

---

[1] Line-Reflect-Match, advanced.
[2] Thru-Reflect-Line

identifying setup drift, as well as verifying calibration standards [13].

In this work, NIST multiline TRL was selected as the benchmark calibration. In conjunction with methods proposed in [14, 15], this procedure allows accurate setting of the measurement system reference impedance to 50 Ohms as well as a precise definition of the measurement reference plane for both semi-insulating and conductive wafers. Thus, it can be successfully applied for verifying the accuracy of a lumped calibration with a wafer-embedded customized set of standards.

III. On-Wafer Calibration Standards

The customized calibration set included all standards required for the benchmark TRL and lumped LRM+ calibrations. The standards were fabricated within a typical 2-port AC GSG on-wafer padset. The ground pads, signal pad, and signal launch lines were placed at a high-level metal layer.

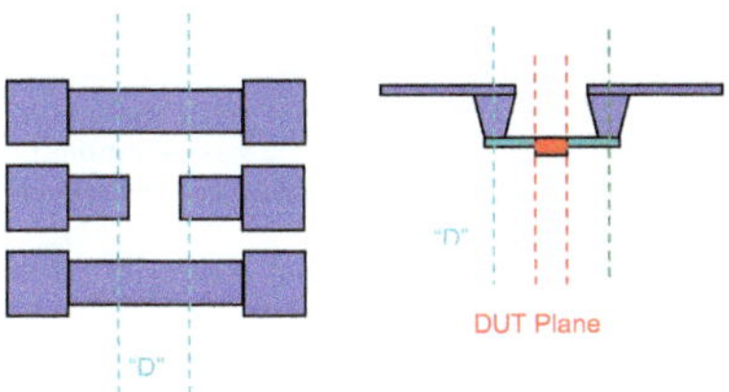

Fig. 1. Design of the customized calibration set.

The transmission lines were fabricated as extensions of the signal launch lines at this high-metal level. Line lengths were optimized to cover a frequency band from 1 GHz to 110 GHz. Lumped load, short, and open elements were developed according to [7] and optimized for the specifics of the bulk CMOS process used and the device under test (DUT) design. These elements were fabricated at the bottom of the via stacks, at plane "D" in Fig. 1, closely coinciding with the placement of the DUT. Thus, both the TRL and LRM+ calibrations establish the reference plane at the via stack, close to the on-wafer DUT.

The electrical parameters of TRL lines were simulated with a simplified model using the 3D full-wave electromagnetic field simulation package Ansoft HFSS®[3]. Simulation of the complete physical structure in Fig. 1 with HFSS requires computing resources and solution times that are not practical, so the structure has to be simplified while retaining the correct response. The simplest approach is a microstrip line, a solid reference conductor, and a single dielectric. This structure was created using the physical dimensions of the launch line, reference plane, and total dielectric thickness.

The cumulative phase of the thru and line standards was used to validate that a single value could be chosen for the real part of the dielectric constant and this would predict the correct phase for all transmission line lengths. The losses of the lines in the model are modulated by the conductivity of the return path and the dielectric loss tangent. These parameters are set in the simplified model to provide a good fit to the measured data.

IV. Measurement Setup

The experimental setup for the 110 GHz wafer-level measurements included an Agilent 8510XF VNA, a semi-automated wafer-probe station, 110 GHz wafer probes with a pitch of 100 μm from GGB Industries, and GaAs reference material RM 8130 available from NIST.

The influence of contact repeatability on measurement data, calibration standards, verification elements, and evaluation transistor S-parameters was reduced by acquiring raw data (with error correction turned off) in three measurement runs. Measurement data from the VNA was recorded with proprietary data acquisition software developed by IBM and with commercial software SussCal®[4]. Off-line LRM+ and TRL calibration, error correction, and calibration comparison was performed using SussCal and MultiCal©[5] software packages.

V. Experimental Results

First, the measurement system drift (instrument drift) shown in Fig. 2 was verified within the time interval of the DUT measurements using the calibration comparison technique and a commercially-available alumina calibration substrate. The probe contact repeatability error is significantly less than the measurement instrument instability within the experimental time (about eight hours) due to fabrication of the standards with gold contact pads. Therefore, it was assumed that the error defined from the calibration comparison method represented the instability of the measurement instrument only.

Also, the contact repeatability error of probing on aluminum pads was found from three identical measurement series A, B, and C, of the same wafer-embedded calibration set. Obtained results showed that the error between series AB, BC and AC are comparable and that it is larger than the instrument drift. It follows that the contact repeatability on aluminum pads

---

[3] HFSS is commercially-available from ANSYS Inc.

[4] SussCal is available from SUSS MicroTec.

[5] MultiCal software package is available from NIST.

affects experimental results more than the test instrument instability.

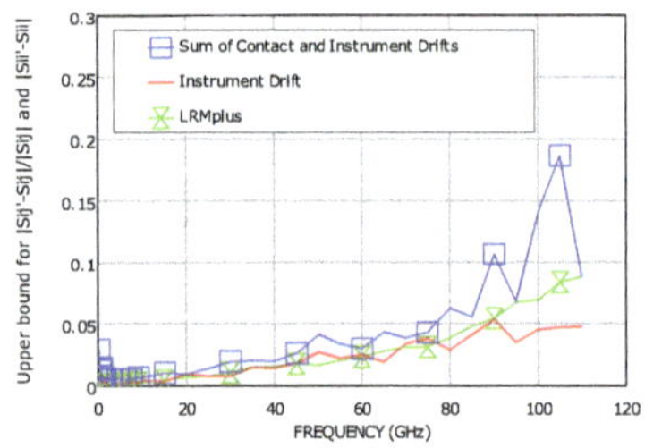

Fig. 2. The accuracy verification of the on-wafer LRM+ calibration.

The line capacitance per unit length was extracted and the characteristic impedance was found using the methods [14, 15]. The extracted value of the line capacitance is 1.5578 pF/cm. The Fig. 3-4 show the measured characteristic impedance, relative phase and attenuation constants of the lines.

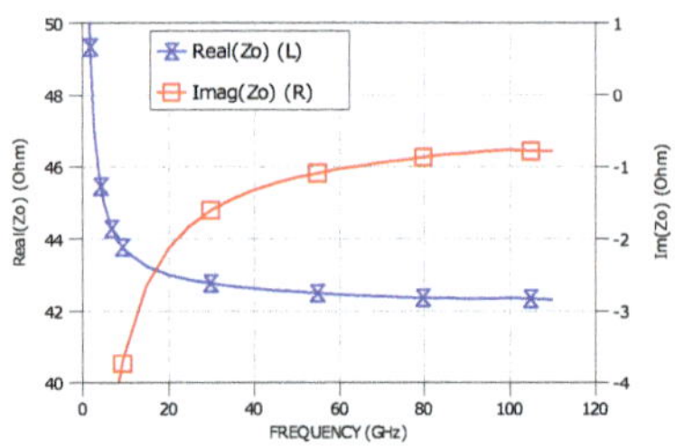

Fig. 3. Measured characteristic impedance of the line standard.

The measured parameters of the customized thru (Fig. 5), load (Fig. 6), and line standards were obtained from the multiline TRL with the reference impedance normalized to 50 Ω. As shown in the figures, the comparison of the measured parameters of standards and the simulation results exhibited good agreement.

Different resistors were compared to evaluate the influence of fabrication tolerances: the load used for the LRM+ calibration and loads from the Open-Resistor, Resistor-Open, Short-Resistor, Resistor-Short structures as illustrated in Fig. 6. All resistor structures had the same layout geometries. It was found that the load impedance variation error was comparable with the contact repeatability error of this experiment.

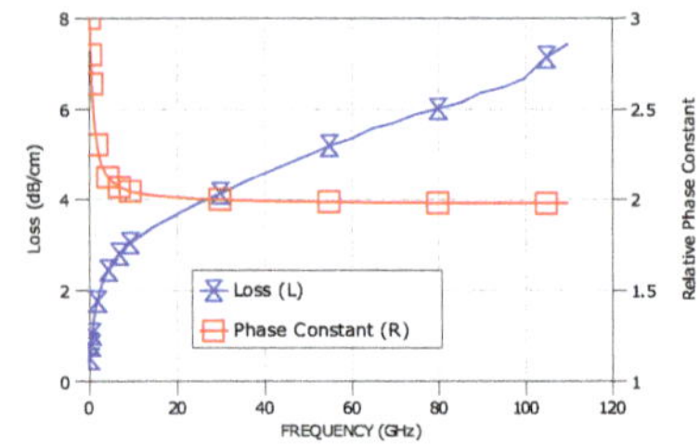

Fig. 4. Measured relative phase and attenuation constants of the line standard.

Next, the on-wafer LRM+ calibration was verified against the benchmark multiline TRL. As shown in Fig. 2, the error of the on-wafer LRM+ is comparable with the system instrument drift and the contact repeatability error for measurements on silicon.

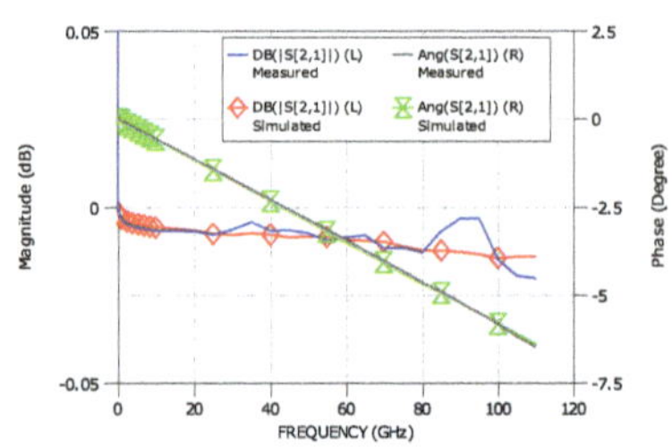

Fig. 5. Comparison of the simulated and measured characteristics of the 24 μm thru standards. The measured parameters were obtained with respect to the multiline TRL calibration method and with the reference impedance normalized to 50 Ω.

Finally, the same NFET transistor with a gate length of 0.12 μm was measured using the conventional two steps probe-tip (off-wafer) SOLT calibration and the open and short DUT contact pads were de-embedded. This data was compared with the data measured using the on-wafer TRL and LRM+ calibration procedures. Key device parameters Cgs and

Cgd were extracted and compared as shown in Fig. 7. These results show both on-wafer calibration methods are comparable and are a significant improvement for critical parameter extraction over the SOLT calibration with open/short de-embedding over the frequency range of 1 to 110 GHz.

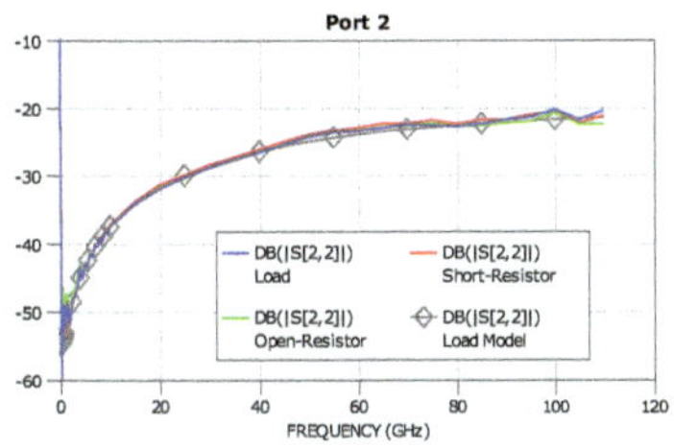

Fig. 6. Comparison of the simulated and measured characteristics of load standard for the second port elements for the LRM+ load, the Open-Resistor, and the Short-Resistor de-embedding structures. The measured parameters were obtained with respect to the multiline TRL calibration method and with the reference impedance normalized to 50 Ω.

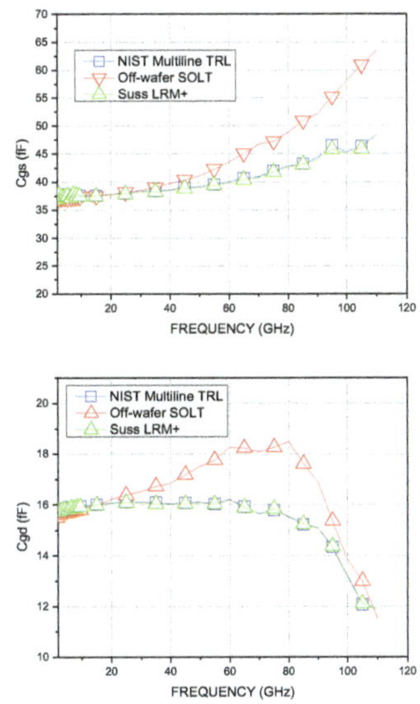

Fig. 7. The test device parameters Cgs and Cgd extracted with respect to the on-wafer multiline TRL and LRM+ calibration procedures and the conventional two-step technique (probe-tip SOLT with the pad de-embedding). The on-wafer TRL and LRM+ shows extremely good agreement over the frequency range and outperforms the conventional method for accuracy of parameter extraction.

VI. CONCLUSION

Summarizing, for the first time the LRM+ calibration was quantitatively verified on a challenging high performance RF silicon CMOS technology. It was demonstrated that the accuracy of the on-wafer multiline TRL and LRM+ calibration methods outperform the conventional off-wafer calibration with the DUT contact pad parasitics de-embedded.

Both the LRM+ and the benchmark NIST multiline TRL method are in very good agreement. This proves LRM+ to be a valuable tool as it overcomes the main drawback of multiline TRL: construction of wafer-embedded transmission lines. LRM+ does not require long transmission line calibration standards but nevertheless provides comparable calibration accuracy. LRM+, therefore, saves wafer space, minimizing the test chip size to only three standards, realized in the same padset as the DUT. Thus, a fully automated calibration is possible even when using a fixed wafer probe configuration. The required determination of the electrical model of the load standards needed in the calibration can be done easily, e.g., by means of the approaches presented in [7].

As shown above, LRM+ can be successfully used for advanced RF CMOS applications up to 110 GHz.

REFERENCES

[1] P. J. van Wijnen and L. C. Smith, "High frequency characterization of small geometry bipolar transistors," in *Bipolar Circuits and Technology Meeting*, 1988, pp. 91-94.

[2] D. Williams, R. Marks, K. Phillips, and T. Miers, "Progress toward MMIC on-wafer standards," in *ARFTG Microwave Measurements Conference-Fall, 36th*, 1990, pp. 73-83.

[3] T. E. Kolding, "A four-step method for de-embedding gigahertz on-wafer CMOS measurements," *IEEE Transactions on Electron Devices*, vol. 47, pp. 734-740, 2000.

[4] T. E. Kolding, "On-wafer calibration techniques for giga-hertz CMOS measurements," in *International Conference on Microelectronic Test Structures*, 1999, pp. 105-110.

[5] L. F. Tiemeijer, R. J. Havens, A. B. M. Jansman, and Y. Bouttement, "Comparison of the "pad-open-short" and "open-short-load" deembedding techniques for accurate on-wafer RF characterization of high-quality passives," *IEEE Transactions on Microwave Theory and Techniques*, vol. 53, pp. 723-729, 2005.

[6] X. Wei, G. Niu, S. Sweeney, Q. Liang, X. Wang, and S. Taylor, "A general 4-port solution for 110 GHz On-wafer transistor measurements with or without impedance standard substrate (ISS) calibration," *Electron Devices, IEEE Transactions on*, vol. 54, pp. 2706-2714, 2007.

[7] R. F. Scholz, F. Korndorfer, B. Senapati, and A. Rumiantsev, "Advanced technique for broadband on-wafer RF device characterization," in *ARFTG Microwave Measurements Conference-Spring, 63rd*, 2004, pp. 83-90.

[8] D. F. Williams, U. Arz, and H. Grabinski, "Characteristic-impedance measurement error on lossy substrates," *Microwave and Wireless Components Letters, IEEE*, vol. 11, pp. 299-301, 2001.

[9] D. F. Williams, U. Arz, and H. Grabinski, "Accurate Characteristic Impedance Measurement on Silicon," in *ARFTG Conference Digest-Spring, 51st*, 1998, pp. 155-158.

[10] J. Leinhos and U. Arz, "Effect of uncertainties in the cross-sectional parameters on the wideband electrical properties of coplanar waveguides," in *Signal Propagation on Interconnects, 2007. SPI 2007. IEEE Workshop on*, 2007, pp. 35-38.

[11] R. B. Marks, "A multiline method of network analyzer calibration," *Microwave Theory and Techniques, IEEE Transactions on*, vol. 39, pp. 1205-1215, 1991.

[12] A. Rumiantsev and N. Ridler, "VNA calibration," *Microwave Magazine, IEEE*, vol. 9, pp. 86-99, 2008.

[13] D. F. Williams, R. B. Marks, and A. Davidson, "Comparison of on-wafer calibrations," in *ARFTG Microwave Measurements Conference-Fall, 38th*. vol. 20, 1991, pp. 68-81.

[14] D. F. Williams and R. B. Marks, "Transmission line capacitance measurement," *Microwave and Guided Wave Letters, IEEE*, vol. 1, pp. 243-245, 1991.

[15] R. B. Marks and D. F. Williams, "Characteristic impedance determination using propagation constant measurement," *IEEE Microwave and Guided Wave Letters*, vol. 1, pp. 141-143, June 1991.

15

Attachment to Chapter 7

15.1 Paper [63]: "Influence of Probe Tip Calibration on Measurement Accuracy of Small-Signal Parameters of Advanced BiCMOS HBTs"

A. Rumiantsev, P. Sakalas, N. Derrier, and D. Celi, "Influence of probe tip calibration on measurement accuracy of small-signal parameters of advanced BiCMOS HBTs," in *IEEE Bipolar/BiCMOS Circuits and Technology Meeting (BCTM)*, Atlanta, GA, 2011.

This paper presented investigation results of the probe-tip calibration impact on the BiCMOS HBT small-signal parameter measurement accuracy. Popular calibration procedures were applied on the same data set and followed by the two-step de-embedding from the device dedicated compete open and complete short dummy elements. This paper proved that only the multiline TRL was capable to establish well-defined calibration reference impedance and to place the measurement reference plane to the desired location. Accuracy of other methods was limited due to imperfection of planar calibration standards and probe misplacement errors. Experimental results showed that the observed difference in cold HBT parameters and parameters of passive devices was minimized by the de-embedding step. The f_T and f_{MAX} demonstrated higher sensitivity to the probe-tip calibration residual errors.

In this work, I developed the methodology of the analysis, planned the work, wrote the major part of the paper and presented the results at a conference. I acquired the measurement data together with Paulius Sakalas University of Technology (TUD) (Dresden, Germany) at the measurement system of the TUD CEDIC laboratory. I did parameter extraction together with Nicolas Derrier and Didier Celi from STMicroelectronics (Crolles, France). The data analysis was shared between all co-authors.

Influence of Probe Tip Calibration on Measurement Accuracy of Small-Signal Parameters of Advanced BiCMOS HBTs

A. Rumiantsev[1,2], P. Sakalas[2,3], N. Derrier[4], D. Celi[4], M. Schroter[2,5]

[1]Cascade Microtech GmbH, Suss-Strasse, 1, D-01561 Sacka, Germany.
[2]CEDIC, Dresden University of Technology, D-01062 Dresden, Germany.
[3]State Scientific Research Institute, FRL, Goštauto 11, 01108 Vilnius, Lithuania.
[4]STMicroelectronics, 850 rue Jean Monnet, F-38926 Crolles, France.
[5]UC San Diego, ECE Dept., La Jolla, CA, USA.

***Abstract* — This paper presents investigation results of the probe-tip calibration impact on the BiCMOS HBT small-signal parameter measurement accuracy. Popular calibration procedures were applied on the same data set and followed by the two-step de-embedding from the device dedicated Compete-Open and Complete-Short dummy elements. Experimental results showed that the observed difference in cold HBT parameters and parameters of passive devices was minimized by the de-embedding step. The f_T and f_{MAX} demonstrated higher sensitivity to the probe-tip calibration residual errors.**

***Index Terms* — Silicon-germanium HBT, silicon bipolar/BiCMOS process technology, calibration, S-Parameters, de-embedding.**

I. INTRODUCTION

Pushing device operation frequencies towards THz range [1] causes serious challenges for conventional device characterization techniques. Presently, RF device characterization uses the two-step approach:

1) probe-tip calibration (off-wafer) that is performed on a commercially available alumina calibration substrate and uses well characterized impedance standards (ISS);
2) de-embedding of the silicon backend parasitics using wafer-embedded test structures such as Open, Short, etc. (also called "dummies").

Each dummy is designed to represent a part (series or parallel) of the backend parasitic impedances. After measurement, these impedances are subtracted by a step-wise procedure yielding characteristics of the device under test (DUT).

A good understanding of possible sources of errors and potential room for improvement of each step are the key factors for increasing the accuracy of device characterization. In this work, we evaluate the impact of popular probe-tip calibration procedures on important Figure of Merits (FoM) and parameters of advanced Si/SiGe:C HBTs from ST Microelectronics' BiCMOS9MW process [2].

Section II describes the calibration procedures that were evaluated, and Section III presents the experimental setup and obtained results for some passive devices as well as for a test HBT under cold and hot (S-parameters) operation conditions. Chapter IV summarizes the observations and gives some practical recommendations for further accuracy improvement.

II. PROBE TIP CALIBRATION

For this work, four calibration methods were selected: SOLT[1], LRM+[2], eLRRM[3], and the multiline TRL[4] (or mTRL). These methods have the widest variation from each other in:

1) systematic error models they are built on;
2) types of required calibration standards;
3) definition of calibration reference impedance Z_{REF};
4) sensitivity to standards non-ideality.

A. Calibration Method SOLT

SOLT requires three reflection standards at each VNA measurement port (highly-reflective elements, such as Open and Short, and the well matched 50 Ω Load) and one transmission standard Thru [3]. All electrical characteristics of standards must be fully known. As a result, the calibration accuracy critically depends on the fabrication and characterization of standards. It remains a challenge to achieve reliable SOLT calibration at high frequencies.

B. Self-Calibration Methods: LRM+, LRRM, TRL

The self-calibration methods are based on the seven-term model of systematic measurement errors. They take more measurements of calibration standards than required for calculating error terms. The gained information redundancy enables the use of partly-

[1] Short-Open-Load-Thru
[2] Line-Reflect-Match, advanced
[3] Line-Reflect-Reflect-Match, enhanced
[4] Thru-Reflect-Line

978-1-61284-166-3/11/$26.00 ©2011 IEEE

defined standards. The missing parameters are calculated from within the calibration procedure (also called "self-calibration").

LRM+ uses one transmission (Thru) and two reflection elements: the Load, also called "Match", and the "Reflect" (Open or Short) [3]. The reflection coefficient of the Reflect is a free parameter. This is an important advantage: there is no more need for a well-defined highly reflective element. So far, this has been a challenging task, especially at high frequencies.

Any arbitrary impedance element can be used as the Load. The LRM+ method is capable to set the reference impedance Z_{REF} to 50 Ω when the impedance of the Load standard Z_{LOAD} is known [4].

In contrast to the LRM+, eLRRM uses an additional Reflect avoiding the need for the second port Load [3]. The automatic Load inductance extraction minimizes the calibration error caused by possible probe misplacement on the Load. This algorithm demonstrated reliable results for the well-defined probe-tip calibration conditions (e.g. on alumina ISS).

The multiline TRL was developed at NIST to solve the frequency limitation of the conventional TRL procedure [5]. Operating with many lines, it applies an extensive statistical analysis of the redundant information. In conjunction with the method proposed in [6], this procedure allows precise setting of the calibration reference impedance Z_{REF} to 50 Ω.

Multiline TRL relies on the measurement of sections of transmission lines and does not require any definition of the impedance of the Reflect. Therefore, it became the accuracy benchmark for comparing wafer-level calibrations.

III. Experimental Results

The experimental measurements were carried out on a broadband S-parameter measurement system from Cascade Microtech, consisting of a PM8 manual probe station, 100 micron pitch Infinity probes model SP-i110-A-GSG-03, ISS 104-783A calibration substrate, equipped with the 110 GHz PNA network analyzer from Agilent Technologies. Also, we used the ceramic chuck add-on that carried the ISS and the test wafer to suppress a possible influence of higher order propagation modes and to reduce coupling effects especially at high measurement frequencies [7].

The calibration and error correction were performed for the same data set outside the VNA on a computer. WinCal®[5], MultiCal®[6], and proprietary IC-CAP®[7] script were used for this purpose, while device parameters were extracted using IC-CAP.

It is important to note that due to their size (0.12x14.86 μm^2), the selected test transistors required extremely low input power of about -40 dBm.

The propagation constant γ and the characteristic impedance Z_0 of the ISS lines were extracted using the method from [6]. It resulted in a capacitance per unit length of C=1.481 pF/cm. The reference impedance of the multiline TRL was set back to 50 Ω and the measurement reference plane was moved to the probe tip ends. After that, multiline TRL established well-defined calibration conditions and could be used as the accuracy benchmark.

A. Verification of Passive Elements

The set of dummies included Pad-Open, Pad-Short, and Complete-Open and Complete-Short elements. The Pad-Open and the Pad-Short give the parasitic impedance of the contact pads and can be used in multi-step de-embedding methods or for moving the measurement reference plane to the metal 6 (M6) level (Fig. 1). The pairs of Complete-Open and Complete-Short structures were designed for de-embedding of complete backend parasitics of a specific device geometry and are device optimized.

Raw measurement data of all de-embedding elements were calibrated by the probe-tip SOLT, LRM+, eLRRM and the multiline TRL methods. Assuming the Π-equivalent circuit for the Open and the T-equivalent circuit for the Short dummies, we extracted equivalent capacitances *C1*, *C2* and *C3* and inductances *L1*, *L2*, and *L3* for the Complete-Open and Complete-Short elements respectively (Fig. 2, Table 1). *C3* and *L3* are negligible.

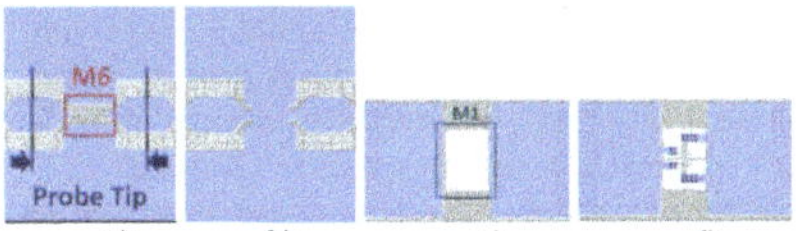

a) b) c) d)

Fig. 1. Pad-Open (a), Pad-Short (b), Complete-Open (c, zoomed), and Complete-Short (d, zoomed) dummies. Location of the measurement plane at the probe tip, M6 (a) and M1 (c).

We observed slight parameter variations that can be attributed to the calibration residual errors. *C1* and *C2* being de-embedded from the Pad Open are about 2 fF. *L1* and *L2* being de-embedded from the Pad Open and the Pad Short are about 7.5 pH for all methods (Fig 2).

[5] WinCal is available from Cascade Microtech
[6] MultiCal is available from NIST
[7] IC-CAP is a product of Agilent Technologies

As de-embedding subtracts impedances, the major part of calibration residual errors vanished.

TABLE I
PARAMETERS OF COMPLETE OPEN AND SHORT

Parameter	Probe Tip Calibration			
	SOLT/ SOLT(m)	mTRL	LRM+/ LRM+(m)	eLRRM
C1, fF	19.7/21.5	21.3	21.6/21.8	21.6
C2, fF	19.6/21.3	22.3	21.4/21.3	22.5
L1, pH	20.1/21.3	24.2	21.7/21.8	21.8
L2, pH	20.5/21.6	24.4	21.9/22.0	22.0

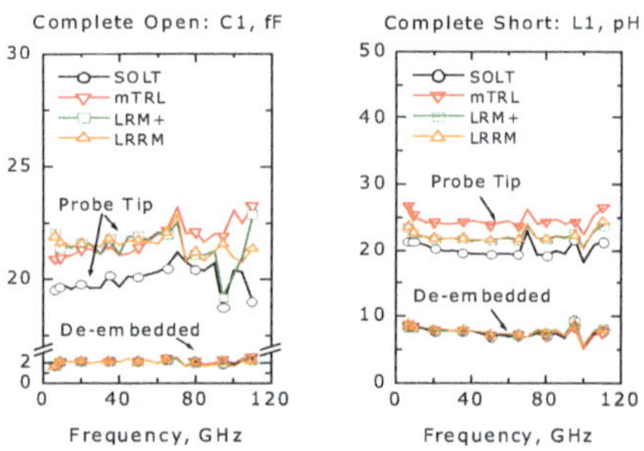

Fig. 2. The capacitance *C1* of the Complete-Open (left) and the inductance *L1* (right) of the Complete-Short elements.

B. Verification for a Transistor

As an active device, we measured S-parameters of a HBT having 14.86 μm length and 0.12 μm emitter stack width at different bias conditions: in a cold-S mode with V_B=-1 V...0.5 V, V_C=0 V and in hot-S (active) mode V_B=0.7 V...1 V and V_{CB}=0 V (emitter grounded in both cases). We extracted some key transistor characteristics: junction capacitances C_{BE} and C_{BC}, f_T and f_{MAX}. f_T was calculated at 20 GHz using the single spot frequency method. The f_{MAX} was defined with Mason's gain. All results were de-embedded from the Complete-Open and Complete-Short dummies. Table II compares obtained parameters.

We found that the probe-tip calibration does not significantly influence de-embedded junction capacitances C_{BE} and C_{BC} up to 67 GHz. f_T is 241 GHz for all methods except SOLT which gave a value lower by 6 GHz. f_{MAX} revealed variations of about 34 GHz in total with the highest value of 329 GHz for SOLT. Obviously, the de-embedding from the Complete-Open and the Complete-Short does not fully compensate for the calibration residual errors in hot-S parameter measurements.

We found that the major portion of the residual errors was caused by the misplacement of probes on calibration standards. SOLT reacts most sensitively, whereas mTRL and eLRRM are significantly less sensitive to probe misplacement [8]. Fig. 4 (left) shows the conductance *G1* of the Pad Open. *G1* is negative for SOLT, which is obviously wrong for any passive element. Such artifacts typically illustrate that the calibration reference impedance was defined incorrectly.

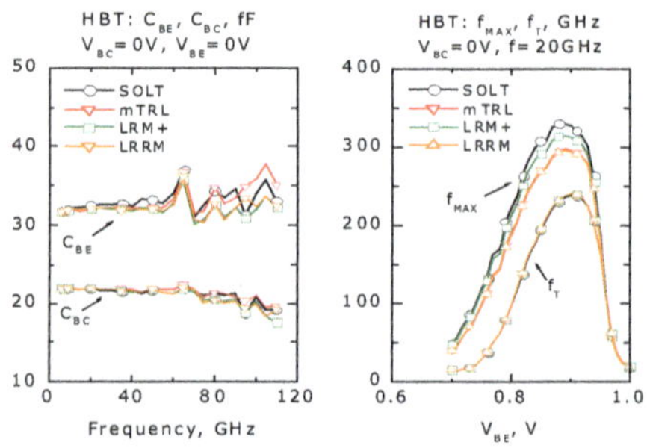

Fig. 3. Extracted junction capacitances C_{BE} and C_{BC} (right), f_T and f_{MAX} of a test HBT (left). Results are de-embedded from the Complete-Open and the Complete-Short dummies.

TABLE II
EXTRACTED PARAMETERS OF TEST TRANSISTOR

Parameter	Probe Tip Calibration			
	SLOT/ SOLT(m)	mTRL	LRM+/ LRM+(m)	eLRRM
C_{BE}, fF[8]	32.2/32.3	32.0	31.9/32.1	32.1
C_{BC}, fF[8]	22.8/22.1	22.8	22.8/22.0	21.9
f_T, GHz[9]	236/242	241	241/241	241
f_{MAX}, GHz[10]	329/302	297	315/298	295

We used the automatic Load inductance extraction algorithm of the eLRRM method to find the actual inductance of the Load [9]. In contrast to the factory value of -3.3 pH, the port 1 *L1* and the port 2 *L2* inductances yielded 0.1 pH and 1 pH respectively (for the frequency range of 20 GHz to 50 GHz). Next, the Open parasitic capacitance and the Short parasitic

[8] V_{BC}=V_{BE}=0V
[9] V_{BC}=0V, V_{BE}=0.9V
[10] V_{BC}=0V, V_{BE}=0.89V

inductance were extracted from the eLRRM-corrected measurements. Then, SOLT was recalculated for the new models of Open, Short and Load. (Fig. 4, legend SOLT(m)). The corrected SOLT(m) was in good agreement with mTRL and eLRRM for both passive and hot-S parameter FoMs (Table 1, Table 2).

Finally, LRM+ was re-calculated for measured *L1* and *L2* of the Load (Table 1, 2, legend LRM+(m)). The variation of f_T and f_{MAX} turned out to be less significant than for SOLT. f_T remained constant, while f_{MAX} decreased by 17 GHz.

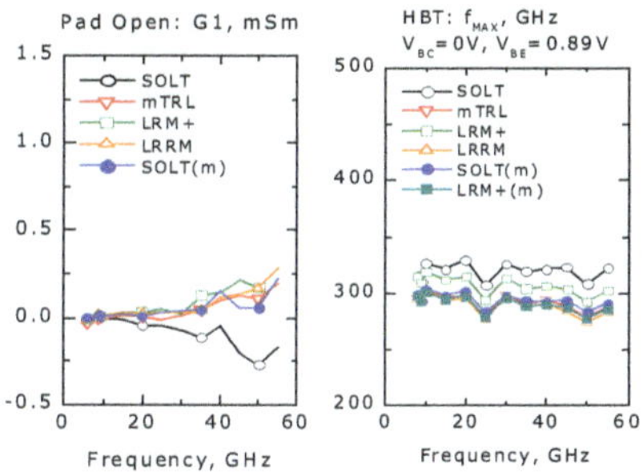

Fig. 4. The conductance *G1* of the Pad Open element (left) and the f_{MAX} of a test transistor (right). The f_{MAX} is de-embedded from the Complete-Open and Complete-Short.

IV. CONCLUSION

In this work, we investigated the impact of the probe-tip calibration on the measurement accuracy of advanced BiCMOS HBT small-signal parameters. We proved that for frequencies below 67 GHz, the measurement accuracy of the junction capacitances is mainly defined by the de-embedding step, as possible calibration errors (e.g., due to probes misplacement) are cancelled. The required low power of the input signal did not allow drawing a clear conclusion for frequencies above 67 GHz due to the impact of the noise.

We found that the inaccurate probe tip calibration affected f_T and f_{MAX}. Due to the probe misplacement error, SOLT overestimated f_{MAX} by 27 GHz (about 10%) and underestimated f_T by 6 GHz (about 3%). LRM+ demonstrated constant f_T and an overestimation of about 17 GHz (or 5%) for f_{MAX}. Both multiline TRL and eLRRM provided the same results given equal conditions.

The undertaken experiments revealed that the measurement accuracy of the small-signal transistor parameters can be improved by:

1) using advanced calibration methods;
2) reducing the equivalent impedances of the silicon backend parasitics as well as the impact of the probe placement calibration error.

The last can be achieved by the in-situ (on-wafer) calibration step.

ACKNOWLEDGEMENT

The EU FP7 IP DOTFIVE project is acknowledged for financial support. Cascade Microtech also acknowledges Agilent Technology EESof division and Dr. Franz Sischka from Agilent for support with IC-CAP. M. Schroter and P. Sakalas are thankful for financial support within the BMBF CoolSilicon project.

REFERENCES

[1] "European dotfive project", *online* http://www.dotfive.eu
[2] P. Chevalier, *et al*, "A conventional double-polysilicon FSA-SEG Si/SiGe:C HBT reaching 400 GHz fmax," in *Bipolar/BiCMOS Circuits and Technology Meeting, 2009. BCTM 2009. IEEE*, Capri, Italy, 2009, pp. 1-4.
[3] A. Rumiantsev and N. Ridler, "VNA calibration," *Microwave Magazine, IEEE,* vol. 9, pp. 86-99, 2008.
[4] R. Doerner and A. Rumiantsev, "Verification of the wafer-level LRM+ calibration technique for GaAs applications up to 110 GHz," in *ARFTG Microwave Measurements Conference-Spring, 65th*, 2005, pp. 15-19.
[5] R. B. Marks, "A multiline method of network analyzer calibration," *Microwave Theory and Techniques, IEEE Transactions on,* vol. 39, pp. 1205-1215, 1991.
[6] D. F. Williams and R. B. Marks, "Transmission line capacitance measurement," *Microwave and Guided Wave Letters, IEEE,* vol. 1, pp. 243-245, 1991.
[7] A. Rumiantsev, *et al*, "Influence of calibration substrate boundary conditions on CPW characteristics and calibration accuracy at mm-wave frequencies," *ARFTG Microwave Measurements Conference-Fall, 72nd,* pp. 168-173, 2008.
[8] A. M. E. Safwat and L. Hayden, "Sensitivity analysis of calibration standards for fixed probe spacing on-wafer calibration techniques," in *Microwave Symposium Digest, 2002 IEEE MTT-S International*, 2002, pp. 2257-2260.
[9] L. Hayden, "An enhanced Line-Reflect-Reflect-Match calibration," in *ARFTG Microwave Measurements Conference-Spring*, 67th, 2006, pp. 143-149.

15.2 Paper [58] (Invited Talk): "State-of-the-art and Future Perspectives in Calibration and De-Embedding Techniques for Characterization of Advanced SiGe HBTs Featuring Sub-THz f_T/f_{MAX}"

N. Derrier, A. Rumiantsev, and D. Celi, "State-of-the-art and future perspectives in calibration and de-embedding techniques for characterization of advanced SiGe HBTs featuring sub-THz f_T/f_{MAX}," in *Bipolar/BiCMOS Circuits and Technology Meeting (BCTM)*, 2012. IEEE, Portland, OR, 2012, pp. 92–99.

This paper presented an overview of RF calibration and pad de-embedding techniques, discusses limitations and demonstrates methods for accuracy improvement applicable for the characterization of advanced BiCMOS HBTs. The impact of the reference plane location was discussed. Numerous experiments with different device geometries showed that the *in-situ* (on-wafer) calibration yields the most accurate results. For a probe-tip calibration, a multiple-dummy de-embedding is crucial to improve measurement accuracy. A comparison with the compact model (HICUM V2.30) confirmed the findings.

This paper was a joint work with Nicolas Derrier and Didier Celi from STMicroelectronics (Crolles, France). In this work, I planned the paper and the measurement comparison campaign, did all measurements, and performed calibration and de-embedding calculations. I also wrote the related sections of the paper and presented the work at the conference. Nicolas Derrier applied the advanced six-step de-embedding method on the probe-tip calibrated data and provided HICUM compact model verification and created the 3D pictures of standards, de-embedding structures and DUTs. The data analysis was shred between all co-authors.

State-of-the-art and Future Perspectives in Calibration and De-Embedding Techniques for Characterization of Advanced SiGe HBTs featuring sub-THz f_T/f_{MAX}

N. Derrier[1], A. Rumiantsev[2], D. Celi[1]

[1]STMicroelectronics, 850 rue Jean Monnet, F-38926 Crolles, France.
[2]Cascade Microtech, Inc. 9100 SW Gemini Drive Beaverton, OR 97008, USA.

***Abstract* — This paper presents an overview of RF calibration and pad de-embedding techniques, discusses limitations and demonstrates methods for accuracy improvement applicable for the characterization of advanced BiCMOS HBTs. The impact of the reference plane location is discussed. Numerous experiments with different device geometries showed that the in-situ (on-wafer) calibration yields the most accurate results. For a probe-tip calibration, a multiple-dummy de-embedding is crucial to improve measurement accuracy. A comparison with the compact model (HICUM V2.30) confirmed the findings.**

***Index Terms* — SiGe HBT, Bipolar modeling and simulation, de-embedding, HF measurements, calibration, S-parameters.**

I. INTRODUCTION

Continuously increasing demand for more content, higher speed and cheaper products pushes development of advanced silicon devices operating near THz range. BiCMOS technologies featuring HBT f_T/f_{MAX} of 220/280 GHz respectively have already become foundry available and address the needs of mass production [1]. Further improvement has recently been achieved for HBTs with 300/500 GHz for f_T/f_{MAX} [2, 3]. Their integration into a 300 mm BiCMOS foundry of 55 nm node is in progress [4].

Though substantial enhancements have already been achieved regarding technology and performance of devices, accurate device characterization and parameter extraction still remain challenging. This originates in the different points of view on the "device" in characterization, modeling, and design engineering communities. Still a subject of debate – the location of the characterization reference plane: at the probe tip, at the lateral device terminals or at one of the levels of the Back End of Line (BEOL). As a result, the following procedure has prevailed as the industry's standard: the RF Figures of Merits (FoM) are extracted at relatively low frequencies (typically at 20 GHz); data are calibrated to the probe tip end by the SOLT method and de-embedded using the OPEN and SHORT contact pad dummy elements [5]. Still, this approach may lead to a serious over/under estimation of FoM of sub-THz f_T/f_{MAX} HBTs [6].

Various investigations were undertaken to improve the accuracy of probe tip calibration and pad de-embedding [7-11]. However, to the authors' best knowledge, a cumulative analysis of calibration and pad de-embedding methods for parameter extraction accuracy across various device geometries is still not available.

This paper presents an overview of RF calibration and pad de-embedding techniques, discusses limitations and demonstrates methods for accuracy improvement applicable for characterization of advanced BiCMOS HBTs. Section II gives a brief summary of the S-parameter calibration background. Section III discusses the optimal location of the measurement reference plane and the proper design of de-embedding elements. It is shown that various transistor geometries may have different requirements. A six-step de-embedding algorithm and the in-situ calibration are proposed as possible solutions to minimize distributive effects in Section IV. Finally, the results of an extensive Design Of Experiment (DOE) are presented in Section V. The DOE was carried out on diverse HBT geometries of ST B5T process transistors and for a wide range of calibration algorithms and de-embedding methods.

II. S-PARAMETER CALIBRATION BACKGROUND

For S-parameter measurements, the systematic measurement errors are represented using the system error model. The purpose of the calibration is to define individual components of this model, the "error terms", and to exclude measurement errors from the DUT data by the error-correction step. Modern VNAs enable application of advanced error models and calibration methods, offering a variety of benefits for measurements at mm-wave frequencies and beyond [12].

978-1-4673-3021-3/12/$31.00 ©2012 IEEE

A. Calibration Method SOLT

SOLT[1] requires three reflection and one transmission measurement conditions: Short, Open, Load, and Thru, respectively. All electrical characteristics of standards must be fully known. Subsequently, calibration accuracy critically depends on the design, fabrication and characterization of standards. Though SOLT is one of the widely used algorithms, it remains a challenge to achieve reliable calibration results at high frequencies.

B Self-calibration: Multiline TRL, LRM+, and LRRM

A number of methods use information redundancy to relax requirements to one (or more) standards. Unknown characteristics are subsequently calculated within the calibration process (e.g. "self-calibration").

The multiline TRL[2] (mTRL) was developed at NIST in the early 1990s [13]. It applies advanced statistical analysis of the redundant information gained from measurements of multiple sections of transmission lines of known length. Furthermore, any symmetrical ($S_{11}=S_{22}$) highly-reflective element can be used as a Reflect. It is sufficient to estimate the phase of its return loss roughly within ± 90 degrees. This is an important advantage for wafer-level application: so far, realizing an ideal or fully-known broadband Open or Short standard has been a challenging task.

Moreover, in conjunction with the method proposed in [14], mTRL calibration reference impedance Z_{REF} can be accurately set to the required value of 50 Ω. Thus, mTRL became the reference method for wafer-level calibration, calibration comparison and accuracy verification.

LRM+[3] uses one transmission (Thru) and two reflection elements: the Reflect (similar to mTRL) and the Load (or "Match"). Any arbitrary impedance element can be used as the Load. The LRM+ method is capable to set the calibration reference impedance Z_{REF} to 50 Ω when the impedance of the Load standard Z_{LOAD} is known [15]. For LRM+, standards can be designed for a constant probe-to-probe distance, significantly reducing the test chip size and enabling an automated calibration.

The enhanced LRRM[4] (or eLRRM) was developed to address the need for accurate broadband probe-tip calibration [16]. The automatic Load inductance extraction embedded in eLRRM minimizes the calibration error caused by probe misplacement on the Load. This algorithm demonstrated reliable results on alumina Impedance Standard Substrate (ISS).

---

[1] Short-Open-Load-Thru
[2] Thru-Reflect-Line
[3] Line-Reflect-Match, advanced
[4] Line-Reflect-Reflect-Match

III. LOCATION OF THE REFERENCE PLANE

The S-parameter reference plane can be located at:

1) The probe tip end.
2) The top metal level, metal6 (M6), and
3) The transistor terminals, metal1 (M1, Fig. 1).

Contact pad de-embedding moves the reference plane from the probe tip towards the lateral Device under Test (DUT). Advantages and drawbacks of M6 and M1 are repeatedly discussed between design and modeling communities.

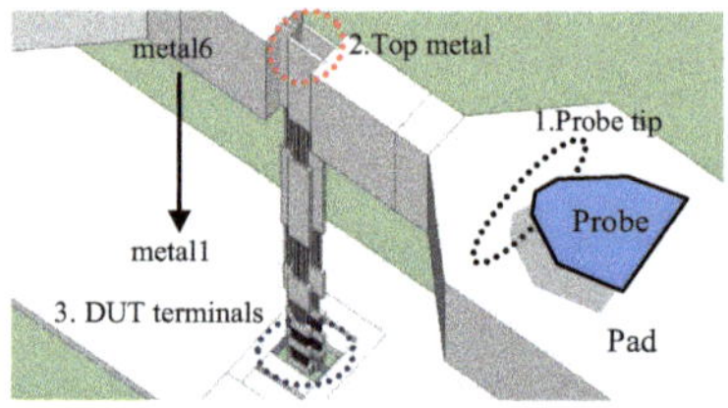

Fig. 1. Three different locations of the reference plane: probe tip, top metal, DUT terminals for a complete-OPEN dummy. Example for the CBEBC HBT, Emitter length of L_E=5 μm.

Design engineers prefer SPICE models that include a minimum of BEOL parasitics and allow realistic pre-layout simulations (schematic mode). Advanced transistors showing very high f_T/f_{MAX}, have junction capacitances of merely several fF. Here, any small parasitics may strongly affect transistor's FoM. Therefore, the top-metal location (e.g. M6) is preferred. For designers, it becomes important that the Process Design Kit (PDK) includes such SPICE libraries.

In contrast, modeling engineers aim to extract "physics-based" parameters of the device compact models. Therefore, the reference plane has to be shifted as close as possible to the intrinsic device terminals (e.g. M1). This paper is mainly focused on device modeling requirements, as they are more challenging.

A. Probe Tip Calibration on ISS

The ISS is an alumina substrate that utilizes several groups of coplanar waveguide elements (CPW): Open, Short, Load, Thru standards as well as transmission Lines of multiple lengths. The equivalent impedances of the Open, Short and Load are represented over a parallel capacitance, serial inductor and a serial connection of an inductor and a resistor respectively. The nominal values depend on the type of ISS, as well as on design, pitch and configuration of a wafer probe and are specified by the vendor. The length of CPW lines is typically optimized to cover a wide frequency range, e.g. from hundreds of MHz up to 110 GHz and beyond. The

Loads are accurately trimmed to R_{LOAD}=50 Ω with typical accuracy better than 0.3 percent. ISS supports various calibration methods, such as SOLT, LRM+, LRRM, mTRL and others.

Due to impact of fabrication inaccuracies as well as a dispersive nature of the CPW lines, additional procedures are required to measure the characteristic impedance Z_{LINE} and the propagation constant γ of the ISS CPW line for accurate TRL calibration[17].

The ISS lumped-standards are significantly better characterized than the custom elements embedded on the test chip. However, their equivalent impedances are sensitive to the probe placement [18]. Inaccurate probe-to-standard alignment may increase calibration residual errors and lead to errors in DUT parameter extraction. As investigated in [6], both mTRL and eLRRM calibration methods are less sensitive to the probe misplacement and are recommended for the probe-tip calibration.

B. Contact Pads and Interconnects Parasitics

A simple method introduced for high-speed bipolar transistors in 1987 allowed characterization of the pad and interconnect parasitics over "dummy" elements [19]. The two-step de-embedding procedure subtracts parallel and serial parasitic impedances, captured by the OPEN (Fig. 2) and the SHORT (Fig. 3) dummies, from the DUT.

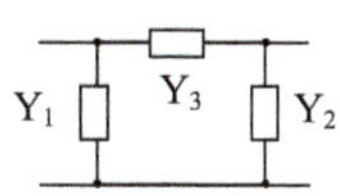

$$Y = \begin{bmatrix} Y_1 + Y_3 & -Y_3 \\ -Y_3 & Y_2 + Y_3 \end{bmatrix}$$

Fig. 2. Π-Equivalent circuit of an OPEN dummy and the Y matrix of parallel parasitics.

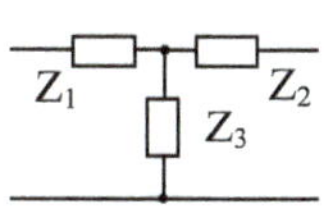

$$Z = \begin{bmatrix} Z_1 + Z_3 & Z_3 \\ Z_3 & Z_2 + Z_3 \end{bmatrix}$$

Fig. 3. T-Equivalent circuit of a SHORT dummy and the Z matrix of serial parasitics.

With the increase of the measurement frequency to the mm-wave range, the complexity of the parasitic equivalent circuit increased (e.g. [8, 20]). Furthermore, the distributed effects exhibit stronger impact on de-embedding accuracy. As these effects cannot be captured by the impedance equivalent circuit model, the cascade-matrix based methods and the general four-port de-embedding algorithms were introduced [21, 22]. As a result, the de-embedding process became cumbersome, accumulating a significant area of silicon, and requesting structures which are difficult to be realized (e.g. known Loads). Therefore, such methods are not widely used, making it necessary to develop alternative solutions for accurate measurements at high frequencies.

IV. METHODS FOR ACCURACY IMPROVEMENT

A. The BEOL Parasitics above the Transistor

Previous technologies often used OPEN and SHORT dummies that did not include the via stack and located the reference plane at the top metal (ref. plane n°2, Fig. 1)[5]. Due to the relatively large size of HBTs and the resulting junction capacitances, the coupling capacitances between the emitter, base and collector metal accesses from the top layer down to the first metal layer were negligible.

Advanced mmW technologies made it necessary to review the layout of the OPEN and SHORT dummies: transistor dimensions had shrunk drastically (e.g., the smallest ST B5T HBT has the effective emitter width of W_E=100 nm and the emitter length of L_E=0.520 μm). The BEOL top layers had become thicker reducing transmission line losses. For modeling purposes, it became crucial to shift the de-embedding reference plane down to the first metal layer.

i. The complete-OPEN

To move the reference plane to the DUT terminals, the dummy elements must repeat the geometry of each transistor[6]. The complete-OPEN is a copy of the associated DUT. While excluding the contact vias, it includes all metals and the via stack between the end of the top layer line and the first metal above the DUT (Fig. 1). Fig. 4 shows a magnified view around the first two metals.

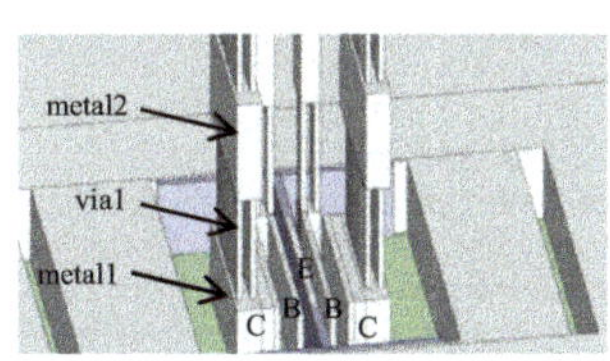

Fig. 4. Magnified view of the area of the M1 and M2 layers of the complete-OPEN. Example for the CBEBC HBT with Emitter length of L_E=5μm.

[5] Such elements will be further called "regular" OPEN and SHORT.
[6] Such elements will be further called "complete" OPEN and SHORT.

Fig. 5 shows f_T de-embedded results (SOLT ISS and OPEN de-embedding) for ST B5T process HBT transistors (section V.*A*) of 2 different lengths (a small one and a medium one). Three different OPEN dummies were used: a regular OPEN (white triangle line, reference plane at top layer M6), a complete-OPEN M1 (black circle line, reference plane at M1, Fig. 4) and a complete-OPEN M2 (white circle line, reference plane at M2). For the small transistor, a difference of 88 GHz on the f_T peak is observed when moving the reference plane from M6 to M1. De-embedding M1 removes additional 1 fF of parasitic capacitance for C_{BE}. The value of the M6-M1 coupling capacitance seems to be small. However, it is in order of the intrinsic capacitance for a small transistor. This ratio reduces with increase of the transistor length.

A prior work attributed the decrease of the f_T for short transistors to the 3D effects [20] whereas it is just a matter of accurate de-embedding and location of the reference plane. Indeed, the f_T FoM should almost be constant for the intrinsic device regardless of its length.

The complete-Open is essential for accurate extraction of FoM for small devices. The obvious drawback is the significant increase of the test chip size.

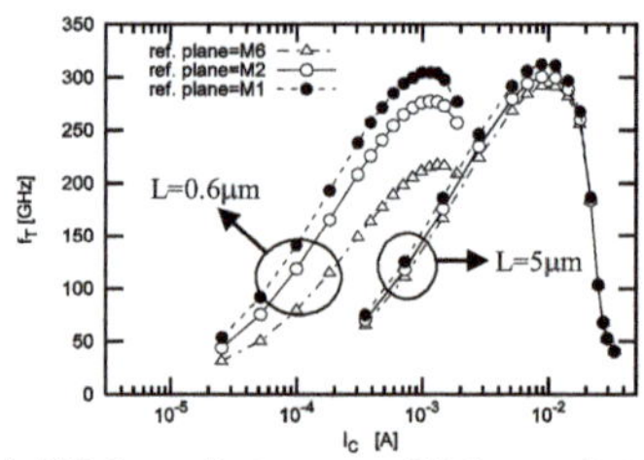

Fig. 5. f_T [GHz] vs collector current [A] (extracted at 20 GHz and V_{CB}=0 V), for two HBT of L_E=0.6 μm and L_E=5 μm, with OPEN de-embedding using three different reference plane locations: Metal6 (white triangle line), Metal2 (white circle line), Metal1 (black circle line).

ii. The complete- SHORT

The complete-SHORT is drawn from the DUT test structure by removing the device and shortening the access to the transistor pins with the M1 plate (Fig. 6).

Similar to the complete-OPEN, the benefit of the complete-SHORT strongly depends on the device geometry. The SHORT de-embedding is more relevant for long or multi-fingers transistors (Fig. 7): due to their small resistances they are very sensitive even to minor series parasitics.

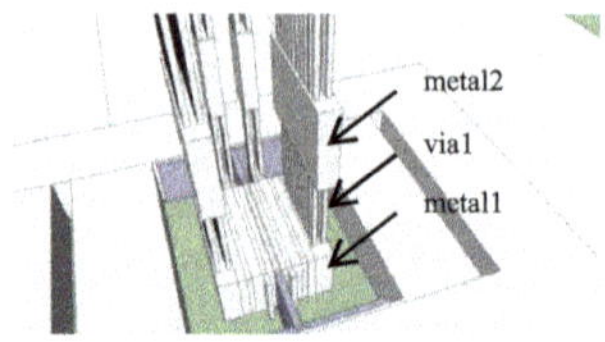

Fig. 6. Magnified area around Metal1 and Metal2 layers of the complete-SHORT.

Fig. 7 clearly demonstrates a resonance around 60 GHz (left plot, circle line) of the inductance parasitics on the C_{BE} RF signature. The right plot shows an improvement of 78 GHz for f_T peak due to accurate de-embedding of the base (Port 1) and the collector (Port 2) resistances by the complete-SHORT.

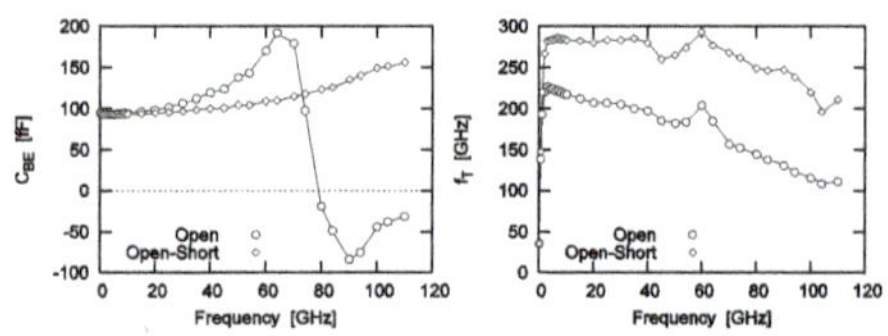

Fig. 7. The Base/Emitter capacitance C_{BE} [fF] @V_{BE}=0 V (left plot) and peak f_T [GHz] (for J_C=15 mA/μm^2) @V_{BC}=0 V (right plot) vs frequency [GHz], for a five Emitter finger HBT of L_E=10 μm after a complete-OPEN (circle) and complete-OPEN/SHORT (diamond) de-embedding.

B. The Importance of the Distributed Effect

Another subject of discussion is the lumped or distributed nature of the de-embedding method. The introduction of the THRU to the OPEN/SHORT sequence did not show a noticeable improvement up to 170 GHz [23].

An advanced de-embedding approach was proposed in [20, 23]. It successively captures the wafer parasitics from the probe tip down to the DUT terminals. Though it implements six steps, all dummy elements are easy to be realized, offering a significant advantage over the four-port method. Extending this methodology with the complete-OPEN and the complete-SHORT, the de-embedding sequence follows as:

1) Probe-SHORT (reduces calibration residuals);
2) Pad-OPEN (parallel impedance of the pad);
3) Pad-SHORT (series impedance of the pad);
4) THRU (distributed parallel impedance of the line);
5) complete-OPEN (parallel impedance of the metal/via access above the transistor);

6) complete-SHORT (series impedance of the metal/via access above the transistor);

The equivalent circuit of this method is shown in Fig. 8.

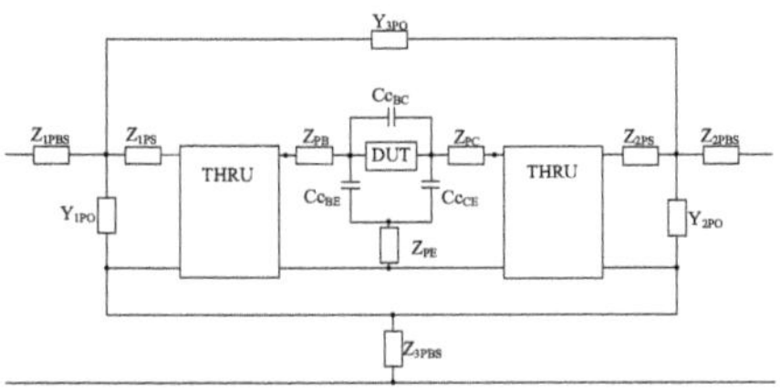

Fig. 8. The equivalent circuit of the 6-dummy de-embedding

Fig. 9 compares the complete OPEN/SHORT (circle lines) against the six-dummy method (diamond lines) for the C_{BE} and f_T FoM. Both are based on the ISS SOLT calibration. The 6-dummy method significantly outperforms above 40 GHz.

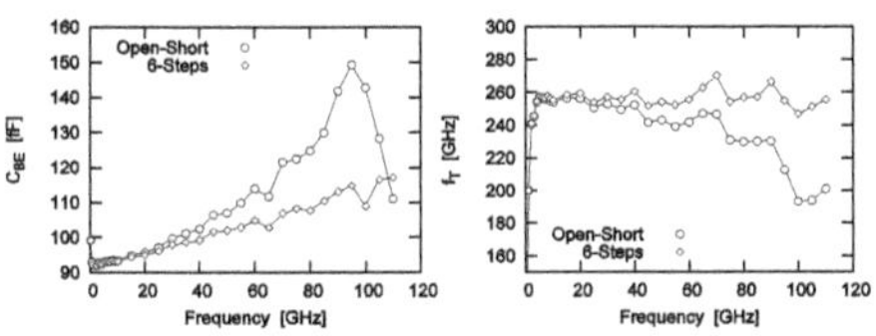

Fig. 9. Base/Emitter capacitance C_{BE} [fF] @V_{BE}=0 V (left plot) and peak f_T [GHz] (for J_C=15 mA/μm^2) @V_{BC}=0 V (right plot) vs. frequency [GHz], for a five Emitter finger transistor L_E=10 μm after the complete-OPEN/SHORT (circle) and the six-steps (diamond) de-embedding.

C. In-situ (On-Wafer) Calibration

Still, the advanced de-embedding method presented above suffers from its complexity [20, 23]: a measurement campaign targeting a scalable library of 30-40 transistors and several temperature conditions will require a tedious and lengthy post-processing of data.

In opposite, the in-situ calibration (with standards implements on a test chip) allows moving the reference plane close to the DUT terminals in one step. The remaining BEOL parasitics are significantly minimized, so that a simple two-step de-embedding method can be used to capture them.

The first in-situ calibration attempts were made in the late 1990s [24]. It was found that the fabrication inaccuracy did not allow implementation of the conventional SOLT procedure on the Si wafer. Over the past years, substantial achievements have been made in fabricating and characterizing planar standards in processes with a conductive substrate (e.g. [10, 25, 26]). LRM+ and mTRL turned out to be the most appropriate methods for in-situ calibration on conductive wafers.

The required calibration elements utilize the same contact pad design as the DUT. The Thru and the Lines are designed as M1-shielded 50 Ω top metal grounded coplanar waveguide. The effective Line length is: 51, 549, 1545, and 3039 μm respectively, covering the frequency range from 1 GHz to 110 GHz. The symmetrical Reflect is a CPW short located at the M1 level. The Load is a single (per port) symmetrical 50 Ω N+ salicided poly resistor located at the M1 level. The calibration reference plane is set at the top metal level.

i. Notes about In-Situ LRM+

The Load element takes a crucial position as it defines the LRM+ calibration reference impedance. The resistance of the in-situ Load is affected by the fabrication variability and may differ from chip to chip. That is why it should be measured during the calibration process. The source monitor unit (SMU), that is used to bias the DUT, provides sufficient accuracy for the measurement of the Load resistance.

The Load reactance is mainly represented by the equivalent reactance of the via stack (the distance between M6 and M1). This parameter can be simulated during the design step as it is less sensitive to the fabrication process instability. Alternatively, it can be measured with respect to the probe-tip calibration and the two-step de-embedding [26].

ii. Notes about In-Situ mTRL

The TRL sets the calibration reference impedance to the Line characteristic impedance Z_0 and the measurement reference position to the center of the Thru. Both the propagation constant γ and the Z_0 of the Line are required to shift the measurement reference plane to an arbitrary position and to transform the calibration results to the 50 Ω system reference impedance. The calibration comparison [27] or the lumped load method [14] can be used to extract the Z_0 of the in-situ lines.

V. EXPERIMENTAL RESULTS

A. The Design Of Experiments (DOE)

The objective of the DOE was to answer the following questions:

1) What is the most suitable calibration method?
2) What is the advantage of in-situ over probe-tip calibration?
3) Which type of de-embedding dummies gives the best results?

STMicroelectronics B5T process featuring f_T/f_{MAX} of 300/400 GHz at collector current density J_C of

15 mA/μm² was used for the experiment [2, 28]. Seven characterization strategies (TABLE I) were applied to the measurement of 4 HBT geometries (TABLE II).

TABLE I
LIST OF CALIBRATION AND DE-EMBEDDING EXPERIMENTS

Experiment	Calibration standards	Calibration algorithm	De-Embedding dummies
n°1	ISS	SOLT	O+S*
n°2	ISS	LRRM	O+S*
n°3	ISS	LRM+	O+S*
n°4	ISS	mTRL	O+S*
n°5	in-situ	LRM+	O+S*
n°6	in-situ	mTRL	O+S*
n°7	ISS	SOLT	6-dummies

* complete Open + complete Short

TABLE II
LIST OF MEASURED TRANSISTORS

Process	Geometry	Emitter width*, μm	Emitter length, μm
ST B5T	CBEBC	0.18	0.6
	CBEBC	0.18	5
	CBEBC	0.18	15
	C-5x(BEBC)	0.18	10

* drawn dimensions

B. The Measurement Setup

The experiment was carried out on a broadband S-parameter measurement system from Cascade Microtech, consisting of a semi-automated wafer-probe system Elite300™-AP, a pair of 110 GHz GSG Infinity Probes® of 100 μm pitch and a fitting alumina calibration[7] substrate (ISS), equipped with broadband 110 GHz PNA-X VNA and 4142B SMU from Agilent Technologies. The DUT input and output power level (port 1 and port 2) was set to -35 dBm and -15 dBm to protect the DUT from overload.

The measured data of the ISS and in-situ standards, the de-embedding structures and the test transistors were acquired in raw format (with the S-parameter calibration turned off). The calibration and error correction were performed by WinCal XE™, and proprietary IC-CAP script outside of the VNA on a computer and for the same raw data set. This approach ensured a minimal impact of contact repeatability and instrument drift on the accuracy of experimental results. Device parameters were extracted using IC-CAP.

C. Results on the Transistors

i. Cold [S] Parameter Results

Fig. 10 shows the Base/Collector capacitance for long multi-fingers transistors extracted from cold [S] measurements. Only the in-situ calibration (both LRM+ and mTRL) gives the expected RF signature of the π-equivalent circuit, which corresponds to the DUT in cold-[S] mode (0 V bias): the capacitances remain constant up to 110 GHz. The in-situ calibrations outperform probe-tip methods from 40 GHz for multi-finger and from 70 GHz for small transistors.

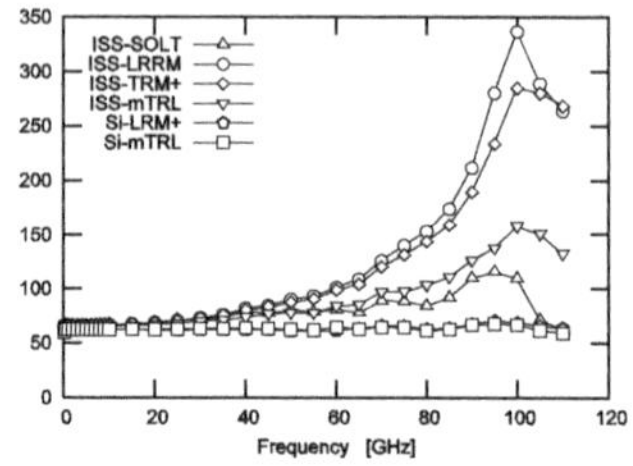

Fig. 10. Base/Collector capacitance C_{BC} [fF] vs frequency [GHz] @V_{BC}=0 V, for a five Emitter finger HBT of L_E=10 μm.

ii. Forward mode [S] Parameter Results

Figure 11 shows peak f_T FoM for a multi-fingers device. Again, only the in-situ calibration methods (LRM+ or mTRL) give expected results for constant f_T up to 110 GHz. For [Y]-parameters in forward mode, the comparison leaded to the same conclusion (not shown here).

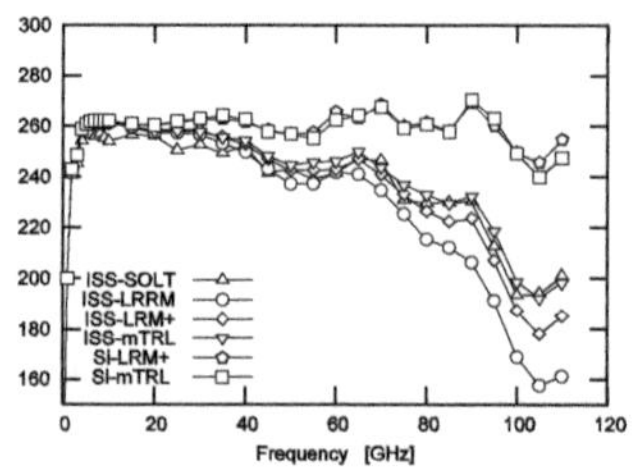

Fig. 11. Peak f_T [GHz] (for J_C=15 mA/μm²) vs. frequency [GHz] @V_{BC}=0 V, for a five Emitter finger HBT of L_E= 10 μm.

D. Probe Tip Calibration and Six-Dummy De-Embedding vs. in-situ Calibration and Open/Short De-embedding

Results presented in section V.*C* are based on the two-step de-embedding method, which is widely accepted as industry standard (TABLE I), [5, 19]. As proved above, a simple two-steps de-embedding yields sufficient results for the in-situ calibration.

As explained in IV.B, the six-dummies de-embedding equivalent circuit accurately captures parasitics remaining after the probe-tip calibration. Consequently,

[7] IC-CAP is a product of Agilent Technologies, Inc.

the next experiment explored in how far this method improves the accuracy of the probe-tip calibration.

Fig. 12 compares the f_T FoMs for the probe-tip SOLT followed by the Open/Short de-embedding versus the six-dummies de-embedding, as well as for the in-situ LRM+ and the in-situ mTRL+, both followed by the Open/Short de-embedding. The implementation of the six-steps de-embedding for the probe-tip calibration yielded results comparable to the in-situ methods for the investigated frequency range. [Y]-parameters showed up the same trend (not shown here).

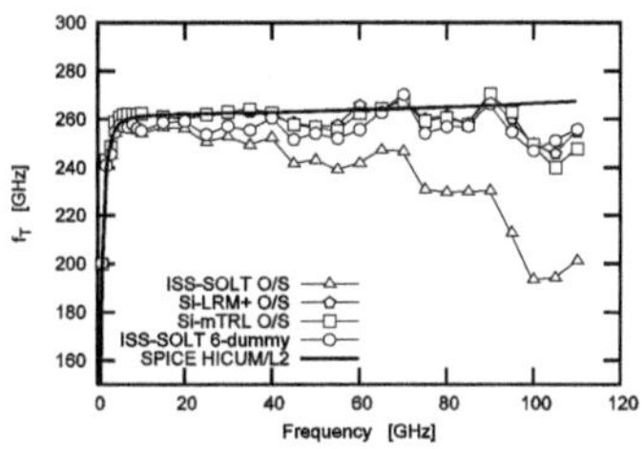

Fig. 12. Peak f_T (for J_C=15 mA/μm^2) [GHz] vs. frequency [GHz] @V_{BC}=0 V, for a five emitter finger HBT of L_E=10 μm, after the SOLT probe-tip calibration + complete-OPEN/SHORT (white triangle) versus six-dummies de-embedding (white circle), in-situ calibrations (white diamond and square), and the HICUM V2.30 compact model simulated with ADS software (solid line).

E. Comparison with a Compact Model

Parameters of the compact model HICUM Level2 (L2) v2.30 [29, 30], for transistors described in TABLE II, were extracted from DC and [S] measurements data, calibrated by the probe-tip SOLT and Open/Short de-embedded. RF FoM like f_T or f_{MAX} were computed at 20 GHz and used inside the extraction sequence, in order to avoid parasitic effects. ADS[8] 2011.10 software was used to simulate the obtained model cards, and compare the simulation with the different calibration/deembedding strategies. Fig. 12 shows that the in-situ calibration and the advanced six-dummies de-embedding validate the model for the entire frequency range of the experiment. The [Y]-parameters demonstrated the same trend (not shown here). Because the HICUM compact model is based on the device's physics, this comparison is an alternative proof of the quality of in-situ methods and the multiple-dummies de-embedding.

VI. CONCLUSION

This work investigated the impact of common calibration methods, de-embedding sequences, and design of de-embedding dummies on the accuracy of S-parameter measurement and FoM extraction of advanced BiCMOS HBTs, featuring f_T/f_{MAX} of 300/400 GHz. It was shown that it is crucial to locate the reference plane close to the device terminals yielding an increase of up to 30 percent of f_T/f_{MAX} for small transistors. For these elements, the M6-M1 coupling capacitance is in order of the intrinsic capacitance for a transistor. It can be completely removed by the introduced complete-OPEN/SHORT dummy elements.

Because the impact of the distributed parasitic effects increases with frequency, a simple two-steps de-embedding failed above 40 GHz for long or multi-finger devices. That is why, an advanced de-embedding, such as the six-steps method, is crucial for measurement accuracy improvement for the probe-tip calibration. The choice of the advanced calibration methods (e.g. mTRL and eLRRM) was not decisive.

The numerous experiments on several ST B5T HBT geometries revealed a substantial accuracy improvement when using the in-situ calibration (mTRL or LRM+). The in-situ standards are optimized to calibrate out the major part of the backend parasitics and to move the measurement reference plane to the M6 level in one step. The remaining M6-M1 parasitics can be removed by a simple two-steps de-embedding. A comparison with the compact model (HICUM V2.30) confirmed the findings.

Though both methods (the probe-tip calibration with six-steps de-embedding and the in-situ calibration with two-steps de-embedding) yield close results, in practice, the choice of the method depends on the available technology and individual preferences.

The in-situ calibration requires preparation work to characterize several electrical parameters of standards. In return, a simple two-step de-embedding can be applied.

The probe-tip calibration demands for an advanced six-steps de-embedding with dummies easy to be realized. However, its application is cumbersome and demands additional silicon area.

ACKNOWLEDGEMENT

A part of the work presented here has been performed in the RF2THZ SiSoC project of the EUREKA program CATRENE, in which the French partners are funded by the ministry of economy, finance and industry. Cascade Microtech acknowledges Giovanni D'Amore from Agilent EMEA for excellent support with the 110 GHz PNA-X system as well as Agilent EEsof division,

[8] ADS is a product of Agilent Technologies, Inc.

particular Dr. Franz Sischka for support with IC-CAP and valuable discussions.

REFERENCES

[1] G. Avenier, M. Diop, P. Chevalier, *et al.*, "0.13 um SiGe BiCMOS technology fully dedicated to mm-wave applications," *Solid-State Circuits, IEEE Journal of*, vol. 44, pp. 2312-2321, 2009.

[2] P. Chevalier, T. F. Meister, B. Heinemann, *et al.*, "Towards THz SiGe HBTs," in *Bipolar/BiCMOS Circuits and Technology Meeting (BCTM), 2011 IEEE*, 2011, pp. 57-65.

[3] "European dotfive project "; http://www.dotfive.eu/.

[4] "European RF2THZ SiSoc project, CATRENE."

[5] "Radio frequency and analog/mixed-signal technologies for wireless communications," *International Technology Roadmap for Semiconductors, ITRS,* p. 36, 2011.

[6] A. Rumiantsev, P. Sakalas, N. Derrier, *et al.*, "Influence of probe tip calibration on measurement accuracy of small-signal parameters of advanced BiCMOS HBTs," in *Bipolar/BiCMOS Circuits and Technology Meeting (BCTM), 2011 IEEE*, Atlanta, GA, 2011.

[7] E. P. Vandamme, D. Schreurs, and G. Van Dinther, "Improved three-step de-embedding method to accurately account for the influence of pad parasitics in silicon on-wafer RF test-structures," *Electron Devices, IEEE Transactions on,* vol. 48, pp. 737-742, 2001.

[8] I. M. Kang, S.-J. Jung, T.-H. Choi, *et al.*, "Five-step (Pad-Pad Short-Pad Open-Short-Open) de-embedding method and its verification," *Electron Device Letters, IEEE,* vol. 30, pp. 398-400, 2009.

[9] R. F. Scholz, F. Korndorfer, B. Senapati, *et al.*, "Advanced technique for broadband on-wafer RF device characterization," in *ARFTG Microwave Measurements Conference-Spring, 63rd*, 2004, pp. 83-90.

[10] A. Rumiantsev, S. L. Sweeney, and P. L. Corson, "Comparison of on-wafer multiline TRL and LRM+ calibrations for RF CMOS applications," *ARFTG Microwave Measurements Conference-Fall, 72nd,* pp. 132-135, 2008.

[11] Q. Liang, J. D. Cressler, G. Niu, *et al.*, "A simple four-port parasitic deembedding methodology for high-frequency scattering parameter and noise characterization of SiGe HBTs," *Microwave Theory and Techniques, IEEE Transactions on,* vol. 51, pp. 2165-2174, 2003.

[12] A. Rumiantsev and N. Ridler, "VNA calibration," *Microwave Magazine, IEEE,* vol. 9, pp. 86-99, 2008.

[13] R. B. Marks, "A multiline method of network analyzer calibration," *Microwave Theory and Techniques, IEEE Transactions on,* vol. 39, pp. 1205-1215, 1991.

[14] D. F. Williams and R. B. Marks, "Transmission line capacitance measurement," *Microwave and Guided Wave Letters, IEEE,* vol. 1, pp. 243-245, 1991.

[15] R. Doerner and A. Rumiantsev, "Verification of the wafer-level LRM+ calibration technique for GaAs applications up to 110 GHz," in *ARFTG Microwave Measurements Conference-Spring, 65th*, 2005, pp. 15-19.

[16] L. Hayden, "An enhanced Line-Reflect-Reflect-Match calibration," in *ARFTG Microwave Measurements Conference-Spring, 67th*, 2006, pp. 143-149.

[17] A. Rumiantsev, R. Doerner, and S. Thies, "Calibration standards verification procedure using the calibration comparison technique," in *36th European Microwave Conference*, 2006, pp. 489-491.

[18] A. M. E. Safwat and L. Hayden, "Sensitivity analysis of calibration standards for fixed probe spacing on-wafer calibration techniques," in *Microwave Symposium Digest, 2002 IEEE MTT-S International*, 2002, pp. 2257-2260.

[19] P. J. van Wijnen, H. R. Claessen, and E. A. Wolsheimer, "A new straight forward calibration and correction procedure for on-wafer high frequency S-parameter measurements (45 MHz–18 GHz)," in *Bipolar Circuits and Technology Meeting*, 1987, pp. 70-73.

[20] F. Pourchon, C. Raya, N. Derrier, *et al.*, "From measurement to intrinsic device characteristics: Test structures and parasitic determination," in *Bipolar/BiCMOS Circuits and Technology Meeting, 2008. BCTM 2008. IEEE*, 2008, pp. 232-239.

[21] M.-H. Cho, G.-W. Huang, C.-S. Chiu, *et al.*, "A cascade Open-Short-Thru (COST) de-embedding method for microwave on-wafer characterization and automatic measurement," *IEICE Trans Electron,* vol. E88-C, pp. 845-850, May 1, 2005 2005.

[22] X. Wei, G. Niu, S. Sweeney, *et al.*, "A general 4-port solution for 110 GHz on-wafer transistor measurements with or without impedance standard substrate (ISS) calibration," *Electron Devices, IEEE Transactions on,* vol. 54, pp. 2706-2714, 2007.

[23] C. Raya, "Modélisation et optimisation de transistors bipolaires à hétérojonction Si/SiGeC ultra rapides pour applications millimétriques," Ph.D., University of Bordeaux, Bordeaux, 2008.

[24] A. Fraser, R. Gleason, and E. W. Strid, "GHz on-silicon-wafer probing calibration methods," in *Bipolar Circuits and Technology Meeting, Proceedings of the*, 1988, pp. 154-157.

[25] A. Rumiantsev, P. L. Corson, S. L. Sweeney, *et al.*, "Applying the calibration comparison technique for verification of transmission line standards on silicon up to 110 GHz," in *ARFTG Microwave Measurements Conference-Spring, 73rd*, Boston, MA, 2009.

[26] A. Rumiantsev, F. Pourchon, N. Derrier, *et al.*, "Designing on-wafer calibration standards for advanced high-speed BiCMOS technology," in *Microwave Integrated Circuits Conference (EuMIC), 2011 European*, 2011, pp. 57-60.

[27] U. Arz, D. F. Williams, and H. Grabinski, "Characteristic impedance measurement of planar transmission lines," in *URSI-GA*, 2002.

[28] P. Chevalier, T. Lacave, E. Canderle, *et al.*, "Scaling of SiGe BiCMOS Technologies for Applications above 100 GHz," in *IEEE CSICS*, 2012.

[29] A. Pawlak, M. Schroter, J. Krause, *et al.*, "A HICUM extension for medium current densities," in *9th European HICUM Workshop*, Würzburg, 2009.

[30] F. Stein, Z. Huszhka, N. Derrier, *et al.*, "Extraction of the emitter related space charge weighting factor parameters of HICUM L2.30 using the Lambert W function," *to be published.*

16

Attachment to Chapter 8

16.1 Paper [87]: "Verification of Wafer-Level Calibration Accuracy at Cryogenic Temperatures"

A. Rumiantsev, R. Doerner, and P. Sakalas, "Verification of wafer-level calibration accuracy at cryogenic temperatures," in *ARFTG Microwave Measurements Conference-Fall, 68th*, 2006, pp. 134–140.

This article presented the results of accuracy verification of wafer-level calibration at cryogenic temperatures based on coplanar calibration standards. For the first time, the electrical characteristics of commercially available coplanar calibration lines were extracted at the temperature of liquid helium. It was demonstrated that the temperature dependent variation of the characteristic impedance of the tested lines is within $\pm 1\%$ tolerance of the nominal value of 50 Ω for a temperature range from room temperature down to 4 K. Finally, the accuracy of the transfer TMR calibration method at cryogenic temperatures was verified by definition of the worst case error bounds for the measurement of passive devices and compared to the reference NIST multiline TRL.

It was a collaborative work with Ralf Doerner from Ferdinand-Braun-Institut (FBH), Leibniz-Institut fuer Hoechstfrequenztechnik (Berlin, Germany) and Paulius Sakalas from Dresden University of Technology (TUD) (Dresden, Germany). My contribution to this work was the idea, planning of the work, development of the test and verification methodology, data acquisition, some calculations and data analysis, as well as writing the paper. Ralf Doerner contributed to the calculation and data analysis of this work while Paulis Sakalas assisted the measurement session.

Verification of Wafer-Level Calibration Accuracy at Cryogenic Temperatures

Andrej Rumiantsev [1], Ralf Doerner [2], Paulius Sakalas [3,4]

[1] SUSS MicroTec Test Systems GmbH, Suss-Str. 1, Sacka, D-01561, Germany.
[2] Ferdinand-Braun-Institut fuer Hoechstfrequenztechnik (FBH), Gustav-Kirchhoff-Str. 4, Berlin, D-12489, Germany.
[3] Dresden University of Technology, CEDIC, 01062 Dresden, Germany.
[4] Semiconductor Physics Institute, Fluctuation Phenomena Lab., 01108 Vilnius, Lithuania.

***Abstract* — This article presents the results of accuracy verification of wafer level calibration at cryogenic temperatures based on coplanar calibration standards. For the first time, the electrical characteristics of commercially available coplanar calibration lines were extracted at the temperature of liquid helium. It was demonstrated that the temperature dependent variation of the characteristic impedance of the tested lines is within ±1% tolerance of the nominal value of 50 Ω for a temperature range from room temperature down to 4 K. Finally, the accuracy of the LRM+ calibration method at cryogenic temperatures was verified by definition of the worst case error bounds for the measurement of passive devices and compared to the reference NIST multiline TRL.**

***Index Terms* — cryogenic, calibration, error correction, calibration comparison, scattering parameters measurement.**

I. Introduction

Accurate calibration is crucial for device measurement, characterization and modeling. Fabrication uncertainty, temperature instability, and mismatch in modeling of calibration standards significantly reduce the final calibration accuracy. Difficulties in traceability of wafer level (planar) calibration standards challenge alternative methods of their verification [1]. These difficulties increase if standards operate in extreme conditions like high vacuum and at the temperature of liquid helium.

The research undertaken by the National Institute of Standards and Technology (NIST) provides a technique for characterizing coplanar lines [2, 3], accurate system calibration [4] and accuracy verification of different calibration methods [5]. This technique is well-accepted in engineering practice and can be successfully used for calibration verification purposes [6]. However, traceable verification results can be achieved only at room temperature and using the reference material RM 8130: the NIST fabricated and characterized GaAs calibration elements [7].

Previously published results of calibration at low temperatures do not consider the temperature influence on the electrical characteristics of thru and line standards [8–11]. Investigations were primarily focused on the temperature stability of thin-film resistors (used as the load standard) as well as on comparing different calibration methods. However, as it was recently demonstrated in [1], even small variations of the electrical parameters of thru and line standards can lead to significant calibration and finally measurement errors. This paper will present the results of extracting electrical characteristics (capacitance per unit length, characteristic impedance Z_0, attenuation and relative phase constants) of commercially available alumina coplanar lines at temperatures down to 4 K, temperature stability of the thin-film load resistance, and finally the accuracy of the LRM+ calibration method at these temperatures.

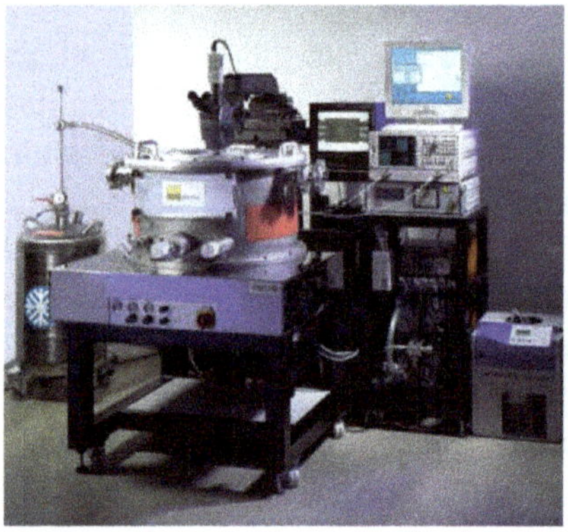

Fig. 1. The PMC200 cryogenic measurement system used for experiments.

II. Verification Method

The calibration comparison technique [5] was used for the evaluation of time and temperature drift of the cryogenic wafer-level RF measurement system during the experiment and for the verification of the LRM+ calibration method. This technique provides the worst-case deviations of the measured S-parameters of passive devices for an examined (first-tier) calibration with respect to a benchmark (second-tier) calibration. Deviations are treated as $|S_{ij} - S_{ij}'|$, for $ij \in \{11, 12, 21, 22\}$, where S_{ij}' is the S-parameter measured by the calibration to be tested, and S_{ij} is the S-parameter measured by the benchmark calibration.

The system drift can be quantitatively defined using the same calibration method with measurements of identical standards once, as the examined calibration, at the beginning of the experiment and again, as the benchmark calibration, at the end. In the ideal case, both calibrations, examined and benchmark, are equal and the error bounds $|S_{ij} - S_{ij}'|$ are zero. Remaining differences can be addressed mainly to system drift and contact repeatability.

Using the calibration comparison technique for the verification of the LRM+ calibration method requires a benchmark calibration. The multiline TRL [1, 4] is well-suited for this purpose and enables an accurate setting of reference plane and reference impedance if the parameters of the line standards are known.

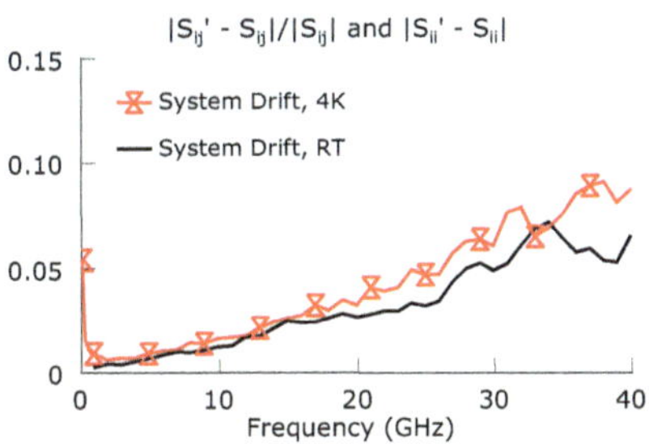

Fig. 2. The system drift of the experimental setup at different temperatures.

To characterize calibration lines at different temperatures the procedure proposed in [2] and evaluated in [1] was used. The resistance R of the load standard can be measured using the 4-terminal method. Then, a first-tier multiline TRL can be performed setting the calibration reference impedance to the characteristic impedance Z_0 of the alumina lines used. The propagation constant known from multiline TRL can be used to extract the value of the line capacitance per unit length C in the frequency range where the tested lines are no longer dispersive and the load reactance is negligible [2, 3]. Once the line capacitance per unit length is found, the characteristic impedance can be obtained from the propagation constants with the help of the MultiCal®[1] software package. The exact determination of the load is essential for accurate line parameter extraction.

[1] Available from NIST, USA

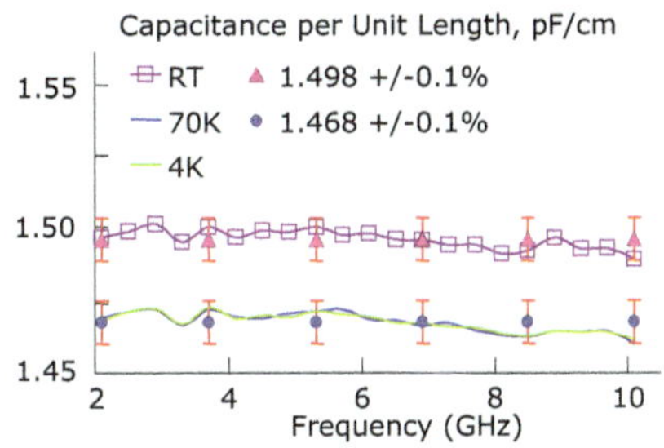

Fig. 3. The results of the line capacitance extraction for different temperatures: room temperature (RT), 70 K, and 4 K. The mean values of 1.498 pF/cm for room temperature and 1.468 pF/cm for 70 K and 4 K data are calculated with maximum error bounds of ±0.1%.

III. Experimental Setup

The experimental setup (Fig. 1) included a high-frequency, manual cryogenic probe system PMC200, 40 GHz GSG |Z| Probes with 150 μm pitch, CSR-8 calibration substrate, SussCal® calibration software (all available from SUSS MicroTec), the Agilent 8722ES opt. 400 40 GHz vector network analyzer (VNA) and the NIST MultiCal® software package. To avoid additional uncertainty due to contact repeatability, all required data were acquired in one measurement series in raw-data format with the help of the calibration software SussCal® and saved externally for further analysis.

Three measurement experiments were performed: one each at room temperature, 70 K and 4 K. The system drift was defined for two extremely different conditions: experiments performed at room temperature and at 4 K and over a time period of 3.5 hours and 4 hours respectively (Fig. 2).

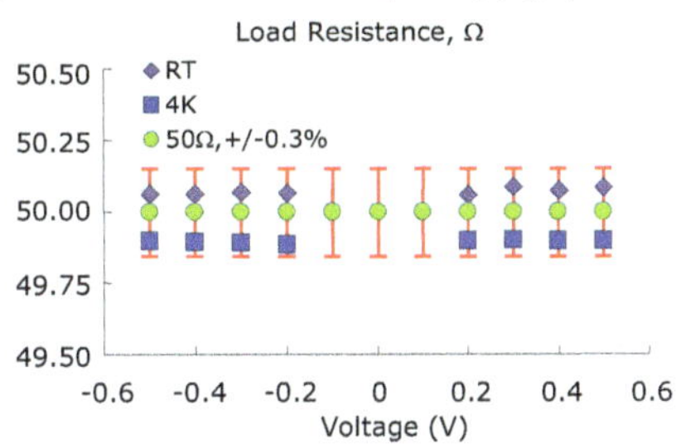

Fig. 4. Load resistance measured over applied voltage of -0.6...+0.6 V at room temperature (RT) and 4 K. The values within -0.1...+0.1 V does not provide acceptable measurement accuracy and are not shown in the graph.

It was found that, considering the 30-minutes difference in experiment time, the drift at 4 K is comparable with the drift at room temperature. It proves that the test system reached a stable condition and validates the measurement data acquired at cryogenic temperatures.

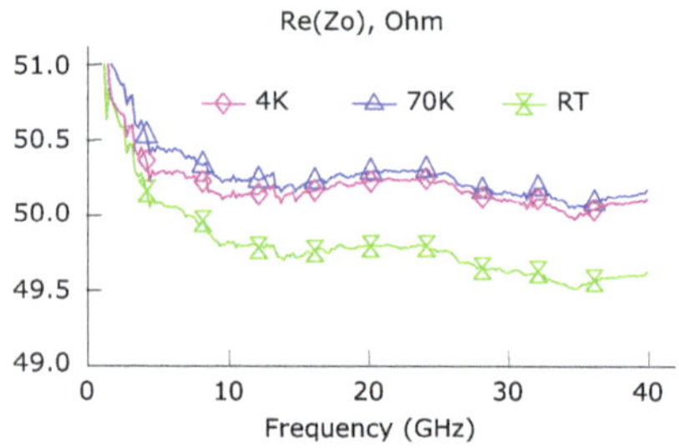

Fig. 5. The extraction results of the line characteristic impedance for the CSR-8 substrate at different temperatures. The extracted values are within the error bounds of ±1% over the range from room temperature down to 4 K.

IV. Results

The line capacitance of the tested CSR-8 alumina substrate was extracted at room temperature, 70 K and 4 K (Fig. 3). To guarantee accurate extraction results, the load resistance was measured at required temperatures. Fig. 4 shows the value of the load resistance provided by the four-terminal method and used for definition of the CSR-8 line capacitance per unit length. The capacitance values are: 1.498 pF/cm at room temperature and 1.468 pF/cm at 70 K and 4 K with maximum error bounds of ±0.1%. The line capacitance per unit length was also extracted at 70 K. No significant difference was observed between the results at 4 K and 70 K.

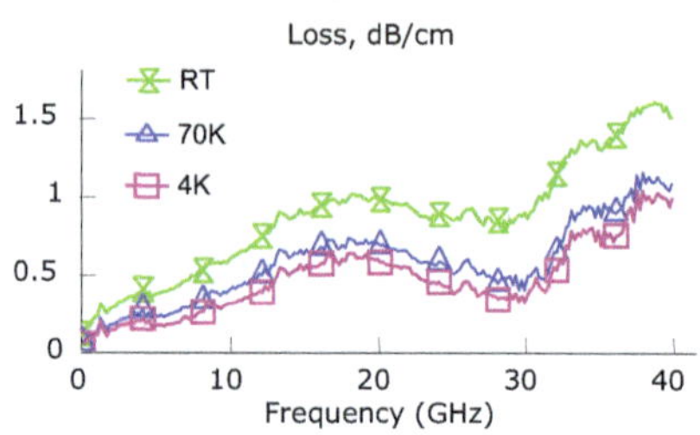

Fig. 6. The extraction results of the attenuation constant at different temperatures.

Next, the characteristic impedance (Fig. 5) was obtained from the measured propagation constant (Fig. 6 and 7). The temperature stability of the characteristic impedance is within ±1% over the range from room temperature down to 4 K. Additionally, the room temperature results demonstrate very good agreement with those extracted using a different setup [1].

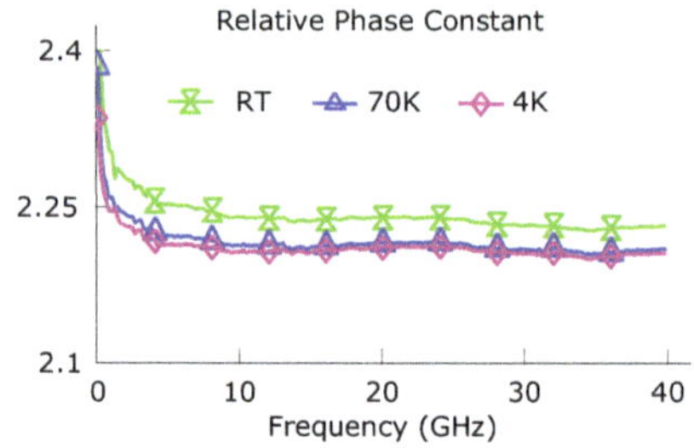

Fig. 7. The extraction results of the relative phase constant at different temperatures.

The temperature dependence of line attenuation and relative phase constants demonstrated expected behavior: decreasing loss and electrical lengths with decreasing temperature. If these effects are not considered when calibrating at different temperatures, they decrease calibration and, finally, measurement accuracy.

Finally, the accuracy of the lumped LRM+ calibration was examined and compared to the multiline TRL in different conditions: at room temperature and 4 K as well as with a full and simplified description of the thru standard. The full description considered the extracted characteristic impedance and propagation constant of the thru standard. The simplified description assumed that the thru standard was lossless, its characteristic impedance was 50 Ω, and the electrical length was known (e.g. 1.16 ps for the used CSR-8 calibration substrate).

Fig. 8 and 9 demonstrate the verification results of the LRM+ calibration with full and simplified characterization of the thru standard at RT and 4 K. The error due to the simplified thru description is negligible for 4 K and slightly decreases calibration accuracy at room temperature from approximately 35 GHz. This effect can be explained by the fact that the LRM+ calibration defines the system reference impedance from the impedance of the load standard. As demonstrated, the electrical characteristics of the used thin-film load element remain stable and ensure the quality of the cryogenic LRM+ calibration over the whole temperature range of interest.

The mismatch between the simplified model of the thru standard and its real characteristics can lead to the error in definition of the measurement reference plane. According to the experimental results, this error is marginal at cryogenic temperatures and at room temperature up to 40 GHz. Therefore, the simple model of the CSR-8 thru standard can be used at low-temperature calibration and will not lead to significant calibration error.

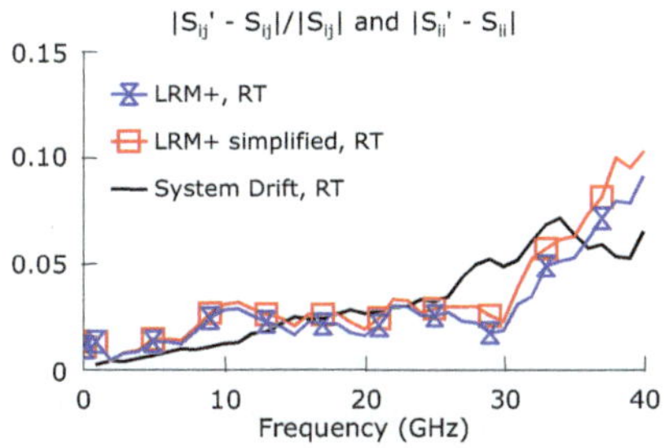

Fig. 8. Verification of the LRM+ calibration method at room temperature using a full and simplified thru description.

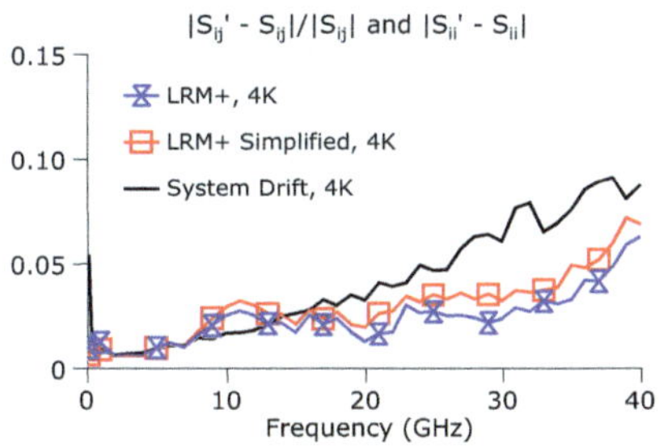

Fig. 9. Verification of the LRM+ calibration method at 4K using a full and simplified thru description.

V. Conclusion

Summarizing, the proven worst-case deviations of the measured S-parameters of passive devices at the temperature of 4 K for the examined LRM+ calibration with respect to the benchmark NIST multiline TRL are less than 0.1 up to 40 GHz and they are less then the measurement system drift within the experiment time. Therefore, the LRM+ method provides calibration accuracy comparable to the NIST reference multiline TRL for the extremely wide temperature range: from room temperature down to 4 K.

The demonstrated temperature stability of the thin-film resistor of the CSR-8 load standard is ±0.3% for the evaluated temperature range. The full characterization of the CSR-8 thru standard is not necessary at low-temperatures. Therefore no additional efforts in characterizing calibration standards are required for accurate LRM+ calibration of a cryogenic wafer-level measurement system.

References

[1] A. Rumiantsev, R. Doerner, S. Thies, "Calibration Standards Verification Procedure Using the Calibration Comparison Technique", *36th European Microwave Conference Digest*, September 2006.

[2] D. Williams, R. Marks, "Transmission Line Capacitance Measurement", *IEEE Microwave and Guided Wave Lett.*, vol. 1, pp. 243–245, September 1991.

[3] R. Marks, D. Williams, "Characteristic Impedance Determination Using Propagation Constant Measurements", *IEEE Microwave and Guided Wave Lett.*, vol. 1, pp. 141–143, June 1991.

[4] R. Marks, "A Multiline Method of Network Analyzer Calibration", *IEEE Trans. Microwave Theory Tech.*, vol. 39, no. 7, pp. 1205–1215, July 1991.

[5] D. Williams, R. Marks, A. Davidson, "Comparison of On-Wafer Calibrations", *38th ARFTG Conference Digest*, pp. 68–81, December 1991.

[6] R. Doerner, A. Rumiantsev, "Verification of the Wafer-Level LRM+ Calibration Technique for GaAs Applications up to 110 GHz", *65th ARFTG Conference Digest*, June 2005.

[7] "Proposed Procedures for Verifying Probe Station Integrity and On-Wafer Measurement Accuracy", *NIST/Industrial MMIC Consortium*, NIST.

[8] S. Delcourt, G. Dambrine, N.E. Bourzgui, S. Lepilliert, C. Laporte, J-P. Fraysse, M. Maignan, "A Non Uniform Thermal De-embedding Approch for Cryogenic On-Wafer High-Frequency Noise Measurements", *IEEE MTT-S Digest*, pp. 1809–1812, 2004.

[9] J. Laskar, J.J. Bautista, M. Nishimoto, M. Hamai, and R. Lai, "Development of Accurate On-wafer, Cryogenic Characterization Techniques", *IEEE Trans. Microwave Theory and Tech.*, vol. 44, no. 7, pp. 1178–1183, July 1996.

[10] M. Nishimoto, M. Hamai, and J. Laskar, "Study and Development of On-Wafer Cryogenic Calibration Techniques", *44th ARFTG Conference Digest*, December 1994.

[11] V.M. Hietala, M.S. Housel, R.B. Caldwell, "Network Analyzer Calibration for Cryogenic On-Wafer Measurements", *43rd ARFTG Conference Digest*, May 1994.

16.2 Paper [88]: "Verification of Wafer-Level Calibration Accuracy at High Temperatures"

A. Rumiantsev and R. Doerner, "Verification of wafer-level calibration accuracy at high temperatures," in *ARFTG Microwave Measurements Conference-Spring, 71st*, 2008, pp. 103–106.

This article presented the results of accuracy verification of wafer-level calibration at high temperatures based on coplanar calibration standards. The electrical characteristics of different commercially available coplanar calibration lines were extracted and compared at different temperatures. Finally, the accuracy of lumped calibrations at variable temperatures was verified by definition of the worst-case error bounds for the measurement of passive devices and compared to the reference NIST multiline TRL.

For this work, I developed the idea, the test and the verification methodology, planned the work, and wrote and presented the paper. Ralf Doerner from Ferdinand-Braun-Institut (FBH), Leibniz-Institut fuer Hoechstfrequenztechnik (Berlin, Germany) did all experimental measurements at the FBH facility. The calculation and the data analysis were shared.

This paper received the Best Interactive Forum Paper Award of the ARFTG-71st Spring Conference.

Best Interactive Forum Paper Award, ARFTG-71st

Verification of Wafer-Level Calibration Accuracy at High Temperatures

Andrej Rumiantsev #, Ralf Doerner *

SUSS MicroTec Test Systems GmbH, Suss-Str. 1, Sacka, D-01561, Germany, a.rumiantsev@ieee.org
* Ferdinand-Braun-Institut fuer Hoechstfrequenztechnik (FBH), Gustav-Kirchhoff-Str. 4, Berlin, D-12489, Germany, ralf.doerner@ieee.org

***Abstract* — This article presents the results of accuracy verification of wafer-level calibration at high temperatures based on coplanar calibration standards. The electrical characteristics of different commercially available coplanar calibration lines were extracted and compared at different temperatures. Finally, the accuracy of lumped calibrations at variable temperatures was verified by definition of the worst-case error bounds for the measurement of passive devices and compared to the reference NIST multiline TRL.**

***Index Terms* — calibration, error correction, calibration comparison, scattering parameters measurement.**

I. Introduction

Accurate calibration is crucial for device measurement, characterization and modeling. Fabrication uncertainty, temperature instability, and mismatch in modeling of calibration standards significantly reduce the final calibration accuracy. Difficulties in the traceability of wafer-level (planar) calibration standards challenge alternative methods of their verification [1, 2]. These challenges increase if standards operate in extreme conditions such as very low or high temperatures.

Previous research (i.e. [3, 4]) has already addressed the subject of high-temperature calibration. Investigations focused on the temperature stability of the thin-film resistors (used as the load calibration standard) [3] or on the compensation for the temperature drift of the measurement system [4]. The temperature influence on the electrical characteristics of calibration thru and line standards was, however, not considered. Recently, as demonstrated in [1], even small variations of the electrical parameters of thru and line standards can lead to significant calibration and thus measurement errors.

II. Verification Method

The research undertaken by the National Institute of Standards and Technology (NIST) provided a technique for characterizing coplanar lines [5], [6] accurate system calibration [7] and accuracy verification of different calibration methods [8]. This technique is well accepted in the engineering practice and can be successfully used for calibration verification purposes [9, 10]. On the other hand, traceable verification results can be achieved only at room temperature and using the reference material RM8130: the NIST-fabricated and characterized GaAs calibration elements.

This paper demonstrates the results when extracting electrical characteristics (attenuation and relative phase constants) of commercially available alumina coplanar lines and the accuracy verification of different calibration procedures at temperatures up to 150°C using the method reported in [6].

To define the time and temperature drift of the high-temperature wafer-level RF measurement system during the experiment, the calibration comparison technique was used [8]. This technique provides the worst-case deviations of the measured S-parameters of passive devices for the examined (first-tier) calibration with respect to the benchmark (second-tier) calibration. Deviations are treated as $|S_{ij}-S_{ij}'|$, for $ij\in\{11, 12, 21, 22\}$, where S_{ij}' is the S-parameter measured by the calibration to be tested, and S_{ij} is the S-parameter measured by the benchmark calibration. Performing the benchmark calibration at the beginning and the examined calibration at the end of the experiment, the system drift can be quantitatively defined. In the ideal case, both calibrations benchmark and examined, are equal and the error bounds $|S_{ij}-S_{ij}'|$ are zero.

To characterize calibration lines, the procedure proposed in [6] was used: the resistance R of the load standard can be measured using the 4-terminal method. Then, the first-tier multiline thru-reflect-line (TRL) [7] can be performed, setting the calibration reference impedance to the characteristic impedance Z_0 of the alumina lines used. The frequency range where the tested lines are no longer dispersive and, on the other hand, the load reactance is negligible can be used to extract the value of the line capacitance C. Once the line capacitance C is found, the characteristic impedance, the attenuation and relative phase constants can be extracted with the help of the MultiCal® software package.

Knowing the electrical parameters of the calibration thru and line standards, it is possible to set the position of the measurement reference plane and the measured reference impedance accurately with the help of the multiline TRL. Therefore, the benchmark calibration can be defined and verification of accuracy of the tested lumped calibration can be performed at different temperatures [2].

III. Experimental Results

The experimental setup included a thermal probe system, an Agilent 8510C 50 GHz vector network analyzer (VNA), 50GHz GSG |**Z**| Probes with 150 µm pitch from SUSS MicroTec and 50 GHz GSG 100 µm pitch probe from another vendor (probe 'P'), and SussCal® and MultiCal® software packages. Commercially available alumina calibration substrates CSR-8 from SUSS MicroTec and substrate 'I' from another vendor were examined.

TABLE I
Measured Load Resistance

Temperature, °C	R, Ω (CSR-8)	R, Ω (Sub. 'I')
RT	49.99	49.96
50	49.99	50.02
100	50.05	50.29
150	50.09	50.54

To avoid additional uncertainty due to contact repeatability, all data were acquired in one measurement series in raw format with the help of the calibration software SussCal® and saved externally for the further analysis.

The experimental data included measurement results of all calibration standards that are required for the lumped (i.e. SOLT[1] or LRM+[2]) as well as for the multiline TRL calibrations. The measurement experiments were repeated at four different temperatures: at room temperature (RT), 50°C, 100°C, and 150°C for both calibration substrates and with the use of two different wafer probes (|**Z**| Probe and probe 'P'). The experiment took approximately four hours for the entire measurement cycle.

First, the drift of the experimental setup was determined using the procedure [10].

Next, the temperature stability of the load standard resistance was measured. As it was discussed in [2], the precise value of the load resistance is required for the accurate characterization of the line standard as well as for the lumped calibration at different temperatures. It was found that the temperature stability of the CSR-8 load resistance is better than 0.18%. The load of the substrate 'I' is less stable and showed a variation in its resistance within 1.08% (see Table I, Fig. 1).

The temperature dependence of the attenuation and relative phase constants of the line standards were extracted for each calibration substrate using the method [6]. Fig. 2 shows the loss and the relative phase constant of the CSR-8 line standard extracted at room temperature and 150°C. As it was expected, these parameters did not depend on the type of the wafer probe used. Fig. 3 and Fig. 4 demonstrate results for the whole temperature range for CSR-8 and substrate 'I' respectively.

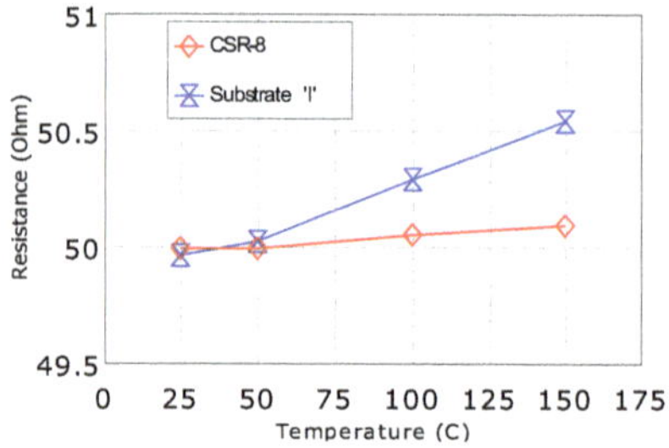

Fig. 1. Temperature stability of the load resistance of the CSR-8 and substrate 'I'. The load of the CSR-8 demonstrated stability better than 0.18% compared to 1.08% for the load of substrate 'I'.

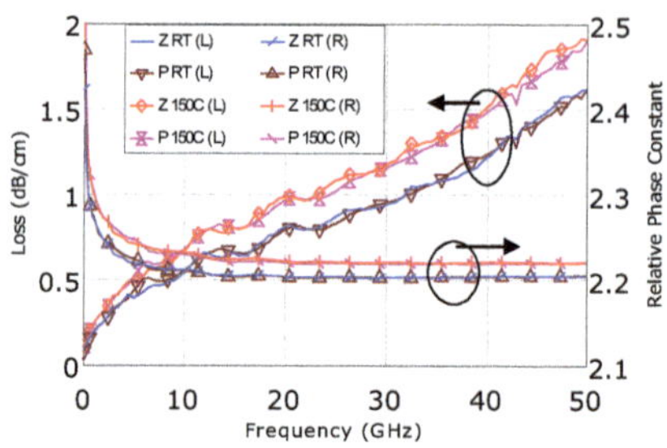

Fig. 2. Temperature stability of the line loss and relative phase constant of the CSR-8 calibration substrate measured with |**Z**| Probes and alternative probes 'P'. Comparison results are presented at room temperate (RT) and 150°C.

Observed changes in the line's loss and phase constant can lead to significant calibration and finally measurement errors of the lumped calibration if no additional correction procedure is applied.

To evaluate the worst-case influence of the line thermal instability on the lumped calibration, the following procedure was used. The equivalent model of the load on substrate "I" was extracted with respect to the multiline TRL calibration. The load series inductance was L=7.8 pH.

[1] Short-Open-Load-Thru

[2] Line-Reflect-Match, advanced

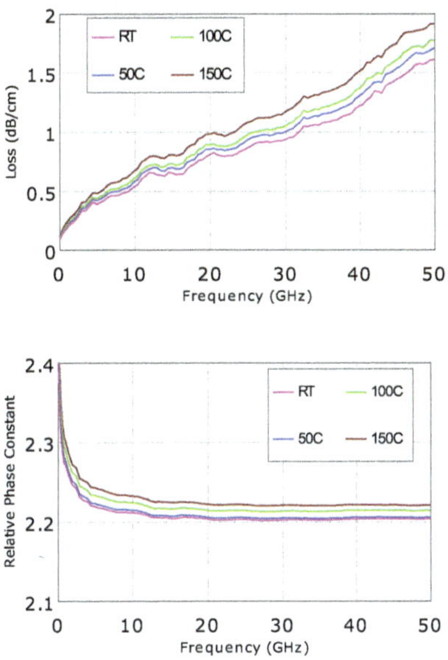

Fig. 3. Temperature stability of the line loss and relative phase constant of the CSR-8 calibration substrate as measured with |**Z**| Probes.

Next, the examined ("working") lumped LRM+ calibration was performed for the data series measured at room temperature and 150°C for both calibration substrates. For this calculation run, the thru standard was assumed to be lossless and perfectly matched. This is a conventional assumption used very often in practice. The lumped calibration accuracy was compared to the benchmark multiline TRL [2, 10]. The measurement reference plane was defined at the middle of the thru standard.

Finally, the temperature variation of the attenuation and the relative phase constants was included in the line model and the same calculations were repeated. Fig. 5 shows selected results from the experiment.

The worst-case 150°C experiment proved that the error of the high-temperature lumped calibration can be significantly decreased by improving the model of the used line standard and by taking its temperature instability into account. Therefore, the error in the calibration at 150°C was able to be corrected to that of the calibration at room temperature. In addition, it was observed that the results obtained from the CSR substrate were almost matched to the system drift. This was, however, not the case for the alternative substrate. Other factors must be taken into account in order to correct for this difference. One of these factors could be the difference in the line's characteristic impedance, which was not considered in this experiment.

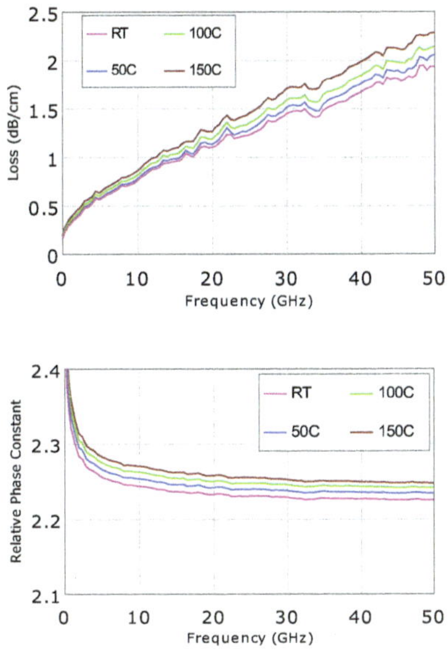

Fig. 4. Temperature stability of the line loss and relative phase constant of the alternative calibration substrate 'I' measured with |**Z**| Probes.

IV. Conclusion

Summarizing, for the first time a quantitative verification of the calibration and measurement accuracy at high temperatures was done. It was demonstrated that the temperature changes in the measurement setup leads to an increased measurement error. Obviously, the system should be re-calibrated at each temperature in order to achieve high measurement accuracy.

A practical method for characterizing the coplanar calibration thru and line standards at high temperatures was presented and verified for different commercially available alumina substrates up to 150°C. Temperature dependent line, attenuation and relative phase constants were extracted for each substrate. Comparison of the lumped and distributed calibration was done with and without compensation for the temperature variation of calibration standards.

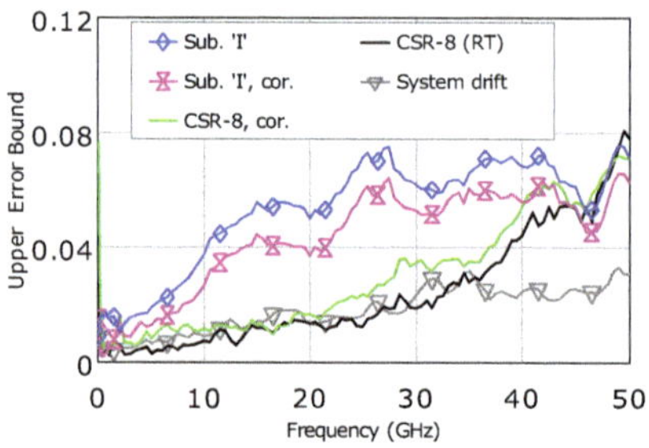

Fig. 5. Accuracy verification of the 150°C lumped calibration for both substrates "I" and CSR-8 with (red and green) and without (blue) correction for line temperature instability. The accuracy of room temperature calibration (black) on the CSR-8 is also shown.

REFERENCES

[1] A. Rumiantsev, R. Doerner, and S. Thies, "Calibration standards verification procedure using the calibration comparison technique," in *36th European Microwave Conference*, 2006, pp. 489-491.

[2] A. Rumiantsev, R. Doerner, and P. Sakalas, "Verification of wafer-level calibration accuracy at cryogenic temperatures," in *ARFTG Microwave Measurements Conference-Fall, 68th*, 2006.

[3] J. E. Pence and R. Anholt, "Calibration and measurement considerations for deriving accurate temperature dependent equivalent circuits," in *ARFTG Microwave Measurements Conference-Spring, 41st*, 1993, pp. 85-92.

[4] Z. D. Schwartz, A. N. Downey, S. A. Alterovitz, and G. E. Ponchak, "High-temperature RF probe station for device characterization through 500/spl deg/C and 50 GHz," *IEEE Transactions on Instrumentation and Measurement*, vol. 54, pp. 369-376, 2005.

[5] D. F. Williams and R. B. Marks, "Transmission line capacitance measurement," *IEEE Microwave and Guided Wave Letters*, vol. 1, pp. 243-245, 1991.

[6] R. B. Marks and D. F. Williams, "Characteristic Impedance Determination using Propagation Constant Measurement," *IEEE Microwave and Guided Wave Letters*, vol. 1, pp. 141-143, June 1991.

[7] R. B. Marks, "A multiline method of network analyzer calibration," *Microwave Theory and Techniques, IEEE Transactions on*, vol. 39, pp. 1205-1215, 1991.

[8] D. F. Williams, R. B. Marks, and A. Davidson, "Comparison of on-wafer calibrations," in *ARFTG Microwave Measurements Conference-Fall, 38th*, 1991, pp. 68-81.

[9] R. Doerner and A. Rumiantsev, "Verification of the wafer-level LRM+ calibration technique for GaAs applications up to 110 GHz," in *ARFTG Microwave Measurements Conference-Spring, 65th*, 2005.

[10] "Proposed procedures for verifying probe station integrity and on-wafer measurement accuracy," NIST/Industrial MMIC Consortium.

16.3 Paper [89]: "Sensitivity Analysis of Wafer-Level Over-Temperature RF Calibration"

A. Rumiantsev, G. Fisher, and R. Doerner, "Sensitivity analysis of wafer-level over-temperature RF calibration," in *Microwave Measurement Symposium (ARFTG), 80th*, San Diego, CA, USA, 2012, pp. 1–3.

This paper analyzed the commonly used wafer-level S-parameter calibration methods LRRM, multiline TRL and LRM+ for the sensitivity to the thermal variation of electrical characteristics of planar calibration standards. It was demonstrated that the calibration error of lumped-standard based methods could be significantly reduced by taking into account the variation of load standard resistance over the temperature. The obtained results proved that for the evaluated commercially available calibration standards and for a given frequency range, the overall calibration error due to the temperature variation was in order of magnitude of repeatability of the manual system calibration. The proposed method can be successfully applied for different calibration substrates, temperature and frequency ranges, as well as to the *in-situ* calibration element.

For this paper, I defined the scope of the work, develop the concept of the investigation, did several calculations, performed the data analysis, and wrote the major part of the paper. Ralf Doerner from Ferdinand-Braun-Institut (FBH), Leibniz-Institut fuer Hoechstfrequenztechnik (Berlin, Germany FBH, Berlin) acquired all measurement results at the FBH facility, did the major part of calculation as well as data analysis. Gavin Fisher from Cascade Microtech (Beaverton, OR, USA) contributed to some calculation tasks as well as to the preparing the final manuscript.

Sensitivity Analysis of Wafer-Level Over-Temperature RF Calibration

A. Rumiantsev[1], G. Fisher[1] and R. Doerner[2]

[1]Cascade Microtech, Inc., Beaverton, OR, USA

[2]Ferdinand-Braun-Institut, Leibniz-Institut fuer Hoechstfrequenztechnik, Berlin, Germany

Abstract — **This paper analyzes the commonly used wafer-level S-parameter calibration methods LRRM, multiline TRL and LRM+ for the sensitivity to the thermal variation of electrical characteristics of planar calibration standards. We demonstrate that the calibration error of lumped-standard based methods can be significantly reduced by taking into account the variation of Load standard resistance over the temperature. The obtained results proved that for the evaluated commercially available calibration standards and for a given frequency range, the overall calibration error due to the temperature variation is in order of magnitude of repeatability of the manual system calibration. The proposed method can be successfully applied for different calibration substrates, temperature and frequency ranges, as well as to the *in-situ* calibration element.**

Index Terms — **S-parameter calibration, device characterization, mm-wave measurements, on-wafer calibration.**

I. INTRODUCTION

Continually increasing demand for more data, more content and high speed pushes operation frequencies of modern RF circuits towards mm-wave and sub-THz ranges. Circuit designers challenge modeling engineers for more accurate device models, optimized and verified at extremely wide frequency range, across multiple temperatures and with the highest level of accuracy. Development of accurate and consistent wafer-level S-parameter calibration methods covering wide frequency and temperature ranges becomes the critical success factor.

While some work has been already done in this area (e.g. [1]), to the best of author's knowledge, the quantitative analysis of contribution of a selected calibration standard to the overall accuracy of over-temperature S-parameter calibration methods is not yet available. Practical recommendations on how to address the variation of electrical properties of standards across wide temperature range are still missing.

In this work, we evaluated the sensitivity of the most popular wafer-level self-calibration procedures, such as multiline TRL[1] [2], enhanced LRRM[2] [3] and LRM+[3] [4] to the variation of electrical characteristics of coplanar calibration standards due to the temperature change. The objective of this work was to identify the method that provides the smallest calibration residual errors for a given set of conditions.

II. ANALYSIS METHOD

Multiline TRL, eLRRM and LRM+ calibration methods differently define calibration reference impedance Z_{REF}. For the multiline TRL, the reference impedance is set to the characteristic impedance Z_0 of the Line standard. Assuming that the dielectric loss of the evaluated alumina calibration substrate is negligible in the frequency range of interest, we calculated Z_0 from the measured capacitance per unit length C' and propagation constant γ, as proposed in [5, 6]. Once Z_0 is known, the calibration reference impedance can be transformed to the desired value of $Z_{REF} = 50\ \Omega$.

The Z_{REF} for eLRRM and LRM+ depends on the impedance of the Load Z_{LOAD} as well as Z_0 and γ of the Thru. Therefore, multiline TRL, eLRRM and LRM+ may show different calibration residual error across the temperature range.

We chose the maximum error bounds for measured S-parameters of a passive device as the figure of merit (FoM) of the calibration accuracy, calculated by the comparison technique [7]. As demonstrated in several previous works (e.g. [8, 9]), the resistance of a coplanar Load standard R_{LOAD} as well as the propagation constant γ are affected by the variation of the Impedance Standard Substrate (ISS) temperature. As a result, the accuracy of the probe tip calibration varies across the temperature range.

To quantify this variation, we measured the resistance of the Load standard R_{LOAD} and extracted the propagation constant γ as well as the capacitance per unit length C' of the Line at each temperature point, including room temperature.

The attenuation coefficient α_{REF} used for the Thru model in lumped-standard based calibration methods, such as eLRRM and LRM+ was extracted from propagation constant:

$$\alpha(f) = \alpha_{REF} \frac{\tau}{\tau_{REF}} \sqrt{\frac{f}{f_{REF}}} \qquad (1)$$

where: α_{REF} is the attenuation of the reference element, dB extracted at the reference frequency f_{REF}, Hz; τ is the electrical delay of the Thru standard, sec; τ_{REF} is the delay of the reference element, sec; f is the measured frequency, Hz.

---

[1] Thru-Reflect-Line

[2] Line-Reflect-Reflect-Match

[3] Line-Reflect-Match, advanced

978-1-4673-4818-8/12/$31.00 ©2012 IEEE

For instance, the reference element for the evaluated ISS model 104-783A is the 27 pico-second long Line.

To avoid additional uncertainty due to contact repeatability, all data were acquired in one measurement series and saved for the further analysis.

Next, different sets of error terms were calculated for $R_{LOAD}(T)$, $\alpha_{REF}(T)$, $Z_0(T)$ (for eLRRM and LRM+) and $C'(T)$ (for multiline TRL) for measurement data taken at a certain temperature as:

1) *Benchmark.* T is the measurement temperature;
2) *Worst case.* $T = 25°C$ (room temperature).

Further, we calculated error sets for configurations where only a single standard parameter was corrected for the temperature impact. This should show the proportional impact of the selected parameter on the overall calibration accuracy.

III. EXPERIMENTAL RESULTS

The experimental setup included a manual thermal probe system from Cascade Microtech, and Agilent 8510C 50 GHz vector network analyzer (VNA). WinCal XE calibration software from Cascade Microtech was used for measuring S-parameters of standards, calibration and analysis.

Extracted parameters of calibration standards are summarized in Table 1. Fig 1 shows the attenuation constant α extracted from the multiline TRL as well as its model approximated by (1). α_{REF} increases with the temperature by 0.1 dB (or 18%) at $T = 150°C$ compared to nominal (room temperature) value.

C' also increases by 0.025 pF/cm resulting in a decrease of Z_0 by 0.38 Ω (or 0.8%) and 0.07 Ω (or 15%) for its real and imaginary parts respectively. Fig. 2 shows Z_0 calculated from a known C' and γ.

TABLE I
STANDARD PARAMETERS VARIATION OVER TEMPERATURE

Parameter	Temperature, C			
	+25	+50	+100	+150
α_{REF} @ 40 GHz, dB	0.56	0.59	0.62	0.66
C', pF/cm	1.560	1.565	1.575	1.585
$\Re(Z_0)$ @ 40 GHz, Ω	47.71	47.66	47.47	47.33
$\Im(Z_0)$ @ 40 GHz, Ω	-0.46	-0.47	-0.50	-0.53
R_{LOAD}, Ω	49.94	50.07	50.32	50.55

Once $Z_0(T)$, $\alpha_{REF}(T)$, and $R_{LOAD}(T)$ are defined, the impact of each of them on the calibration accuracy can be calculated.

The maximum error bounds for the multiline TRL, eLRRM and LRM+ were calculated at three temperature points: $T = 50°C$, $T = 100°C$, and $T = 150°C$ (Fig. 3). The maximum error of 2.4% at 50 GHz corresponds to the worst-case multiline TRL for $C'(25°C)$ and $T = 150°C$. Both eLRRM and LRM+ have comparable errors of about 2.2% for the worst-case scenario of $T = 150°C$, $R_{LOAD}(25°C)$, $Z_0(25°C)$, $\alpha_{REF}(25°C)$.

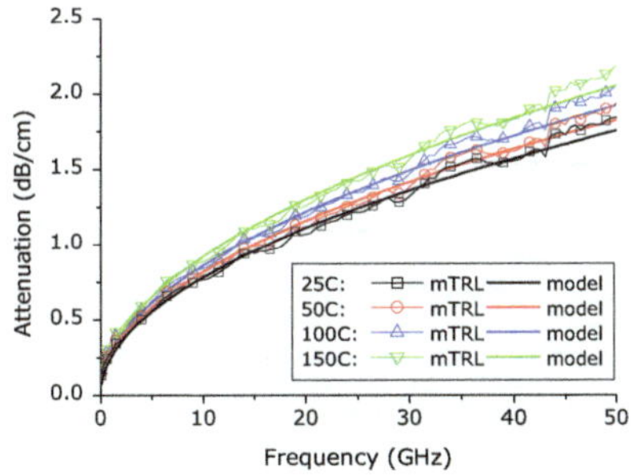

Fig. 1. Attenuation constant of the CPW Line standard measured across the temperature and approximated by (1).

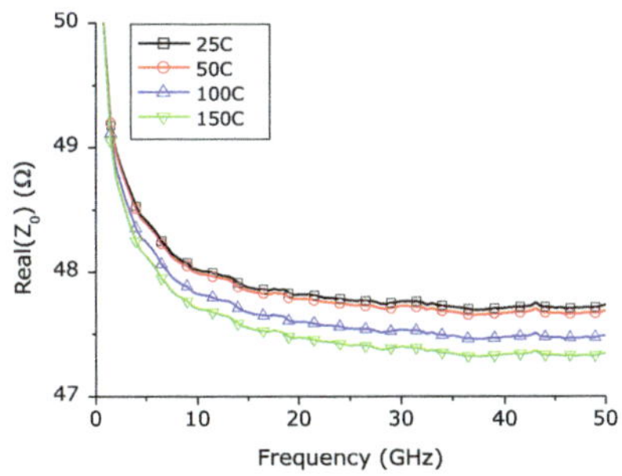

Fig. 2. Real part of the characteristic impedance Z_0 of the CPW Line standard measured across the temperature.

Last, we calculated the impact of each standard parameter on the accuracy of eLRRM (Fig. 4). Results showed that the resistance of the Load R_{LOAD} is the most influencing factor for the evaluated ISS. Thus, addressing its temperature variation reduced overall calibration error from 2.2% down to negligible value of 0.6%. Correction for α_{REF} and Z_0 had marginal effect.

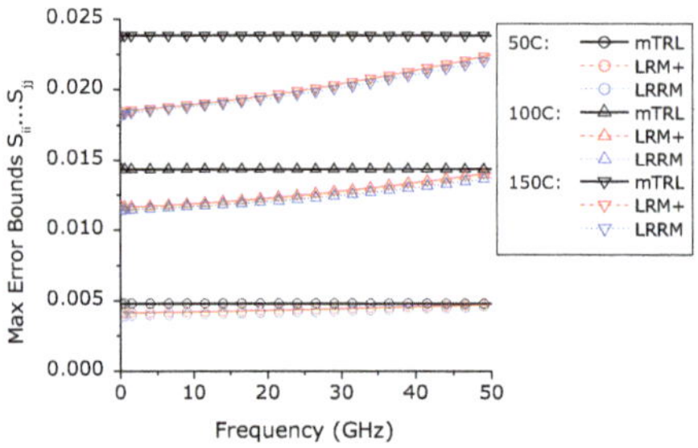

Fig. 3. Maximum error bounds calculated for multiline TRL, eLRRM+ and LRM+ for $T_1 = 50°C$, $T_2 = 100°C$, and $T_3 = 150°C$ with benchmark and worst case definition of standard properties.

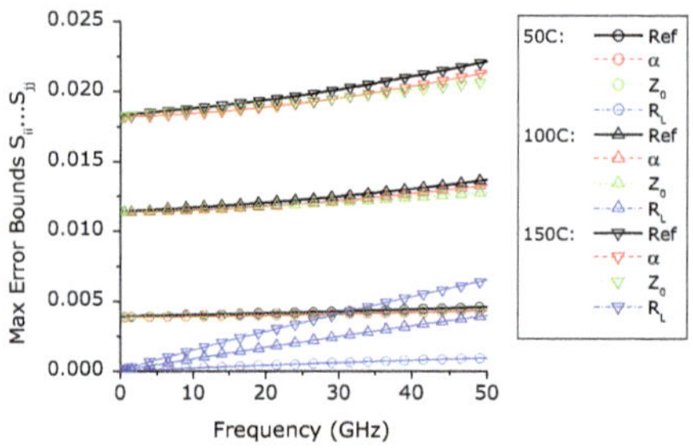

Fig. 4. Maximum error bounds for eLRRM calibration procedure for $T_1 = 50°C$, $T_2 = 100°C$, and $T_3 = 150°C$ calculated individually for the benchmark definitions of the Load resistance R_{LOAD}, the loss α_{REF} or the characteristic impedance Z_0.

IV. Conclusion

Sensitivity analysis of the wafer-level over-temperature calibration was performed for commonly-used methods: multiline TRL, eLRRM, and LRM+. This investigation was focused on a commercially available alumina Impedance Standard Substrate.

The obtained results demonstrated that temperature variation of the electrical characteristics of Load and Thru standards leads to different calibration residual errors depending on the calibration method used.

For the considered experimental setup, we found that the lumped-standard based calibration methods were less sensitive to temperature. In addition, the Load resistance R_{LOAD} is the main influencing factor for both eLRRM and LRM+ methods. Therefore, a simple correction for $R_{LOAD}(T)$ can significantly improve calibration accuracy.

The proposed method can be successfully applied for other types of calibration substrates, temperature and frequency ranges, as well as to the *in-situ* (on-wafer) calibration elements.

References

[1] L. Pattison, N. Buchanan, and D. Linton, "New techniques for on-wafer calibration over temperature," in *European Microwave Conference, 29th*, 1999, pp. 236-239.

[2] R. B. Marks, "A multiline method of network analyzer calibration," *Microwave Theory and Techniques, IEEE Transactions on,* vol. 39, pp. 1205-1215, 1991.

[3] L. Hayden, "An enhanced Line-Reflect-Reflect-Match calibration," in *ARFTG Microwave Measurements Conference-Spring, 67th*, 2006, pp. 143-149.

[4] R. Doerner and A. Rumiantsev, "Verification of the wafer-level LRM+ calibration technique for GaAs applications up to 110 GHz," in *ARFTG Microwave Measurements Conference-Spring, 65th*, 2005, pp. 15-19.

[5] R. B. Marks and D. F. Williams, "Characteristic impedance determination using propagation constant measurement," *IEEE Microwave and Guided Wave Letters,* vol. 1, pp. 141-143, June 1991.

[6] D. F. Williams and R. B. Marks, "Transmission line capacitance measurement," *Microwave and Guided Wave Letters, IEEE,* vol. 1, pp. 243-245, 1991.

[7] D. F. Williams, R. B. Marks, and A. Davidson, "Comparison of on-wafer calibrations," in *ARFTG Microwave Measurements Conference-Fall, 38th*, 1991, pp. 68-81.

[8] A. Rumiantsev, R. Doerner, and P. Sakalas, "Verification of wafer-level calibration accuracy at cryogenic temperatures," in *ARFTG Microwave Measurements Conference-Fall, 68th*, 2006, pp. 134-140.

[9] A. Rumiantsev and R. Doerner, "Verification of wafer-level calibration accuracy at high temperatures," in *ARFTG Microwave Measurements Conference-Spring, 71st*, 2008, pp. 103-106.

16.4 Paper [92]: "Impact of in-situ TRL Reference Impedance Determination on Parameter Extraction"

A. Rumiantsev, R. Doerner, and F. Lenk, "Impact of in-situ TRL reference impedance determination on parameter extraction," in *European Microwave Conference, 42nd*, Amsterdam, NL, 2012, pp. 593–596.

This paper investigated the impact of possible parameter extraction errors caused by inaccurate definition of the calibration reference impedance of *in-situ* multiline TRL. Two calibration sets implemented on GaAs and Si/SiGe:C wafer processes were quantitatively analyzed. Obtained results demonstrated that for most practical cases, the desired 5%-level of confidence of extracted parameters of high-reflective devices can easily be achieved without additional efforts. Thus, implementation of the *in-situ* TRL into a characterization workflow of high-performance microwave devices can be significantly simplified.

For this paper, I developed the analysis methodology, planned the work, acquired data for the BiCMOS process, did a part of calculations and parameter extraction, and wrote the major part of the paper. I performed the data analysis together with Ralf Doerner from Ferdinand-Braun-Institut (FBH), Leibniz-Institut fuer Hoechstfrequenztechnik (Berlin, Germany). Ralf Doerner acquired data for GaAs process and did the major part of calculations. The access to the GaAs test structures was provided by Friedrich Lenk from Hochschule Lausitz (FH) (Senftenberg, Germany).

Proceedings of the 42nd European Microwave Conference

Impact of *in-situ* TRL Reference Impedance Determination on Parameter Extraction

Andrej Rumiantsev
Cascade Microtech GmbH
Sacka, Germany
a.rumiantsev@ieee.org

Ralf Doerner
Ferdinand-Braun-Institut (FBH)
Berlin, Germany
ralf.doerner@ieee.org

Friedrich Lenk
Hochschule Lausitz (FH)
Senftenberg, Germany
friedrich.lenk@hs-lausitz.de

***Abstract*—This paper investigates the impact of possible parameter extraction errors caused by inaccurate definition of the calibration reference impedance of *in-situ* multiline TRL. Two calibration sets implemented on GaAs and Si/SiGe:C wafer processes were quantitatively analyzed. Obtained results demonstrated that for most practical cases, the desired 5%-level of confidence of extracted parameters of high-reflective devices can easily be achieved without additional efforts. Thus, implementation of the *in-situ* TRL into a characterization workflow of high-performance microwave devices can be significantly simplified.**

Keywords*—component; S-parameter; calibration; on-wafer measurements; HBT

I. INTRODUCTION

The continuous increase of operation frequencies of advanced microwave devices (e.g. [1, 2]) causes serious device characterization challenges. Wafer-level measurement systems must provide more accurate, reliable and repeatable calibration and measurement results at mm and sub-mm frequencies.

Wafer-embedded (*in-situ*) calibration techniques have already demonstrated substantial advantages over conventional probe pad de-embedding at mm-wave frequencies [3-6]. Implementing customized standards on-wafer allows for capturing the major part of device backend parasitics and for moving the measurement reference plane close to the DUT terminals in one step [7]. Therefore, with increasing of measurement frequency, *in-situ* calibration becomes an effective tool.

However, its practical realization suffers from some difficulties:

- A certain number of electrical parameters of calibration standards must be known, such as characteristic impedance Z_{LINE} of the Line or the impedance Z_{LOAD} of the Load [3].
- Process instability and fabrication tolerances affect the accuracy of standards. Therefore, each calibration chip should be characterized before use.
- Conventional S-parameter measurement systems are not necessarily equipped with instruments that may be required for accurate characterization of customized standards (e.g. precision impedance analyzers).

That is why *in-situ* calibration is still not widely implemented into a device characterization workflow.

This work investigates the impact of characteristic impedance Z_{LINE} of *in-situ* multiline TRL[1] [8] on parameter extraction accuracy of passive and active devices. The test chips were implemented in a passive GaAs process from FBH and in advanced Si/SiGe:C ST Microelectronics' BiCMOS9MW process technologies [1]. Z_{LINE} was characterized over measured capacitance per unit length C' and propagation constant γ of the Line. Obtained results demonstrate that, in most cases, the desired 5%-level of confidence in parameter extraction of passive and active devices can be easily achieved on conventional S-parameter characterization setup without additional efforts.

II. DESIGN AND IMPLEMENTATION

A. GaAs Calibration Standards

The achievements made over the past years in fabricating and characterizing custom calibration standards on semi-insulating and conductive wafer processes established a solid background for accurate *in-situ* calibration [7, 9].

All GaAs calibration standards and DUTs utilize the same ground-signal-ground (GSG) RF contact pad design. They are realized on semi-isolated 625 μm GaAs substrate with 3 μm electroplated Au metallization as top metal layer in coplanar

[1] Thru-Reflect-Line

configuration. The process also includes NiCr thin-film resistors and air-bridges.

The contact pad is realized by a 50 Ω coplanar line with a center and slot width of 20 µm and 15 µm, respectively. The half-length of the Thru offset l_{OFF} of 100 µm (Fig. 1, a) is used for Short, Open, and Load. The offset ensures a sufficient separation of the probes and a well-guided coplanar mode. A nonius is added to every contact pad for precise definition of the probe overlap of 25 µm and for repeatable probe placement. The Line length varies from 480 µm to 9400 µm to cover the frequency range from 500 MHz to 110 GHz.

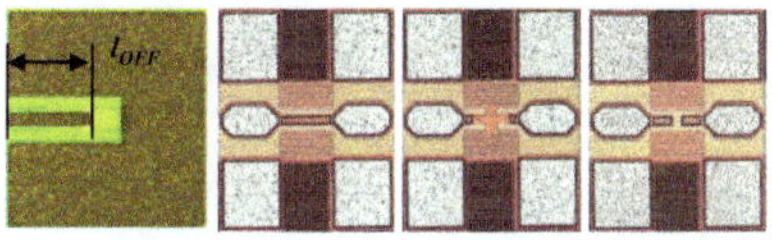

a) b) c) d)

Figure 1. The offset Load implemented on GaAs test chip (a). Thru, Short and Open implemented on the SiGe test chip (b)-(d).

B. SiGe Calibration Standards

SiGe standards and DUT utilize the same GSG RF contact pad configuration as the GaAs test chip. Metal 1 (M1) ground shield eliminates the losses in a resistive substrate and makes the parasitics of the pads purely capacitive. The signal pads consist only of the upper Metal 6 (M6) minimizing the parasitic pad capacitances to the ground. The pad size is reduced to the smallest possible dimension.

The calibration standards are designed as a M1-shielded 50 Ω top metal grounded coplanar waveguide. The effective length was optimized to cover the frequency range from 1 GHz to 110 GHz and yielded 51, 549, 1545 and 3039 µm for the Thru and Lines, respectively. The symmetrical Short element is located at M6. The Open and the Thru have the same offset (Fig. 1).

C. Calibration Reference Impedance

As shown in [10], multiline TRL sets the calibration reference impedance Z_{REF} to the characteristic impedance of the Line Z_{LINE}. Because planar waveguides are dispersive and affected by fabrication inaccuracies, it is required to measure Z_{LINE} accurately and to transfer the TRL calibration reference impedance to the desired value of $Z_{REF} = 50\,\Omega$. The simple "lumped load" method of Z_{LINE} measurement over the line capacitance per unit length C' for semi-insulating wafers was introduced in [11]. For proper designed standards, it can also be applied for conductive wafers, as shown in [12]. Thus, C' can be extracted from:

$$C' \approx \Re\left(\frac{\gamma}{j\omega R_{LOAD}} \frac{1+\Gamma_{LOAD,Z_{LINE}}}{1-\Gamma_{LOAD,Z_{LINE}}}\right) \tag{1}$$

where γ is the propagation constant of the Line. It can be directly extracted from the TRL algorithm. $\Gamma_{LOAD,Z_{LINE}}$ is the reflection coefficient of the Load, measured with respect to the calibration reference impedance Z_{LINE}. R_{LOAD} is Load resistance.

From:

$$\frac{\gamma}{Z_{LINE}} = j\omega C' + G' \tag{2}$$

and, assuming that $G/\omega C$ is negligible:

$$Z_{LINE} \approx \frac{\gamma}{j\omega C'} \tag{3}$$

Once Z_{LINE} is defined, the measurement results can be transformed to any desired reference impedance Z^{n}_{ref} (which is, in general, complex) over T-parameters [10] by:

$$T^{nn} = Q^{nm} T^{mm} Q^{mn} = \frac{1}{1-\Gamma^2_{nm}} \times \\ \times \begin{pmatrix} 1 & \Gamma_{nm} \\ \Gamma_{nm} & 1 \end{pmatrix} T^{mm} \begin{pmatrix} 1 & -\Gamma_{nm} \\ -\Gamma_{nm} & 1 \end{pmatrix}, \tag{4}$$

where T^{mm} and T^{nn} are T-parameter matrices of the DUT for reference impedances Z^{m}_{ref} (source) and Z^{n}_{ref} (target), respectively. Γ_{nm} is defined as:

$$\Gamma_{nm} \equiv \frac{Z^{m}_{ref} - Z^{n}_{ref}}{Z^{m}_{ref} + Z^{n}_{ref}}. \tag{5}$$

III. Experimental Results

A. Measurement Setup

The experimental measurements were carried out on a broadband S-parameter measurement system from Cascade Microtech, consisting of a manual PM8 probe station, 100 µm pitch Infinity probes model SP-i110-A-GSG-03, and equipped with the 110 GHz PNA network analyzer from Agilent Technologies and the source monitor unit (SMU). The measured data are acquired in raw format (with the S-parameter calibration turned off). Calibration and error correction were performed for the same data set outside of the VNA on a computer. We used WinCal XE®[2], MultiCal®[3], and proprietary MATLAB®[4] script for this purpose, while device parameters were extracted using MATLAB.

B. Measurement of Line Characteristic Impedance

The multiline TRL was performed on both SiGe and GaAs test chips for $Z_{REF} = Z_{LINE}$ and the reflection coefficient $\Gamma_{LOAD,Z_{LINE}}$ was calculated. The SMU (connected to the mm-wave test heads bias-Ts) was used to measure R_{LOAD} through RF probes. The test port resistance was defined using the Short standard and subtracted from the measured R_{LOAD}.

[2] Commercially available from Cascade Microtech, Inc.
[3] Available from NIST
[4] Commercially available from MathWorks, Inc.

While the Load was designed for $R_{LOAD} = 50\,\Omega$, process fabrication inaccuracies caused a deviation of 2.4 Ω for SiGe and 11.1 Ω for GaAs Load (Table 1).

(1)-(3) show that ill-defined R_{LOAD} leads to inaccurate C' and Z_{LINE}. With the help of a mathematical model, error bounds for C', Z_{LINE} as well as for the extracted parameters of DUT were calculated for a given worst-case error in R_{LOAD} of 10%. (Fig. 2, Table 1). Table 1 gives the real and imaginary parts of the Z_{LINE} extracted at 10 GHz.

TABLE I. EXTRACTED PARAMETERS OF LINES

	Measured/ Extracted Parameter		
	Nominal value	***Lower bound***	***Upper bound***
Si/SiGe: C Process			
R_{LOAD}, Ω	47.6	42.84	52.36
C_{LINE}, pF/cm	1.259	1.399	1.144
Re(Z_{LINE}), Ω @10 GHz	51.12	46.00	56.23
Im(Z_{LINE}), Ω @10 GHz	-3.07	-2.86	-3.39
GaAs Process			
R_{LOAD}, Ω	61.1	55.0	67.2
C_{LINE}, pF/cm	1.725	1.917	1.568
Re(Z_{LINE}), Ω @10 GHz	47.69	42.92	52.46
Im(Z_{LINE}), Ω @10 GHz	-1.34	-1.20	-1.47

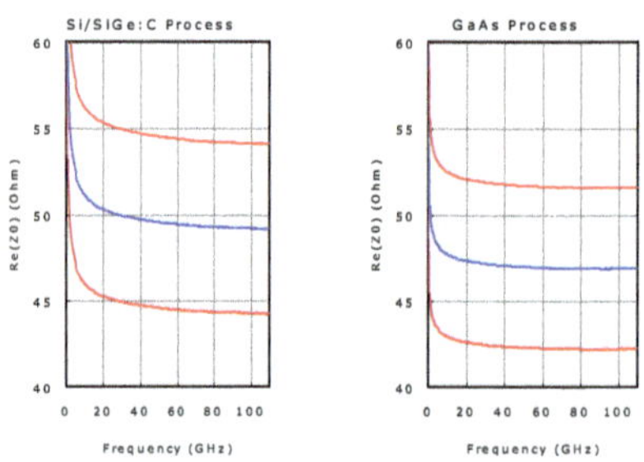

Figure 2. Real part of the extracted characteristic impedance Z_{LINE} (blue) and measurement error bounds (red) from Si/SiGe:C (left) and GaAs (right) processes.

C. *Results for Passives*

The passive GaAs test chip included the offset Short with a serial inductance $L_{SHORT,GaAs}$ of about 31.5 pH and the Open exhibiting a parallel capacitance $C_{OPEN,GaAs}$ of about 14 fF. The measured S-parameters of these elements were corrected with respect to multiline TRL for $Z_{REF} = Z_{LINE}$. Next, the results were transformed to $Z_{REF} = 50\,\Omega$ using (4) for nominal, upper and lower bounds of Z_{LINE} (±10% error for R_{LOAD}). Equivalent capacitances of the Open and inductances of the Short where extracted from the Π- and the T-equivalent circuits, respectively. Figure 3 shows the port 1 capacitance $C1$ of the Open and Port 1 inductance $L1$ of the Short, normalized to nominal values. Even for worst-case bounds for R_{LOAD} of 10% from the nominal value of 61.1 Ω, the calculated extraction errors are within -9%...+11% for capacitance $C1$ and +/-10% for inductance $L1$ (Fig. 3, red lines). Parameters that were extracted without re-normalization of the reference impedance from Z_{LINE} to $Z_{REF} = 50\,\Omega$ are also within this region: -4.5% @ 10 GHz for $C1$ and +4.9% @ 10 GHz for $L1$ (Fig. 3, green diamonds). Results for other elements are comparable.

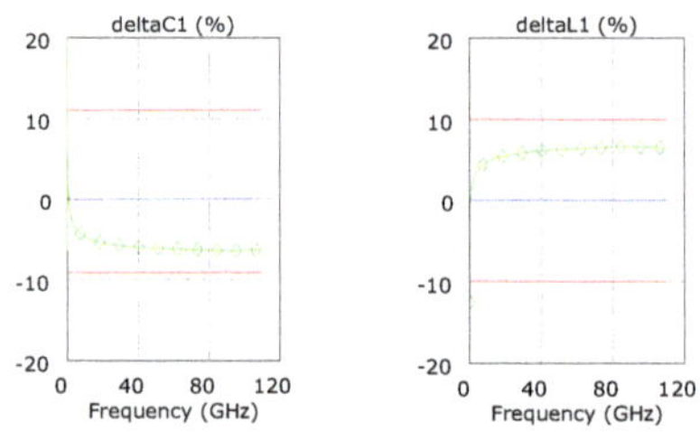

Figure 3. Error bounds for Port 1 capacitance *C*1 of the Open (left) and Port 1 inductance *L*1 of the Short (right) of GaAs test chip, normalized to nominal values. Nominal values are extracted from $Z_{REF} = 50\,\Omega$ (blue). Green diamond curve corresponds to $Z_{REF} = Z_{LINE}$.

D. *Active DUT Results*

An HBT with 14.86 μm length and 0.12 μm emitter stack available from BiCMOS9MW process was measured in a cold-*S* mode with $V_B = 0$ V...+1 V and $V_C = 0$ V and in hot-*S* (active) mode with $V_B = 0.7$ V...+1 V and $V_{CB} = 0$ V (emitter grounded in both cases). Important figures of merits (FoM), such as C_{BC}, C_{BE}, f_T and f_{MAX} were extracted. For this experiment, the calibration reference plane was kept at the M6 (top metal) level.

The extracted nominal value of C_{BE} is 26.5 fF[5]. For the worst-case error bounds of 10% for measured R_{LOAD} of 47.6 Ω, the errors for C_{BE} are within -9%...+11% (Fig. 4, red lines). It is only 2% @10 GHz without re-normalization of the reference impedance, e.g. for $Z_{REF} = Z_{LINE}$ (Fig. 4, green diamonds).

[5] Because the calibration reference plane is set to M6, C_{BE} also includes via parasitic inductance and M6-M1 parasitic capacitance.

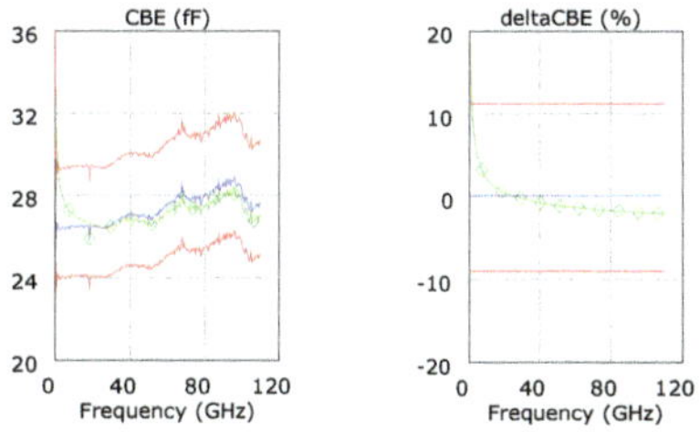

Figure 4. The absolute(left) and normalized (right) results for base-emitter capacitance of a test Si/SiGe:C HBT at V_{BE} = 0V for Z_{REF} = 50 Ω (blue), $Z_{REF} = Z_{LINE}$ (green diamonds) and estimated error bounds (red). The measurement reference plane is set to the M6 (top metal).

IV. Conclusion

A quantitative accuracy analysis of *in-situ* multiline TRL calibration implemented in both semi-insulating and conductive wafers was performed. The impact of possible measurement errors of R_{LOAD} on extraction of C' and Z_{LINE} was estimated. The measurement error bounds for extracted parameters of passive and active DUTs were calculated for a worst-case error in R_{LOAD}. Obtained results showed that the I-V measurement capability (provided by SMU) of a conventional S-parameter characterization setup is sufficient for accurate extraction of C' and Z_{LINE} of customized Line standards.

It was also shown, that Z_{LINE} stays within the desired level of confidence even at low frequencies for proper layout of the Line. For such cases and when characterizing high-reflective DUTs, the extraction of Z_{LINE} and the re-normalization of the TRL reference impedance to 50 Ω may not be required. Thus, implementation of the *in-situ* TRL into a characterization workflow of high-performance microwave devices can be significantly simplified.

References

[1] A. Chantre, P. Chevalier, T. Lacave, G. Avenier, M. Buczko, Y. Campidelli, L. Depoyan, L. Berthier, and C. Gacquiere, "Pushing conventional SiGe HBT technology towards "Dotfive" terahertz," in Microwave Integrated Circuits Conference (EuMIC), 2010 European, 2010, pp. 21-24.

[2] S. Munkyo, M. Urteaga, J. Hacker, A. Young, Z. Griffith, V. Jain, R. Pierson, P. Rowell, A. Skalare, A. Peralta, R. Lin, D. Pukala, and M. Rodwell, "InP HBT IC technology for terahertz frequencies: fundamental oscillators up to 0.57 THz," Solid-State Circuits, IEEE Journal of, vol. 46, pp. 2203-2214, 2011.

[3] C. Huang, M. Chang-Chien, Y. Hsu, and K. Ku, "Applying line, offset-series, offset-shunt calibration for on-wafer CMOS device characterizations," in Asia-Passific Microwave Conference Singapore, 2009.

[4] I. Martinez, "In-situ calibration and verification techniques for the characterization of microwave circuits and devices," in 25th IEEE International Conference on Microelectronic Test Structures San Diego, CA, USA, 2012.

[5] A. Rumiantsev, P. Sakalas, F. Pourchon, P. Chevalier, N. Derrier, and M. Schroter, "Application of on-wafer calibration techniques for advanced high-speed BiCMOS technology," in Bipolar/BiCMOS Circuits and Technology Meeting, 2010. BCTM '10. IEEE Austin, TX, 2010, pp. 98-101.

[6] A. Rumiantsev, S. L. Sweeney, and P. L. Corson, "Comparison of on-wafer multiline TRL and LRM+ calibrations for RF CMOS applications," ARFTG Microwave Measurements Conference-Fall, 72nd, pp. 132-135, 2008.

[7] A. Rumiantsev, F. Pourchon, N. Derrier, P. Sakalas, and D. Celi, "Designing on-wafer calibration standards for advanced high-speed BiCMOS technology," in Microwave Integrated Circuits Conference (EuMIC), 2011 European, 2011, pp. 57-60.

[8] R. B. Marks, "A multiline method of network analyzer calibration," Microwave Theory and Techniques, IEEE Transactions on, vol. 39, pp. 1205-1215, 1991.

[9] D. Williams, R. Marks, K. Phillips, and T. Miers, "Progress toward MMIC on-wafer standards," in ARFTG Microwave Measurements Conference-Fall, 36th, 1990, pp. 73-83.

[10] R. B. Marks and D. F. Williams, "A general waveguide circuit theory," Journal of research of the National Institute of Standards and Technology, vol. 97, pp. 533-562, September-October 1992.

[11] D. F. Williams and R. B. Marks, "Transmission line capacitance measurement," Microwave and Guided Wave Letters, IEEE, vol. 1, pp. 243-245, 1991.

[12] A. Rumiantsev, P. L. Corson, S. L. Sweeney, and U. Arz, "Applying the calibration comparison technique for verification of transmission line standards on silicon up to 110 GHz," in *ARFTG Microwave Measurements Conference-Spring, 73rd* Boston, MA, 2009.

17

Attachment to Chapter 9

17.1 Paper [108]: "The Modifed Ripple Test for On-Wafer S-Parameter Measurements"

H. Heuermann and A. Rumiantsev, "The modifed ripple test for on-wafer S-parameter measurements," in *ARFTG Microwave Measurements Conference-Spring, 69th*, 2007, pp. 1–5.

This paper addressed the background, implementation, and the on-wafer measurement results of and automatic analysis of the accuracy of S-parameters. The classic ripple-test was automated and expanded to the use of a dispersive transmission line as a reference. For the first time a dispersive coplanar line standard was used for the ripple-test. The novel technique was demonstrated through a number of examples. Numerical and measurement results verified the proposed method.

In this work, I provided a list of requirements, planned the work, and made all measurements. I also calculated electrical characteristics of the reference coplanar line required for the algorithm. Prof. Holger Heuermann from Fachhochschule Aachen (Aachen, Germany) implemented the impedance transformation step into the automated ripple test, did all calculations and wrote the major part of the paper. I presented this work at the ARFTG-69th conference.

The Modified Ripple Test for On-Wafer S-Parameter Measurements

Holger Heuermann* and Andrej Rumiantsev**
* Univ. of Applied Sciences Aachen, Institute of High Frequency Tech., Germany, 52066 Aachen, Eupener Str. 70
** SUSS MicroTec Test Systems GmbH, Germany, 01561 Sacka, Sussstr. 1

Abstract— **This paper addresses the background, implementation, and the on-wafer measurement results of an automatic analysis of the accuracy of S-parameters. The classic ripple-test is automated and expanded to the use of a dispersive transmission line as a reference. Consequently, it was the first time a dispersive coplanar line was used for the ripple-test. The novel technique is illustrated through a number of examples. Numerical and measurement results have verified the proposed method.** [1]

I. Introduction

Many users of vector network analysers (VNAs) do not question their S-parameter measurements. Easily identifiable stochastic error is intrinsic because of the good repeatability and reproducibility of the VNA measurements the user often believes the results. However, every VNA-measurement includes errors. The non-ideal calibration standards (especially matched loads and opens) and the corresponding calibration method have strong influence on the measurement accuracy.

Experienced users are aware of the classic ripple test and apply it for evaluating the measurement accuracy of coaxial systems using an air-line as a reference, as shown in [1], [2]. More complex ways to estimate measurement errors are given in [3].

The ripple test is a simple calculation of the worst case estimates of S-parameter measurement errors. However, for on-wafer measurements as well as for waveguide measurements, the classic ripple test is not practical: the reference transmission line is not ideal; it is lossy and typically dispersive.

This paper will demonstrate the modification of the classic ripple test for wafer-level applications and the measurement results obtained from a 110 GHz setup.

The new algorithm can handle dispersive and lossy lines. It also implements the source match estimation from [2] to automate the procedure and to calculate the error bars for reflection S-parameter measurements.

This paper is organized as follows: first, the classic ripple test is presented. Then, the automation of the ripple test is explained. Next, the implementation of the dispersive reference lines is presented. Then, experiments with synthetic measurements will demonstrate the precision of the algorithm. Finally, experimental measurement results verify the proposed method and is illustrated in a number of examples.

[1]©ARFTG-spring-conference, Honolulu 2007

II. Sources of Measurement Errors

For every measurement, users would like to see the uncertainty in the measurement results approximated with the error bars, as shown in Fig. 1.

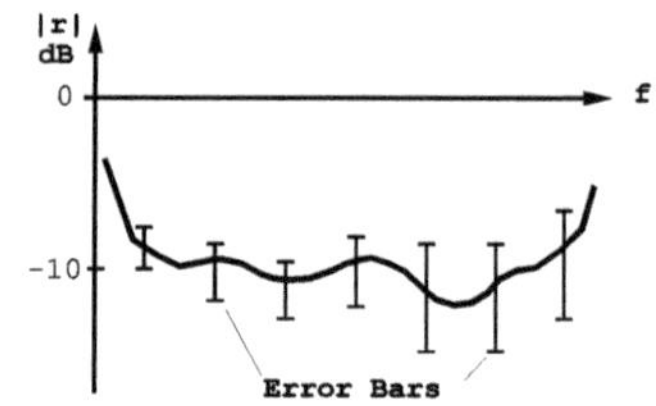

Fig. 1. Reflection measurement over frequency with error bars

The measurement errors of scattering parameter measurements with modern VNAs can be identified into two types: random errors and systematic errors.

Random errors over frequency can be identified very easily. The finite measurement dynamic of the VNA is an example. The repeatability of connections introduces errors in every measurement.

Often, the largest errors are systematic ones. In contrast to random errors, systematic errors can not be identified easily. Here the non-ideal calibration standards have the largest influence. Most problems come from matched loads and opens. Also the calibration method and/or its implementation can have an influence. An example is the old and new TRL-application. If the equations from [4] or [5] are used, error can be observed around the frequencies $n \cdot \lambda/2$ ($n = 0, 1, 2, ...$). By using [6], the error-areas are minimized. A comparison is illustrated in Fig. 2 with $n = 0, 1$.

Other sources of measurement errors are cable phase dispersion and temperature drift, [7].

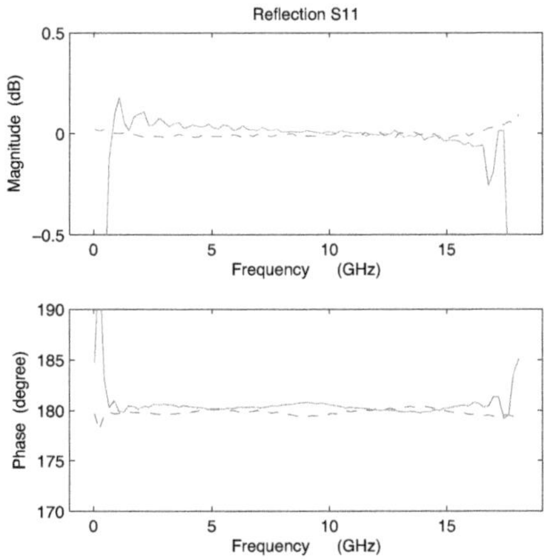

Fig. 2. Reflection coefficient of a short on a PTFE-substrate by using TRL from [5] (solid) and [6] (dashed)

III. The Classic Ripple Test

The classic ripple test can be used for small reflection measurement errors. It does not use the error coefficient of classic calibration procedures and is based on a simpler model.

The error of a VNA-measurement corresponds to the so-called source match with the error vector r_q, which is the vector between the real reflection vector r_w and the measured reflection vector r_m, as shown in Fig. 3.

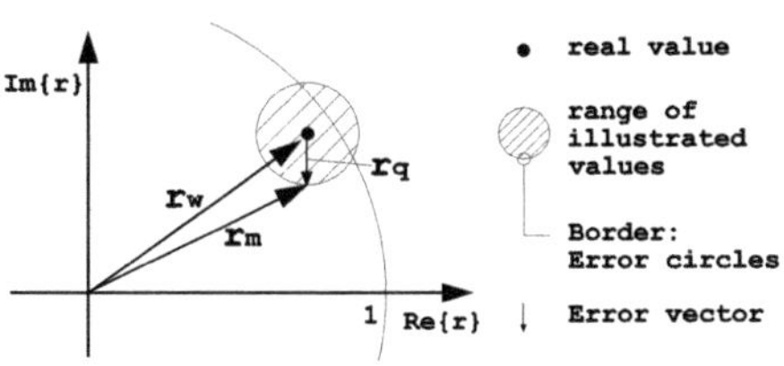

Fig. 3. Model: emergence of ripples in the r-plane

All errors are included in the final source match with the error vector r_q, the value to be identified. Fig. 4 shows the error model for the classic ripple test.

A measurement of a long, matched transmission line loaded with the short is needed to calculate the source match Z_q, which corresponds to the error vector r_q by:

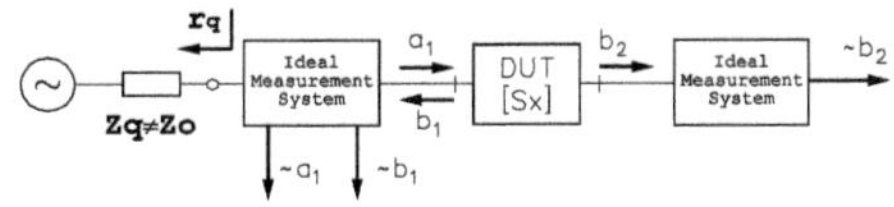

Fig. 4. VNA error model for the classic ripple test

$$r_q = \frac{Z_q + Z_0}{Z_q - Z_0} \quad . \tag{1}$$

Fig. 5 illustrates the measurement set-up for the classic ripple test to calculate the source match.

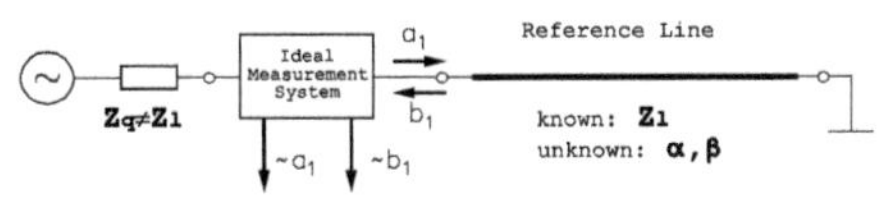

Fig. 5. Set-up to perform the r_q-calculation for the classic ripple test

In practice, the source impedance Z_q and the system impedance Z_0 are not the same for a real VNA measurement. The measurement of the reflection values of the long transmission line is shown in Fig. 6. Instead of

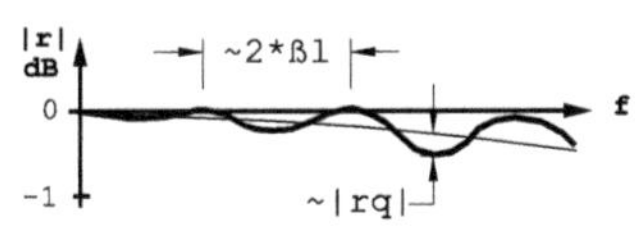

Fig. 6. Reflection results of a matched line loaded with the short for a real VNA measurement

showing an increase in the magnitude of reflection losses over the frequency sweep it shows a ripple response. The results should show only an increase of the losses over the frequency as an effect of the skin deep.

The mathematical approximation for the ripple test is based on Fig. 3. It can be approximated as:

$$r_m \simeq r_w + r_q \quad . \tag{2}$$

This approximation can be proved by the following equations:

$$r_w = \frac{Z_w + Z_0}{Z_w - Z_0} \quad , \tag{3}$$

$$r_m = \frac{Z_m + Z_0}{Z_m - Z_0} \quad , \tag{4}$$

with the approximation: $Z_q \simeq Z_0$.

The practical ripple test is performed in this way: first, measure the peak-to-peak ripple by using a long, matched line loaded with the short. Next, use a table (see [2]) to find the corresponding source match. Then, calculate the so-called delta-match between the reflection value of the DUT and source match. Finally, use the delta-match from the table to find the corresponding magnitude and phase error.

In [2] a practical example is given along with the modified ripple test, which is necessary to automate the ripple test.

IV. Automation of the Ripple Test

The modified ripple test uses the model illustrated in Fig. 7 to measure the source match with the error vector r_q. Only the reflection value S_{11} is measured.

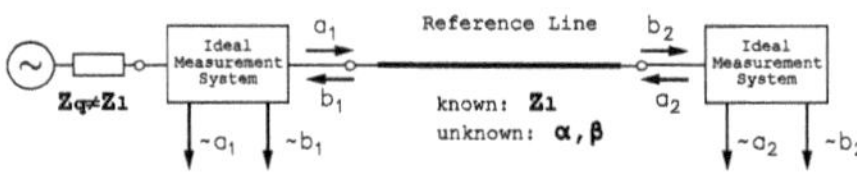

Fig. 7. Setup to measure the source match for the modified ripple test

Using the classic ripple test, it was only possible to estimate the source match for a few frequency points. The results are the average value for a frequency band of one ripple. So, it was necessary to use a very long line to have a lot of ripples.

The test of Fig. 7 works for every frequency point, except the areas around $n \cdot \lambda/2$ for $n = 0, 1, 2, \ldots$. To overcome this problem, a multi-line solution can be chosen.

However, the line impedance Z_l of reference line must be similar to the system impedance Z_0, often $50\,\Omega$. Due to this fact, the classic or modified ripple tests can only be used in a line system with real TEM wave guides, e.g. coaxial lines.

V. Implementation of a Dispersive Reference Line

The dispersive characteristic of a coplanar line for on-wafer an calibration substrate[2] is shown in Fig. 8.

The line impedance was extracted using the procedure introduced in [8] and verified for commercial calibration substrates in [9].

In order to use a dispersive transmission line to estimate the source match, renormalization is necessary.

The Z-parameters were used for the renormalization. The so-called port impedance matrix

$$[\mathbf{Z_{La}}] = \begin{pmatrix} 50 & 0 \\ 0 & 50 \end{pmatrix} \tag{5}$$

is used for a reference to $50\,\Omega$. Port impedance matrices have the real or complex port impedance value in the trace. All other elements are zero.

[2]The CSR-8 calibration substrate from SUSS MicroTec was used.

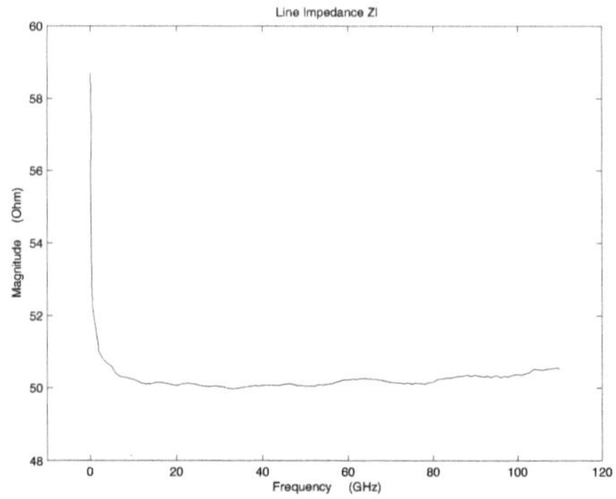

Fig. 8. Magnitude of impedance of the dispersive $50\,\Omega$ coplanar line on Al_2O_3-substrate

$[\mathbf{I}]$ is the unit matrix.

The first step is the calculation of the normalized matrix $[\mathbf{za}]$ from the measured S-parameter of the line:

$$nen = (1 - S_{11}) \cdot (1 - S_{22}) - S_{21} S_{12}\,, \tag{6}$$
$$za_{11} = ((1 + S_{11}) \cdot (1 - S_{22}) + S_{21} S_{12})/nen\,, \tag{7}$$
$$za_{12} = 2 \cdot S_{12}/nen\,, \tag{8}$$
$$za_{22} = ((1 - S_{11}) \cdot (1 + S_{22}) + S_{21} S_{12})/nen\,, \tag{9}$$
$$za_{21} = 2 \cdot S_{21}/nen\,. \tag{10}$$

Now, the absolute Z-parameter can be calculated:

$$[\mathbf{Z}] = [\mathbf{Z_{La}}]^{0.5} \cdot [\mathbf{za}] \cdot [\mathbf{Z_{La}}]^{0.5}\,. \tag{11}$$

Z_w is the complex line impedance from [8]. $[\mathbf{Z_{Lb}}]$ is the port impedance matrix with the complex line impedance.

$$[\mathbf{Z_{Lb}}] = \begin{pmatrix} Zw & 0 \\ 0 & Zw \end{pmatrix} \tag{12}$$

Using the Z-matrix, the Zw normalized Z-parameter can be calculated.

$$[\mathbf{zb}] = [\mathbf{Z_{Lb}}]^{-0.5} \cdot [\mathbf{Z}] \cdot [\mathbf{Z_{Lb}}]^{-0.5} \tag{13}$$

Finally, the renormalized S-matrix is found:

$$[\mathbf{S_{renorm}}] = [\mathbf{I}] - 2 \cdot ([\mathbf{zb}] + [\mathbf{I}])^{-1}\,. \tag{14}$$

The difference between the renormalized and the non-renormalized matching of this line is high, especially up to 20 GHz. Fig. 9 shows both reflection values of the coplanar line.

It is obvious that the source match is better at lower frequencies than higher ones. The renormalization up to 20 GHz helps achieve more realistic values.

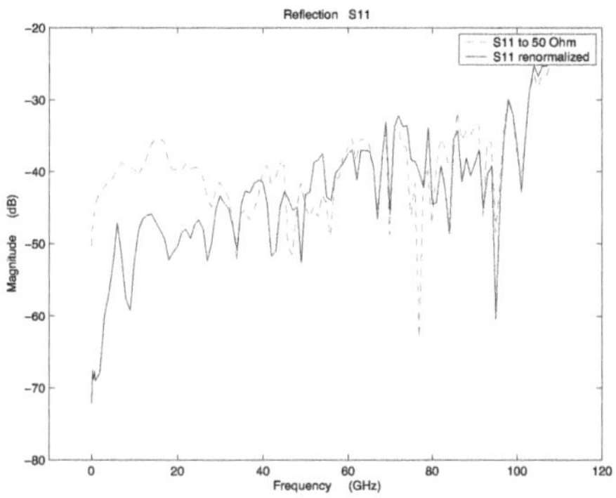

Fig. 9. Magnitude of the source match of dispersive 50 Ω coplanar line on Al_2O_3-substrate with and without renormalization

VI. Experimental Results

<u>Based on the Numerical Simulation</u>

At first, we calculated the S-parameter of a non-50 Ω reference line. The line impedance starts at 0.1 GHz with 60 Ω and decreases to 40 Ω at 10 GHz. The measurement error of this reference line was -65 dB up to 1 GHz and -55 dB up to 10 GHz. Fig. 10 shows both: S_{11} of this line in reference to 50 Ω and the renormalized results.

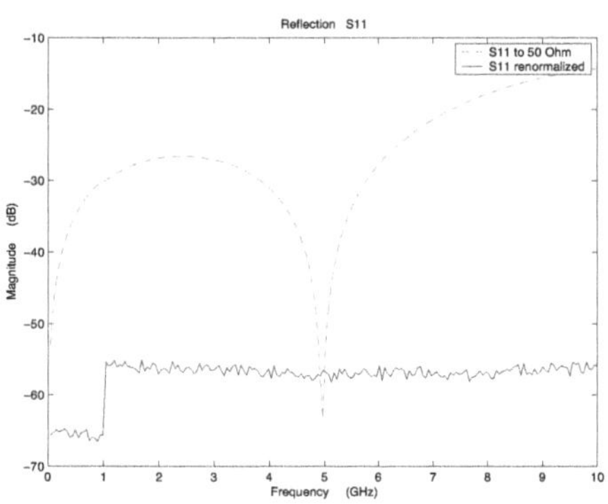

Fig. 10. Magnitude of the source match of dispersive transmission line with and without renormalization

Using this highly-dispersive reference line, the measurement error for a 55 Ω transmission line was calculated. The results in Fig. 11 show a significant decrease in the measurement error over the frequency belonging to the larger delta-match, see [2].

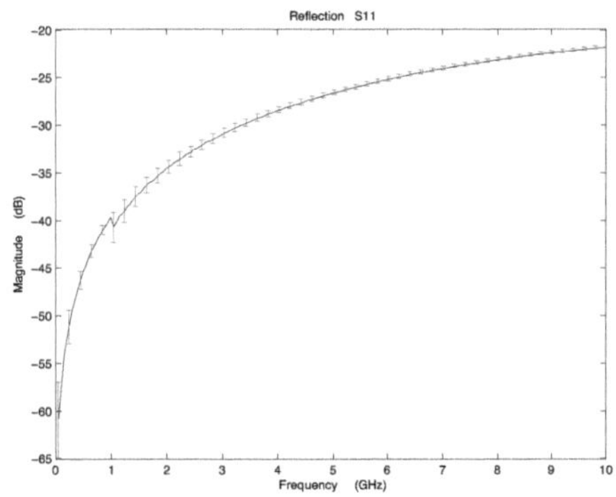

Fig. 11. Magnitude of the reflection of a 55 Ω line and calculated error bars

<u>Based on Measurement Experiments</u>

We used the 110 GHz PNA network analyzer from Agilent, measuring the frequency range from 0.1 GHz to 110 GHz. The calibration was performed using SussCal Professional calibration software from SUSS MicroTec and the LRM+ calibration procedure. The measurement data was exported and processed on an external computer. Additionally, the reproducibility was accepted to be -70 dB for low frequencies and -60 dB for high frequencies.

The first experimental results show the magnitude (Fig. 12) and the phase (Fig. 13) of the reflection coefficient of a 25 Ω load termination.

The final experimental results show the magnitude (Fig. 14) and the phase (Fig. 15) of the reflection coefficient of a 100 Ω load termination.

These experimental results validate the modified ripple test for on-wafer measurements. The error bars are directly generated by the source match values. Additionally, a multi-line procedure can be used to eliminate the $\lambda/2$-problem.

VII. Conclusion

This work addresses users as well as manufacturers of modern network analyzers. The theory gives worst case estimations of the reflection parameter measurement errors. Sources of measurement errors as well as the classic ripple test were explained. The background, implementation, and the on-wafer measurement results of an automatic analysis of the accuracy of S-parameter were presented. The ripple test was automated and expanded to the use of a dispersive transmission line as a reference. Numerical

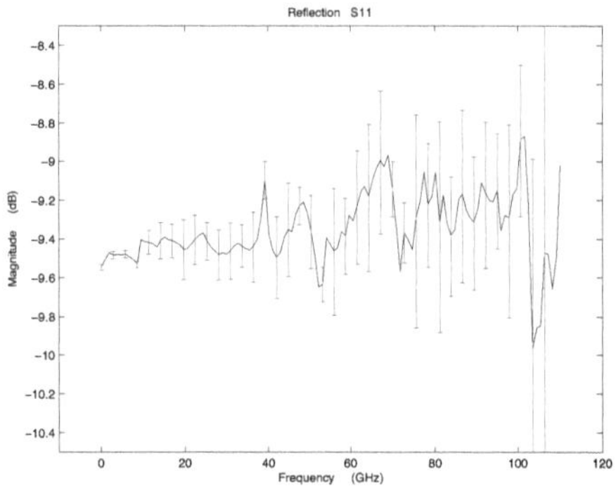

Fig. 12. Magnitude of the reflection coefficient of a 25 Ω load on Al_2O_3-substrate and calculated error bars

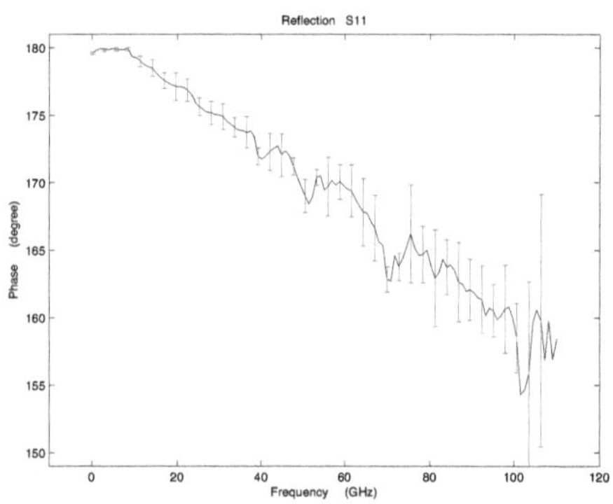

Fig. 13. Phase of the reflection coefficient of a 25 Ω load on Al_2O_3-substrate and calculated error bars

and measurement results verified the proposed automatic error bar calculation for reflection measurements.

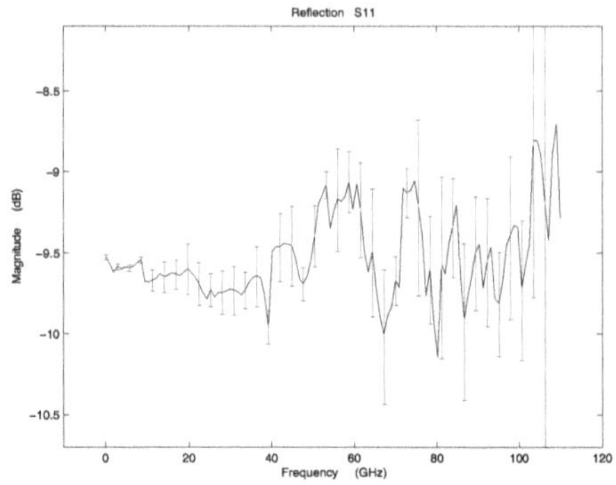

Fig. 14. Magnitude of the reflection coefficient of a 100 Ω load on Al_2O_3-substrate and calculated error bars

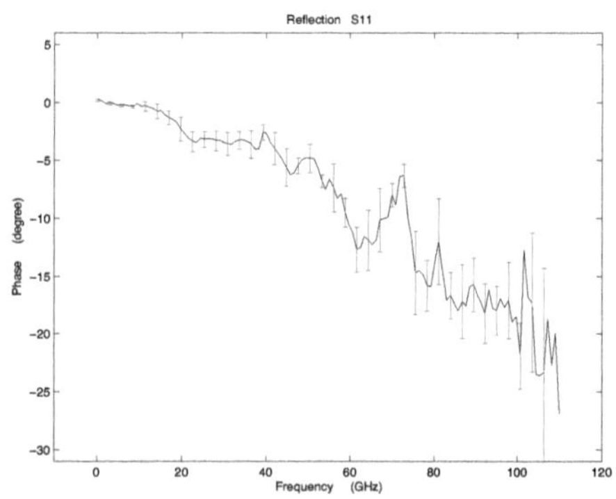

Fig. 15. Phase of the reflection coefficient of a 100 Ω load on Al_2O_3-substrate and calculated error bars

REFERENCES

[1] Maury Microwave, Verifying the Performance of Vector Network Analyzers, Application Note, July 1999.
[2] Heuermann, H., Old and New Accuracy Estimation of S-Parameter Measurements with the Ripple-Test, MTT-S International Microwave Symposium Workshop TMB, San Francisco, Juni 2006.
[3] European co-operation for Accreditation, Expression of the Uncertainty of Measurement in Calibration, EA-4/02, 12.1999.
[4] Engen, G.F., Hoer, C.A., Thru-Reflect-Line: An Improved Technique for Calibrating the Dual Six Port Automatic Network Analyzer, IEEE Trans. Microwave Theory Tech., MTT-27, Dec. 1979, pp. 987-993.
[5] Eul, H.J., Schiek, B., A Generalized Theory and New Calibration Procedures for Network Analyzer Self-Calibration, IEEE Trans. Microwave Theory Tech., MTT-39, Apr. 1991, pp. 724-731.
[6] Heuermann, H., Schiek, B., Robust Algorithms for Txx Network Analyzer Procedures, IEEE Trans. Instrum. Meas., Feb. 1994, pp. 18-23.
[7] Wartenberg, S. A., RF Measurements of Die and Packages, Artech House 2002, ISBN 1-58053-273-x.
[8] Marks, R., Williams, D., Characteristic Impedance Determination Using Propagation Constant Measurements, IEEE Microwave and Guided Wave Lett., vol. 1, pp. 141143, June 1991.
[9] Rumiantsev, A., Doerner, R., Thies, S., Calibration Standards Verification Procedure Using the Calibration Comparison Technique, 36th European Microwave Conference Digest, September 2006.

Index

About the Author

Andrej Rumiantsev received the Diploma-Engineer degree (with highest honors) in Telecommunication systems from the Belarusian State University of Informatics and Radio Electronics (BSUIR), Minsk, Belarus, and the Dr.-Ing. Degree (with *summa cum laude*) in Electrical Engineering from Brandenburg University of Technology (BTU) Cottbus, Germany, in 1994 and 2014, respectively.

He joint SUSS MicroTec Test Systems (from January 2010 Cascade Microtech) in 2001 were he held various engineering product management and marketing positions. He significantly contributed to the development of the RF wafer probe, the |Z| Probe, wafer-level calibration standards, calibration software and probe systems. From 2010 to 2013 he was Product Marketing Manager of Device Characterization for Modeling and Process Development at Cascade Microtech and responsible for Elite300 system product line. In March 2013, he joined Ulrich L. Rohde Chair for RF and Microwave Techniques at Brandenburg University of Technologies (BTU), Cottbus, Germany. Dr. Rumiantsev is currently with MPI Corporation, holding a position of Director of RF Technologies of the Advanced Semiconductor Test Division. His research interests include RF calibration and wafer-level measurement techniques for advanced semiconductor devices.

Dr. Rumiantsev is a member of the IEEE MTT-11 Microwave Measurements Committee and the ExCom member of Automatic RF Techniques Group (ARFTG). He is the past Chair of the Modeling and Simulation Sub-Committee and the ExCom member of IEEE Bipolar/BiCMOS Circuits and Technology Meeting (BCTM). He holds several patents in the area of wafer-level RF calibration and measurements techniques. Dr. Rumiantsev received the ARFTG-71th Best Interactive Forum Paper Award. His doctoral thesis was awarded as “Best Dissertation of 2014 at Brandenburg University of Technologies”.

For Product Safety Concerns and Information please contact our EU
representative GPSR@taylorandfrancis.com
Taylor & Francis Verlag GmbH, Kaufingerstraße 24, 80331 München, Germany

www.ingramcontent.com/pod-product-compliance
Ingram Content Group UK Ltd.
Pitfield, Milton Keynes, MK11 3LW, UK
UKHW022210190926
475702UK00003B/285
* 9 7 8 8 7 7 0 2 2 1 1 2 2 *